History, Philosophy and Theory of the Life Sciences

Volume 23

Editors
Charles T. Wolfe, Ghent University, Belgium
Philippe Huneman, IHPST (CNRS/Université Paris I Panthéon-Sorbonne), France
Thomas A. C. Reydon, Leibniz Universität Hannover, Germany

Editorial Board
Marshall Abrams, University of Alabama at Birmingham
Andre Ariew Missouri
Minus van Baalen UPMC, Paris
Domenico Bertoloni Meli Indiana
Richard Burian Virginia Tech
Pietro Corsi EHESS, Paris
François Duchesneau Université de Montréal
John Dupré Exeter
Paul Farber Oregon State
Lisa Gannett Saint Mary's University, Halifax
Andy Gardner Oxford
Paul Griffiths Sydney
Jean Gayon IHPST Paris
Guido Giglioni Warburg Institute, London
Thomas Heams INRA, AgroParisTech, Paris
James Lennox Pittsburgh
Annick Lesne CNRS, UPMC, Paris
Tim Lewens Cambridge
Edouard Machery Pittsburgh
Alexandre Métraux Archives Poincaré, Nancy
Hans Metz Leiden
Roberta Millstein Davis
Staffan Müller-Wille Exeter
Dominic Murphy Sydney
François Munoz Université Montpellier 2
Stuart Newman New York Medical College
Frederik Nijhout Duke
Samir Okasha Bristol
Susan Oyama CUNY
Kevin Padian Berkeley
David Queller Washington University, St Louis
Stéphane Schmitt SPHERE, CNRS, Paris
Phillip Sloan Notre Dame
Jacqueline Sullivan Western University, London, ON
Giuseppe Testa IFOM-IEA, Milano
J. Scott Turner Syracuse
Denis Walsh Toronto
Marcel Weber Geneva

More information about this series at http://www.springer.com/series/8916

Marta Bertolaso • Silvia Caianiello
Emanuele Serrelli

Editors

Biological Robustness

Emerging Perspectives from within the Life Sciences

 Springer

Editors
Marta Bertolaso
FAST Institute of Philosophy of Scientific
Practice and Faculty of Engineering
University Campus Bio-Medico of Rome
Rome, Italy

Silvia Caianiello
Institute for the History of Philosophy and
Science in Modern Age (ISPF)
Italian National Research Council
Naples, Italy

Emanuele Serrelli
CISEPS - Center for Interdisciplinary
Studies in Economics, Psychology and
Social Sciences
University of Milano Bicocca CISEPS
Brescia, Italy

ISSN 2211-1948   ISSN 2211-1956   (electronic)
History, Philosophy and Theory of the Life Sciences
ISBN 978-3-030-01197-0   ISBN 978-3-030-01198-7   (eBook)
https://doi.org/10.1007/978-3-030-01198-7

Library of Congress Control Number: 2018962881

This Springer imprint is published by the registered company Springer Nature Switzerland AG.
The registered company address is: Gewerbestrasse 11, 6330 Cham, Switzerland

Foreword

The papers collected in this volume are the outcome of a series of workshops organized by the Bio-Techno-Practice group on the different ways in which philosophers, biologists, neuroscientists, and engineers employ the concept of "robustness." The goal, which was successfully realized, was to stimulate an interactive interdisciplinary engagement that would highlight differences and commonalities across disciplines and perspectives. The hope was that, by this type of engagement, confusions would be evaporated and insights from one part of the intellectual landscape could aid those exploring robustness in another part. This volume is evidence of the success of that methodology, as the individual papers reveal the benefits of interactive engagement. Even more important is the reframing of future work by the participants in light of the refraction of individual disciplinary commitments in the context of cross-disciplinary connections.

Robustness is a perfect subject for this type of engagement. It marks the system level property of maintaining system function in response to internal and external perturbations. We see it in evolved systems like organisms in the way body temperatures are maintained, or neurons are reassigned after head trauma. We see it in engineered systems like bridges and buildings or software algorithms whose design aims to preserve performance under a range of expected conditions. But how robustness is achieved varies both within and between types of systems. There is clearly intellectual traffic between disciplines studying robustness. Systems biology employs notions of control networks, feedback, and modularity and reverse engineering. Top down and bottom up approaches converge on how a complex system not only does what it does but continues to do it when there are changes in the external environment and loss or change in internal components.

Philosophy of science can abstract away from the details of any one mechanism for achieving robustness to characterize what it is for a system to be robust. Robustness is always relative, robust with respect to this function, or that equilibrium. Indeed, by evolving or engineering robustness for a particular function in a range of values for internal and external variables, fragility will be introduced for other functional stabilities in other conditions. Redundancy, modularity, and multiple pathways are features that promote robustness.

Biological Robustness: Emerging Perspectives from Within the Life Sciences combines the detailed explorations of robustness by engineers, biologists, neuroscientists, and philosophers, inviting the reader into reflective engagement that the workshops promoted. I have learned much from participating in the project that gave rise to this volume. By bringing together a plurality of perspectives, this volume extends the reach of interdisciplinary engagement.

Distinguished Professor   Sandra D. Mitchell
Department of History and Philosophy of Science
University of Pittsburgh
Pittsburgh, PA, USA

Contents

Chapter 1
Introduction: Issues About Robustness in the Practice of Biological Sciences

Marta Bertolaso, Emanuele Serrelli, and Silvia Caianiello

Abstract Robustness has lately become a bridging notion, in particular across the sciences of the natural and the artificial, crucial for prediction and control of natural and artificial systems in recent scientific practice, in biomedicine, neurobiology and engineering, as well as for risk management, planning and policy in ecology, healthcare, markets and economy. From biological, neurological and societal systems, arising by the interplay of self-organizing dynamics and environmental pressures, to the current sophisticated engineering that aims at artificially reproducing the adaptability and resilience of living systems in front of perturbations in manmade devices, robustness seems to hold the key for orchestrating stability and change. This introduction offers a general survey of the contribution that the notion of robustness is providing to reframing major concepts within the life sciences, such as development, evolution, time and environment, and to reframing the relationship between biology and engineering, as well as between biology and physics.

M. Bertolaso
Departmental Faculty of Engineering and FAST, Institute for Philosophy of Scientific and Technological Practice, University Campus Bio-Medico of Rome, Rome, Italy
e-mail: m.bertolaso@unicampus.it

E. Serrelli (✉)
CISEPS – Center for Interdisciplinary Studies in Economics, Psychology and Social Sciences, University of Milano Bicocca, Brescia, Italy
e-mail: emanuele.serrelli@epistemologia.eu

S. Caianiello
Institute for the History of Philosophy and Science in Modern Age (ISPF), National Research Council, Naples, Italy

Zoological Station Anton Dohrn, Naples, Italy
e-mail: silvia.caianiello@ispf.cnr.it

M. Bertolaso et al. (eds.), *Biological Robustness*, History, Philosophy and Theory of the Life Sciences 23, https://doi.org/10.1007/978-3-030-01198-7_1

Tardigrades, also known as "water bears", count among the most fascinating animals in the world. There are 900 very diverse species of them. They are 1 mm long, segmented animals that can survive the most extreme conditions: extreme temperatures, drought, radiation. They can withstand freezing and overheating, lack of water and air, toxicity, they can even survive in outer space. And they can for hundreds of years. Hence, tardigrades occasionally make it to the media outlet, like in the 2013 catchy Daily Mail article entitled "Meet the toughest animal on the planet: The water bear that can survive being frozen or boiled, float around in space and live for 200 years (shame it isn't much to look at)" (Pow 2013). Tough or *robust*?

Giuliani (this volume) provocatively summarizes the notion of robustness as 'die hard'. Robustness is, in fact, commonly understood as the ability to withstand attacks, perturbations and offences without being disrupted or heavily modified. A robust chair will bear the weight of a person sitting there for many years, its structure remaining largely unchanged with respect to the initial state (van der Krogt et al. 2009; Shahbazi et al. 2015). But the intuitive simplicity of the 'robustness' notion opens the way to many philosophical problems (e.g., Jen 2003) and interesting reflections on the nature of knowledge and on the ontology of the most diverse phenomena, from physical objects to engineered systems, up to living organisms, their components and the communities they form.

This volume is the first outcome of a series of workshops organized by the Bio-Techno-Practice Research Empowering Network (now Hub), coordinated by Marta Bertolaso and based at University Campus Bio-Medico in Rome. Practitioners from different sciences explored robustness as a putative general concept with common epistemological and ontological problems, as well as necessary domain specifications.[1] But this was also a work on the deep entanglement between robustness and Nature. The results were indeed very interesting, providing both definite conclusions and new research questions.

Robustness is a crucial concept in the very definition of an organism, as it reveals its individuality and persistence, its ability to maintain its characteristic functional structure through contingent changes (internal and external perturbations). This link between robustness and the definition and identity of a living being was the main focus of the first workshop, held in 2014.[2] Robustness was

---

[1] A very long list of terms identify, in different disciplines, cognate notions that bear important affinities with robustness. Some terms have to do with the current organization of a system, e.g., resilience, homeostasis/negative feedbacks, dynamical stability, plasticity, functional/functioning. Other terms are more change-related, e.g., homeorhesis, evolutionary stasis, canalization/ entrenchment, evolvability; and more generic terms such as persistence, lawlike/lawful, invariance, entropy. Such linguistic richness and redundancy is both an obstacle and an interesting point of departure for interdiciplinary work on robustness.

[2] "First Interdisciplinary Workshop on Robustness", *Robustness in Biological Systems*, University Campus Bio-Medico, Rome, October 14–15, 2014. A special methodology was experimented: philosophers and scientists gave short, focused talks and then interacted in groups by means of a sound and designed methodology. Group discussions were held in which each participant, being an expert of his/her own field, focusing on examples more than on definitions, then reported their conclusions, disagreements, and collective views.

probed with regards to its relevance in characterizing the peculiarity of living systems dynamics and relationships. Special attention was given to three organismic dimensions: (1) the relationship between autonomy and robustness; (2) the identification of the main organizational principles underlying robustness; (3) the increasingly evident crucial role of robustness in endorsing both evolutionary and developmental changes. These different dimensions shed light both on the ontological closure of living systems and on their peculiar capacity for adaptive transformation.

In biological systems, robustness comes across different scales, from molecular to organismal dimensions, and involves change and developmental aspects, thus becoming a pillar in their dynamics. Is it possible to obtain robustness artificially, or is it a natural property (i.e., is non-living systems' robustness distinct from organismic robustness)? What is the definition of robustness in engineering, and which synthetic models may be inspired by the concept of robustness? Which applications and technologies does robustness inspire? This was the focus of the second workshop, held in 2015.[3]

The third workshop, in the final part of 2015, was focused on the brain, the poster child for plasticity in biology.[4] Neurons and networks constantly rebuild themselves in response to the continual and ongoing change in component ion channels and receptors that are necessary for neuronal signaling. On the other side, external changes drive homeostatic responses. Robust responses can be triggered both ways. The meeting laid special attention on recent modeling and experimental work about the mechanisms, constraints, and outcomes of robust dynamics in the brain.

---

[3] "Second Interdisciplinary Workshop on Robustness", *Robustness – Engineering Science*, University Campus Bio-Medico in Rome, February 5–6, 2015. Goal of the workshop was to hold interdisciplinary discussions on relevant areas such as: (a) *macromolecular robustness*: stability of macromolecules, ability to react to environmental changes without modifying their functionality; (b) *material resistance*: mechanical resistance, brittleness, hardness are all material properties, that reflect the ability of solid objects to resist to deformations; (c) *biological dynamics and robustness*: the analysis of patterns of evolution of biological systems upon perturbation, considered in a theoretical physics and Systems Engineering frame; (d) *autonomous systems*: the ontological definition of autonomy as mirrored in mathematical modelling of systems evolving on their own, on the base of a self-consistent dynamics; autonomy as the basis of robust system design, thought to be resilient towards attacks or faults; (e) *resilience*, as the specific property of systems to return to the previous equilibrium state after perturbation; (f) *environmental robustness*, understood in the objective sense of the sustainability, at the ecological scale, of the interactions between human production systems and environment, and crucial for the assessment of affirmative sustainability principles; (g) *software robustness*, as the ability of an algorithm or of a program to cope with errors or abnormalities during execution, an acceptation strictly related to the management of increasing computational complexity.

[4] "Third Interdisciplinary Workshop on Robustness", *Robustness in Neurological Systems*, held on 2015, November 13–15 at the University of Pittsburgh.

1.1 Biological Robustness

The present volume is not simply about robustness. It is about *biological* robustness. Is there something unique to biological robustness? Robustness has multiple dimensions that must be analyzed and combined by the researchers according to the particular research question to be answered. According to Krakauer (2005) "…as of yet there is no unified theory of biological robustness, only collections of illustrative models". Nonetheless, a taxonomy of robustness cases suggests "some hints of meta-principles of robustness" that, "to a suitably shrewd theorist, might suggest some means and direction of formal unification". Some authors foretell the advent of a unified theory of robustness, or at least the constitution of 'robustness studies' as a field. The convergence among different fields toward "a single… integrated theory of robustness" (Alderson and Doyle 2010; Krakauer 2005) would represent a paradigmatic shift across different disciplinary borders. Robustness notions that emerge from biological and engineering systems tend to part ways from notions of dynamic stability in nonlinear physical systems (Lesne 2008; Carlson and Doyle 2002). Yet, if life sciences and robustness entertain a privileged relationship, it is also true that robustness seems to touch the very heart of scientific practice as such, so that it might become a unifying principle for philosophy of science, starting right from the sciences of the living (Bertolaso 2014).

Leaving aside, for the moment, such epistemological considerations, let us preliminarily delve into the use of robustness that researchers have made in biological research: a fundamental meaning of biological robustness refers to the robustness of the *development* of multicellular organisms. On the other hand, extremely complex and interesting relationships exist between robustness and biological evolution.

1.1.1 Developmental Robustness

Although 'development' is undergoing a deep theoretical revision (Pradeu et al. 2016; Minelli and Pradeu 2014), *developmental robustness* may still be seen as a strong peculiarity of multicellular living beings (Nijhout 2002). During development, many characteristics of organisms are relatively unaffected by substantial perturbation of the environment and by cryptic genetic variation. This phenomenology is, however, only a starting point: developmental robustness is not related to the return of a system to a previous state after perturbation (Allen and Starr 1982), but rather to the preferred trajectories of a morphogenetic process, or to its dynamic repertoire (Goodwin et al. 1993). Dynamic models of development are today possible thanks to the abundance of data on genes expression in development, the huge progress of computational capabilities, and the introduction of new mathematical and statistical methods (Gibson 2002; Levin 2012). Thanks to systems biology, to some extent biological development can also be artificially simulated (Devert et al. 2011; Jin and Meng 2011). The developmental repertoire may include, besides the observed phenomenology, unobserved phenotypic characteristics, alternative routes to the same characters, and trajectories that are differentially stable and reachable. The developing organism is an integrated system. This also implies that the features and dimensions of an organism cannot be dichotomized into plastic and non-plastic (robust) features. Through 'accommodation' (West-Eberhard 2005; Braendle and Flatt 2006; Pigliucci et al. 2006), the individual may develop structures and behaviours that are not seen in other individuals of the same species. Hence, accommodation is today studied for its evolutionary relevance. This logic is carried on through adult life, albeit with more limited flexibility.

The robustness of development is highlighted by probing the outcome of organismal development in non-typical environments. In animals, experimental embryology shows that deviations of development are possible, but they can only happen in limited time frames. Moreover, according to a classic and fortunate term by Waddington, development is *canalized* (Fusco et al. 2014; Siegal and Bergman 2002). In *Organisers and Genes* (1940), Waddington envisioned the development of any 'embryo part' as a cascade bifurcation diagram, where, through a sequence of developmental decisions, the part is driven from an undifferentiated state towards one of its alternative possible fates, represented by the tips of the diagram. The familiar behaviour of water streaming by gravitation provided Waddington with the means of conjugating several ideas, namely that embryo's parts (i) are in dynamic disequilibrium (like water running downstream) with a progressive loss of potential, (ii) follow a developmental track which, as a whole, is more or less stable, and (iii) generally decrease their own sensitivity to disturbances, from periods of high sensitivity where regulation is possible to periods of strong canalization. Waddington also argued for a chemical explanation of development, where concentrations of different chemicals are causally relevant to developmental pathways and decisions. Several authors have thus seen Waddington as a pioneer of the application of the dynamical systems theory. In Waddington's perspective, robustness applies both to the whole embryo and to the many embryo parts, in two different senses that are expressed by the metaphor of canalization: as the reliability of the dynamical system *in reaching a particular end state* (by return 'on track' in face of perturbations), and as the stability *of the 'landscape'* (Waddington's word) of bifurcations and alternative end states.

In organismal development and growth, therefore, plasticity and robustness are faces of the same coin. Moreover, they do not consist in stability of features; on the contrary, they pertain to a dynamic process: morphogenesis. Waddington's landscape is a point of view to look at the robustness of development *through* the variability of characters. Another similar point of view is provided by the reaction norm. The height of a tree heavily depends on the environment in which the seed is planted (humidity, population density, availability of nutrients in the soil, altitude, to name only a few relevant factors). The reaction norm is a mathematical relationship between some particular variables of the environment and some traits achieved by the organism during development. A population with great phenotypic variance can thus be expression of a robust reaction norm in face of environmental conditions that vary considerably across the habitable range. Sometimes phenotypic differences are obtained by simply altering the timing of development. The timing of metamorphosis in the spadefoot toad (*Scaphiopus hammondii*) is accelerated when the desert ponds in which the tadpoles live start to dry up. In response to the evaporation (detected as an increase in population density), the tadpoles undergo earlier metamorphosis and as a result grow into small adults (Bateson and Gluckman 2011, p. 34).

"The resistance of bodies to deviation from the form or forms that are typical for the species is also expressed in behaviour" (Bateson and Gluckman 2011, p. 20). Indeed, the general point that "organisms may reach the same end-point via many different pathways" (*ivi*, p. 25) is exemplified in the highest degree in the domain of stereotyped behaviors of animals. Cats can acquire and improve their adult predatory skills via a number of different developmental routes: by playing with their siblings, by playing at catching prey when young, by watching their mother catch live prey, by practicing catching live prey when young, or by practicing when an adult. Hence a kitten deprived of opportunities for play may still develop into a competent adult predator, but by a different developmental route.

The explanatory appeal of robustness in the post-genomic biological debate is due to its capacity of accounting for the dynamic stability of living systems at disparate organizational levels (including the molecular ones), as the result of complex, sophisticated networks of interactions rather than of the specific properties of the individual components.

1.1.2 Robustness and Biological Evolution

Exploring how biological systems have been 'designed' by evolution to achieve robust behaviours is a subject of increasing research effort, as well as the classification of specific kinds of organism-environment relationships that may correlate to different degrees of robustness (Levy and Siegel 2012). At the evolutionary timescale, robustness may consist in the stability of features over evolutionary time. Incidentally, the robustness of features may coincide with the survival of those biological groups (species, genuses, families) that carry those features. An interesting example of a robust feature is the body temperature of mammals. Across all different environments all over the planet, placental mammals have a body temperature of 37 °C. Relying on several comparative studies, Bokma (2015) observes that 37 °C is the temperature to which most processes are adapted inside all these different species of placental mammals. In environments where 37 °C is a quite extravagant body temperature, mammals develop other compensative characters, for example, a change in color or thickness of the fur. On one hand, the evolutionary robustness of body temperature is believed to be due to the high interconnectedness of this characteristic with many aspects of the organism. On the other hand, the robustness of this character is made possible by change in other characters that "cause less internal disruption than a change in body temperature would" (Bokma 2015, p. 103; see also Jones 2012).

At first sight, robustness and evolvability entertain an antagonistic relationship. Krakauer (2005) even recognizes two largely independent research traditions focused on these two aspects of evolution. Several modeling studies based on networks, however, lead to demonstrations that robustness will promote evolvability in systems characterized by phenotype-genotype distinction (Wagner 2005b, 2008; Félix and Wagner 2008; Rutherford 2000; Bloom et al. 2006). These studies counter the intuition that the more robust a system is, the less phenotypic variation a given number of mutations generate, and hence the less evolvable the system is. They consider mutational robustness (i.e., the robustness of phenotype with respect to genetic mutations) and evolvability (i.e., the ability to produce heritable variation), and conclude that while genotype (sequence) robustness opposes evolvability, phenotype robustness promotes evolvability: "a highly robust RNA genotype has low evolvability. In contrast, a highly robust phenotype has high evolvability" (Wagner 2008, p. 98). Mutationally robust organisms harbour cryptic genetic variation which can become visible in certain environments or genetic backgrounds. There are typically many alternative genotypes that can produce a considered phenotype; but such genotypes are often connected through series of single mutations. Robustness is related to this connectedness, because for a typical genotype some mutations leave the phenotype unchanged, as well as to evolvability, because, if a phenotype is underlain by many different genotypes, new phenotypes might more easily be produced by single mutations. The genotype-phenotype distinction that allows a positive relationship between robustness and evolvability may be taken as a specific case of hierarchical organization with a certain degree of 'disconnect' between the levels (Ereshefsky 2012). In the same vein, the assumption that modularity operates an efficient balance between robustness and evolutionary change is widespread in evolutionary biology (e.g., Hartwell et al. 1999). Functional modularity, by which particular functions are embedded in discrete modules, allows core functions to be robust to change. Such modules are highly conserved in evolution. The evolution of modular systems will thus consist in an alteration of *connections* between different modules, bringing about new properties and higher-level functions in offspring systems. Modularity is thus understood as related to both robustness and evolvability (Thieffry and Romero 1999; Force et al. 2005; Caetano-Anollés et al. 2010).

1.2 The Book

1.2.1 Robustness and Scientific Practice

Robustness is a fundamental notion about how science works. Science is largely based on 'robust methods' for detection of features and phenomena. This is a common underlying theme of the volume, which is nonetheless more directly addressed in the first three chapters, those by **Caianiello**, **Buzzoni**, and **Boone**. Tellingly, however, one of the earliest usages of robustness concerns the robustness of mathematical models with respect to changes in their own assumptions (Weisberg 2006). On *epistemic* robustness, concerned with the conditions ensuring the robustness of knowledge, much work has been carried out in philosophy of science (Soler et al. 2012; Stegenga 2009; Wimsatt 1980, 1981; Morohashi et al. 2002). Epistemic robustness is intertwined with *ontic* robustness, i.e. the robustness of systems and behaviors in the world (see **Silvia Caianiello**, *this volume*). **Marco Buzzoni** discusses two meanings of robustness: robustness as stability against variations in parameters, and robustness as consilience of results from different sources of evidence. He shows their being essentially interconnected to the notion of intersubjective reproducibility. Scientific experiment is, in this light, deeply related to robustness. What any mechanistic approach does, Buzzoni points out, is to explain events as products of robust and regular systems and processes. Mechanisms can be robust, but they are context-dependent both by their identification and by their operation. Buzzoni argues that the concept of robustness of a mechanism, if applied to biological systems, is one-sided and incomplete without a heuristic-methodical reference to final causes. In biology, teleological points of view are widely employed as a counterfactual artifice, "capable of bringing to light causal relations which have a robustly reproducible content". Purposefulness may thus be employed to investigate living beings scientifically, that is, in an intersubjectively testable and reproducible way. Moreover, in biology we have goals and norms that evolve by natural selection: the norms and functions we can allocate to organic parts have been historically determined and consolidated through selective retention of blind variation in the process of survival and reproduction over the generations. Biology assumes that functions, which are a fundamental reference point for biological robustness, owe their existence to history, not to any intelligent design. Teleological and normative dimension both appear as rooted in history.

Robustness might be a key to resolve dilemmas of *multiple realization*, **Trey Boone** argues. How can we conciliate constancy of function with significant change in the parts and mechanisms that realize such function? Focusing on the causal analysis of processes by which regularities arise, Boone proposes better tools to address the multiple realization issue, and in general to integrate mechanistic and functional accounts. To do this, he exploits examples from the neurosciences, but it is biology at large that raises the challenge to integrate robustness into scientific explanation. Some dimensions of robustness seem to be at odds with each other, such as the physico-chemical stability of specific features and the persistence of an organism through constant change. As Boone illustrates in his chapter, *functions* can be robust. Some authors even define functions *as* robust properties of a complex system, as very rarely can they be attributed to an individual molecule (one possible case is the function of haemoglobin to transport gas molecules in the bloodstream). Some biologists emphasize that discrete functions arise from interactions among different kinds of components (e.g., proteins, DNA, RNA and other molecules) and cannot be predicted by studying the properties of isolated components (Kitano 2004; Hartwell et al. 1999).

1.2.2 The Robustness of Living Beings

Right to the point of biological robustness are Leonardo Bich's and Philippe Huneman's essays. **Leonardo Bich** in Chap. 5 points out the crucial biological nexus between *robustness, autonomy and individuality*. Living systems employ several mechanisms and behaviors to achieve robustness and maintain themselves under changing internal and external conditions. Regulatory mechanisms, while enhancing robustness, play a fundamental role in the realization of an autonomous biological organization. Uncoupling of constitutive and second-order control subsystems is a fundamental feature of living beings, in so far as it endows them with additional degrees of freedom for effectively coping with new environmental conditions and internal variation. Specifically, control hierarchies are crucial for the remarkable functional *integration* of biological systems, insofar as they coordinate and modulate the activity of distinct functional subsystems.

Philippe Huneman analyzes several acceptations of robustness according to the specific question that is asked in scientific practice, or, more appropriately according to multiple equally legitimate and interrelated *explanatory projects*. As several other concepts in evolutionary biology, robustness has been questioned from the viewpoint of its *consequences* upon evolution as well as from the side of its *causes* (Ernst Mayr's "ultimate" and "proximate" viewpoints, Mayr 1993). Robustness is thus the *explanandum* for some enquiries in evolution and ecology, and it is the *explanans* for some interesting evolutionary phenomena such as evolvability. Robustness is generally considered as a possible target of natural selection (Kitano 2004; Wilke 2001; de Visser et al. 2003; Hammerstein et al. 2006; Hunter 2009; Klopčič et al. 2009; Delattre and Félix 2009; Muir et al. 2014).[5] At the same time, robustness may be among the features that enable evolutionary change (Kitano 2004), an example of the characteristic coalescence between *explanans* and *explanandum* that Huneman labels "explanatory reversibility". Thus, with respect to evolution, robustness may be seen as an aspect to be explained by (*explanandum*) or as a feature that explains (*explanans*) evolutionary change and/or the particular evolutionary trajectories that are discovered in the history of life. Robustness as an *explanandum* connects with other evolutionary *explananda* such as complexity, modularity or evolvability. "Topological explanations", a class of non-mechanistic

---

[5] To enlarge the context from natural selection only, we should mention that Kitano and Oda (2006) argue that enhancement of robustness in evolution may happen through symbiosis. They refer to major biological innovations such as horizontal gene transfer, serial endosymbiosis, oocytes-mediated vertical infection, and host-symbiont mutualism for bacterial flora. For Kitano and Oda, symbiosis contributes to robustness – in the evolutionary sense – because symbiotic foreign biological entities can enhance the adaptive capacity of a system against environmental perturbations as well as contribute novel functions. The degree of symbiosis achieved can vary from tight integration into the genome (much more frequent in the ancient eras of life) to loose integration as in bacterial flora (a more recent strategy). Again, robustness and plasticity are complementary interpretative lenses of biological evolution. Loose symbioses are highly adaptive, the most dramatic example being immune systems (Feinerman et al. 2008) and bacterial flora in which substantial functions of host defense depends on the proper maintenance of symbionts and their adaptive capability.

explanations based on the formal properties of the system, are for Huneman particularly relevant for addressing robustness: they abstract away from the nature of causal relations in the system of study to highlight the "invariance through continuous transformation" exhibited at the network-level. Ultimately, Huneman's analysis corroborates the fecundity of a *pluralistic* approach to biological causality, where the causal autonomy of topological explanations does not conflict with mechanistic explanation, which can be easily integrated at higher levels of resolution (see also **Giuliani**, Chap. 7).

1.2.3 Systems Biology and Robustness

As **Silvia Caianiello** reconstructs, the first instantiations of the robustness concept in biology are to be found in the 1990s (Endy 2005). Earlier on, the modern robustness notion had emerged in statistics, then it was crucially involved in dynamical systems theory, as well as in control theory in engineering (Kacser and Burns 1973; Sastry and Bodson 1989; cf. Rollins 1999; Safonov 2012), where it contributed to the shift from "modern" to "robust" control theories (Bhattacharyya et al. 1995). Only with systems biology, however, biological robustness came to be identified with the organized complexity common to living and engineering systems, and, unlike unorganized complexity, crucially related to function (Boogerd et al. 2007; Bertolaso et al. 2011; Giuliani et al. 2014). In turn, the global analyses of systems performed by systems biology were urged by the development of high throughput methodologies, which allow high resolution representation at the system scale.

In Chap. 6, **Christian Cherubini, Simonetta Filippi and Alessandro Loppini** consider robustness and stability in biological systems from a biophysical point of view, taking advantage of advanced mathematical and computational tools together with specific physical experimental techniques previously developed in other contexts. In fact, especially in this field, robustness turns out as a territory of interdisciplinary encounter and integration. Cherubini et al.'s approach shows that biological robustness is a very delicate outcome of a dynamical activity of specifically arranged cellular architectures. Stability is crucially relevant to many natural systems, but the relationship between stability and robustness is different if it is framed into a classical deterministic scheme, with respect to nondeterministic systems, where the presence of noise dramatically affects the underlying dynamics. Coordinated behaviours are determined by stochastic noisy dynamical parameters in high-dimensional spaces. The emergence of coordinated patterns in nonlinear dynamics is paradigmatic and can also be interpreted by using tools typical of Quantum Field Theory, i.e. the algebraic approach of coherent state formalism. On one hand, noise is a fundamental ingredient for attaining a robust functional coordination of the network; on the other hand, the change of biochemical parameters in the system dramatically affects the electrochemical patterns. "Nature has found a particular glucose range for which coordinated and robust bioelectrical insulin-producing oscillations of the entire compact cluster are possible". Coherent molecular domains appear in consequence of proper functional conditions in parallel with "phase transition" phenomena.

Phase transitions are also important for **Alessandro Giuliani**, who focuses on robustness in metabolic networks. The network approach allows the scientist to make sense of the intricacies of biological regulation and, for the very mathematical nature of graphs, to obtain a multilevel description linking single node and whole network topological features. This paradigm allows for the detection of a clear "signature of robustness", i.e., the ability of a system to keep different scales of response to environmental stimuli separated. For Giuliani, the biological way to robustness in an ever-changing environment is the presence of a network in which elements self-organize, by the only effect of their location in the network, in differentiated roles, so as to ensure both high sensitivity to environmental stimuli and the maintainance of an invariant structure. The metabolism of an organism (i.e., the set of life-sustaining chemical transformations that take place in its cells) may be represented as a network whose edges stand for the biochemical interactions among the kinds of molecules that are present in the organism. An example of a very robust metabolic network has been found in *Salmonella*. Its metabolic network is very robust thanks to a combination of redundancy and to the richness of host environments that render *Salmonella* partially independent from many biosynthetic and catabolic capabilities (Becker et al. 2006). One way to design antibiotics is to target metabolic enzymes, which have a central role in microbial physiology, and are highly conserved and easily identified by molecular techniques. But the advantages of targeting metabolic enzymes by antibiotics are largely neutralized by the robustness of the metabolic network of *Salmonella* in its usual environments. Robustness makes the large majority of enzymes non-essential for *Salmonella* virulence. Characterizing the behavior and robustness of enzymatic networks by numerous variables and unknown parameter values is a major challenge in biology, especially when some enzymes have counter-intuitive properties or switch-like behavior between activation and inhibition (Donzé et al. 2011). The robustness of metabolism is very relevant in medical research, where the metabolic robustness of pathogens may limit therapeutical interventions (e.g., Blume et al. 2015). One way of looking at diseases such as cancer and diabetes understands them as manifestations of co-opted robustness, in which mechanisms that normally protect our bodies are effectively taken over to sustain and promote the epidemic states (Kitano 2004; Csete and Doyle 2004).

1.2.4 The Relevance of Engineering Principles

Current research in neuroscience elucidates dynamic mechanisms and design principles that enable robust and reliable function in non-trivial – but understandable – ways. **Timothy O'Leary** examines in depth the applicability of engineering principles, such as activity dependent feedback control, the internal model principle, as well as architectural features such as degeneracy, for accounting for the robustness of the nervous system, whose extraordinary degree of integration "of many systems at many levels that can all be considered robust individually" appears still to resist extant mechanistic models. For O'Leary, biological robustness in the nervous system remains a deep scientific puzzle, but not one that demands radically new concepts. O'Leary commits to the view that nervous systems can be approached as formidably complex networks of nonlinear interacting components that self-organise and continually adapt to enable flexible behaviour (cf. Thieffry and Romero 1999; Freeman 2000; Hintze and Adami 2008; Edelman and Gally 2001; Duc-Hau Le and Kwon 2013).

Understanding and controlling the behavior of dynamic distributed systems, especially biological ones, represents a challenging task. This is partly due to the fact that robustness of living beings is related to distributed control and regulation. Regulation is different from control (Pichersky 2005) and implies a specific dimension of causality (Bertolaso 2016). To a certain extent, distributed robustness and control are antithetical: in a robust system, any localized perturbation should have small effects only; robust properties do not depend on many, not one, components and parameters of the system (Gorban and Radulescu 2007). Cells contain from millions to a few copies of each of thousands of different components, each with very specific interactions. In addition, in biology each of the components is often a microscopic device in itself, able to transduce energy and work far from equilibrium. Nonetheless, the programmatic merging of control engineering and Systems Biology, under the banner of 'reverse engineering' (Csete and Doyle 2002; Alderson and Doyle 2010; Carlson and Doyle 2002; Csete and Doyle 2004), discloses new perspectives. Robustness is a fundamental bridging notion in this interdisciplinary development: "it is in the nature of their robustness and complexity that biology and advanced engineering are most alike" (Csete and Doyle 2002 p. 1664). **Menci and Oliva** in Chap. 9 expose the main concepts related to control and robustness of dynamic systems, both traditional and distributed ones, and show their relevance to biological systems. A typical pattern observed in distributed systems is the emergence of complex behaviors, in spite of the local nature of the interaction among elements in close spatial proximity. In such systems, the elements tend to implement feedback control or regulation strategies, where the outputs of a subsystem are fed as inputs to another subsystem and so on, until the first subsystem is eventually itself influenced. Their behavior can be understood only by considering, at the same time, low and high-level perspectives, i.e., by regarding such systems as a collection of systems, and as a whole emerging entity. In particular, dynamic distributed systems show nontrivial robustness properties, which, from one side, are inherent to each subsystem and, from another, depend on the complex web of interactions. Menci and Oliva analyze the relation between robustness, model and control in dynamic distributed systems as a whole, highlighting similarities and differences with other kinds of complex systems. To this aim, they present a case study related to the chemotaxis of a colony of bacteria *E. coli*. Chemotaxis is the movement of bacteria towards a chemical attractant (Alon et al. 1999). Such movement requires the robust detection and amplification of chemical signals in a noisy environment (Hartwell et al. 1999): a few molecules bind to receptors on the cell surface; such event is interpreted as a signal and amplified. Menci and Oliva show that even in the case of the colony – i.e. a distributed complex system – a revised version of the Internal Model Principle, an essential feature of "robust control" (see **Caianiello**), seems to apply.

1.2.5 Robustness, Time and the Environment

An embodied perspective on the robustness of human neurophysiological performances emerges from musical experience. In Chap. 10, **Flavio Keller and Nicola Di Stefano** portray musical language – a species-specific human cultural trait – as a peculiar complex system, issued by the seamless interaction between the realm of machines (the musical instrument) and the realm of biology (the player and the listeners). They discuss some of the properties of music experience in terms of different attributes of robustness, focusing in particular on stability. The music "system" in fact also exhibits the property of maintaining its "function" against a wide range of external and internal changes. Such is the human ability of isolating and maintaining stable information within the perceptual flow and despite changes in the external world that reach our perceptions, that robustness in music perception must refer both to the search for regularities *and* to the range of tolerance. In their conclusion, Keller and Di Stefano highlight the dynamic nature of such system, by distinguishing four multiple interaction cycles that typically characterize music experience and that involve both internal properties of the organism and the environment.

Time is an important dimension related to biological robustness. According to some definitions of cancer, for example, affected relationships among coupled biological rhythms and long-range spatial interactions are a crucial factor in the development of tumours in metazoans (Bertolaso 2016). All forms of life contain circadian clocks, i.e., biochemical oscillators with a robust period of about 24 h (Edery 2000; Leibler and Barkai 2000; Gonze et al. 2002). These clocks are crucial in coordinating many rhythms in the biochemistry, physiology, and behavior with the day and night cycle through a series of output pathways tied to distinct phases of the oscillator. Clocks are robust in that they can maintain oscillation phase, frequency and amplitude for many cycles even in the absence of external cues. The remarkably stable circadian oscillations of single cyanobacteria enable a population of growing cells to maintain synchrony for weeks (Teng et al. 2013). In normal conditions, a series of input pathways to the clock allow correction for systematic small deviations away from 24 h (e.g. human clocks average 24.2 h per day, rather than 24 h exactly, in artificial conditions of stable light).

Nonetheless **Giuseppe Vitiello**, relying on physics and in particular on Quantum Field Theory, shows in Chap. 10 how robustness properties of physical systems, such as dynamical and functional robustness, topological robustness, multilevel and semantic robustness may find their common root in *coherence*. The mechanism of the dynamical rearrangement of symmetry in quantum field theory underlies the phenomenon of coherent boson condensation in the vacuum state. Coherent states appear to be related to fractal self-similarity. The dynamical paradigm of coherence, crucial in the symmetry breaking processes that pave the way to order (including morphogenesis), might thus lead to an integrated vision of natural robust phenomena.

The philosophical perspective on robustness embraced by **Alfredo Marcos** at the end of the volume proposes an ontological framework for dealing with such issues. He argues for embedding robustness in Aristotle's view and in the postmodern continental philosophy of the organism, whose distance is bridged by a fundamental commitment to the notion of *difference*. In the paradigmatic case of living organisms, they can be said to arise precisely by difference-producing processes. The ontology of robustness can therefore be framed within an ontology of difference, whereby the distinction between constitutive and comparative difference becomes crucial. To address the problem of the intelligibility of the organism as the unity of constitutive differences, Marcos proposes an analysis of the ontological and epistemological relationships between difference, similarity and identity.

1.3 Emerging Epistemological Perspectives from Within the Life Sciences

We glimpse in this volume that biological robustness is approached by a great variety of scientific practices and disciplines. We appreciate the importance of analogy, experiment, modeling and simulations. Control theory and network theory are combined with traditional molecular analyses and high-throughput techniques, leading to new hybrid scientific practices (MacLeod and Nersessian 2013). While robust biological modules can be reconstituted *in vitro*, '*in silico* reconstitution' – made possible by the analytical tools of systems biology – is gaining an increasingly important role (Bertolaso and MacLeod 2016). Different spatio-temporal scales are reached by robustness, each one with its appropriate observation and research methods: from gene expression profiles (Hartman et al. 2001; MacNeil and Walhout 2011), through proteins and proteomes (Guo et al. 2004; Bloom et al. 2007), whole metabolisms, cells and multicellular organisms with their organs, phenotypes and their development (Savageau et al. 2009), multi-species aggregations and ecosystems (with their abiotic components, Levin and Lubchenco 2008; Ramos-Jilberto et al. 2012), up to whole phylogenies evolving (Hejnol and Lowe 2015).

Along with the plurality of methods that can be chosen and combined to study biological robustness, we see that robustness is always a relative notion, which implies active choice: scientists cannot but make choices to study robustness 'of' something which is relevant (a system to be identified) 'with respect to' something which is relevant, from practical, intellectual, societal, economic, and very often ethical points of view. The universe of possibilities against which robustness is measured is built by scientists who are humans in a social context.

Robustness is a bridging notion, in particular across the sciences of the natural and the artificial (Tempesti et al 1997; Fernandez-Leon 2011). It has become crucial for prediction and control of natural and artificial systems in recent scientific practice, in biomedicine, neurobiology and engineering, as well as for risk management, planning and policy in ecology, healthcare, markets and economy. From biological, neurological and societal systems, arising by the interplay of self-organizing dynamics and environmental pressures, to the current sophisticated engineering that aims at artificially reproducing the adaptability and resilience of living systems in front of perturbations in man-made devices, robustness seems to hold the key for orchestrating stability and change.

Despite attempts at a non-reductive unification among a wide range of scientific disciplines dealing with functional complex systems, it might be the case that the way to go with robustness will be a pluralistic one. Pluralism is the interaction among different epistemic and causal accounts (Mitchell 2003, 2009). In other words, robust dynamics might admit no unique explanatory tool as either necessary or sufficient, while, instead, encouraging a different *"theory of explanation* that captures several different possibilities" (Woodward 2011). In any of these scenarios, robustness seems to be likely to remain a fertile ground for interdisciplinary dialogue (Serrelli 2016).

A crucial aspect in a reflection devoted to biological robustness is how far is it possible to characterize *specifically biological* dimensions or kinds of robustness (Whitacre 2012). Are biological systems robust in a peculiar way? Can robustness serve as a demarcation criterion of what is biological, and of what is alive? After all, in biology a peculiar kind of control is assumed, one that respects and takes advantage of the system's dynamics, and that, on the other side, can dynamically change its very set points. The extant difference between living and artificial robustness clearly has concrete implications for different sciences in which the ambition of steering the behavior of complex systems is at stake: therapeutic approaches, markets' projections, functional performance of autonomous robots.

Another aspect of specificity which appears to be strictly related to the biological instantiation of robustness is multilevel, hierarchical organization. There are several kinds of hierarchy which are conceptually very different: order, inclusion, control, or level hierarchy (Pumain 2006; Pattee 1973). All of them are relevant in the discourse about robustness. Which kind of philosophical account of hierarchy is more adequate for dealing with organismal robustness?

A nested compositional hierarchy is a pattern of relationships among entities based on the principle of increasing inclusiveness, so that entities at one level are composed of parts at lower levels and are themselves nested within more extensive entities (Tëmkin and Eldredge 2015). The hierarchical assembly of 'parts' and 'devices' is often seen as essential for the construction of complex biological systems (Noman et al. 2015). As formerly recalled, biology routinely studies systems in which many and different parts contribute to system functions across DNA, the genetic code, RNA, proteins, moving to genetic networks (Krakauer and Plotkin 2001) and embryonic development, and whole organisms. Unreliable parts with their details (Kitano 2004) enter in the robustness of collective properties. This is a well studied kind of inter-level robustness (Wagner 2005b, a).

From another point of view, however, the hierarchical organization may itself be examined regarding its robustness. This point of view is particularly relevant in biology. Living part-whole organizations emerge through various evolutionary and physiological processes, they are robustly maintained, and robustness at the organismal level can be lost too, as in in the case of cancer (Hagios et al. 1998; Bissel et al. 2003; Bertolaso 2016). Thus, a promising perspective for framing the robustness of organisms in hierarchical terms is a dynamic and relational view (Bertolaso 2016), which emphasizes the synchronic constitutive relationship between parts and the whole (organism), rather than the "mereological" accounts in terms of parts-whole organization. The life history of the biological entity intrinsically depends on a constitutive and continuous orientation of the parts among themselves and depending on contextual signals. As the biology of cancer shows, the stability of constitutive elements depends on the organization: cells change their behaviour depending on their functional integration in the tissue; alteration in cell communication alters gene expression, and the loss of integration of cells within a functional tissue leads to genetic instability and apoptosis. The collapse of levels, as characterized in cancer, results from the loss of the general functional integration of a biological entity. Interesting research questions concern the emergence of this integration either from a web of pairwise connections or through specific modules (Hartwell et al. 1999).

As for hierarchies of control, we have already seen that in biology the existence, autonomy and causal relevance of single 'controllers' (i.e., parts that are high in a chain of command) is controversial. In fact, many hierarchies of control may be found in living entities (Treviño et al. 2012), but they cannot account for the global system's capability to achieve integration, overall homeostasis nor the kind of biological robustness that keeps the system alive.

Robustness has become one of the pivotal concepts for addressing the urge towards prediction and control of natural and artificial systems in bio-medicine, neurobiology and engineering. As an emerging area of expertise, robustness is perhaps more of a 'trading zone' (Galison 1997; Gorman 2010) than a melting pot. Despite many contact points and reciprocal inspirations, biological robustness and robustness of technological artifacts are still separated by a fundamental difference, which calls for a specific ontology and epistemology of living beings as objects of scientific knowledge. Studies of robustness have elucidated how this feature is related to highly structured, nongeneric, self-dissimilar internal configurations of biological systems, finely tuned by their long evolutionary history. A crucial dimension of biological robustness is 'organized heterogeneity' (Bertolaso and Caianiello 2016), engendered and maintained by the 'relational ontology of levels' which characterises living entities (Bertolaso 2016). Robustness is indeed key to understanding their yet unique capacity to orchestrate stability and change, and to inspire design principles for artificial systems. The more we look at contemporary endeavors in this direction, such as the deep interpenetrations between bio-medicine and technology, the more we become aware of the peculiarity of living beings with respect to technology, but also of how technology itself, in its endeavor to mimic life, is contributing to understanding this difference.

Acknowledgements   So many people must be thanked for this volume. Fist of all we need to thank all the authors who have accepted to contribute a chapter for this volume. But our gratitude needs necessarily to extend to all those who have participated with great enthusiasm and generosity to the three Robustness Workshops held by the Bio-Techno-Practice Research Hub, as well as those who have served ad advisors for the same workshops, along with their institutions: Alessandro Giuliani, Alfred Nordmann, Alfredo Marcos, Alison Barth, Alvaro Moreno, Anna Maria Dieli, Arnon Levy, Dino Accoto, Edwin Morley-Fletcher, Emilio Bizzi, Flavio Keller, Gabriele Oliva, Giuseppe Vitiello, Guido Caniglia, Jane Maienschein, Lorenzo Farina, Luca Valera, Luisa Di Paola, Marcella Trombetta, Marco Buzzoni, Mazviita Chirimuta, Miles MacLeod, Nicola Di Stefano, Philippe Huneman, Raffaella Campaner, Sandra D. Mitchell, Simonetta Filippi, Timothy O'Leary, Trey Boone, Viola Schiaffonati. Very special thanks goes to Sandra Mitchell who supported this initiative since the beginning, being active in all phases of the process, from workshop organization to post-workshop elaboration. For institutional and material support, we are grateful to the Institute for Philosophy of Scientific and Technological Practice (FAST) at University Campus Bio-Medico, Rome and to the Centre for Philosphy of Science at the University of Pittsburgh (PA). For sponsoring the workshops, we must thank Fondazione Cattolica Assicurazioni, M3V ONLUS and the Istituto per la Storia del Pensiero Filosofico e Scientifico Moderno (ISPF) of Italian CNR. Finally, we need to thank Philippe Huneman who, as series editor, believed in publishing this volume. We thank also the other series editors and staff at Springer who have done a patient and wonderful job in all phases of the book production.

References

Alderson, D. L., & Doyle, J. C. (2010). Contrasting views of complexity and their implications for network-centric infrastructures. *IEEE Transactions on Systems, Man, and Cybernetics – Part A: Systems and Humans, 40*(4), 839–852.

Allen, T. F. H., & Starr, T. B. (1982). *Hierarchy: Perspectives for ecological complexity*. Chicago: University of Chicago Press.

Alon, U., et al. (1999). Robustness in bacterial chemotaxis. *Nature, 397*(6715), 168–171.

Bateson, P., & Gluckman, P. (2011). *Plasticity, robustness, development and evolution*. New York: Cambridge University Press.

Becker, D., et al. (2006). Robust Salmonella metabolism limits possibilities for new antimicrobials. *Nature, 440*(7082), 303–307.

Bertolaso, M. (Ed.). (2014). The future of scientific practice: "Bio-Techno-Logos," Pickering & Chatto.

Bertolaso, M. (2016). *Philosophy of cancer: A dynamic and relational view*. Dordrecht: Springer.

Bertolaso, M., & Caianiello, S. (2016). Robustness as organized heterogeneity. *Rivista di Filosofia Neo-Scolastica, CVIII*, 293–303.

Bertolaso, M., & MacLeod, M. (Eds.). (2016). In silico modeling: The human factor, *Humana. Mente Journal of Philsophical Studies 30*.

Bertolaso, M., Giuliani, A., & De Gara, L. (2011). Systems biology reveals biology of systems. *Complexity, 16*(6), 10–16.

Bhattacharyya, S. P., Chapellat, H., & Keel, L. H. (1995). *Robust control: The parametric approach*. Upper Saddle River: Prentice-Hall.

Bissell, M. J., Rizki, A., & Mian, I. S. (2003). Tissue architecture: The ultimate regulator of breast epithelial function. *Current Opinion in Cell Biology, 15*(6), 753–762.

Bloom, J. D., et al. (2006). Protein stability promotes evolvability. *Proceedings of the National Academy of Sciences of the United States of America, 103*(15), 5869–5874.

Bloom, J. D., et al. (2007). Evolution favors protein mutational robustness in sufficiently large populations. *BMC Biology, 5*(1), 29.

Blume, M., et al. (2015). A Toxoplasma gondii gluconeogenic enzyme contributes to robust central carbon metabolism and is essential for replication and virulence. *Cell Host & Microbe, 18*(2), 210–220.

Bokma, F. (2015). Evolution as a largely autonomous process. In E. Serrelli & N. Gontier (Eds.), *Macroevolution. Explanation, interpretation and evidence* (pp. 87–112). Cham/Heidelberg/New York/Dordrecht/London: Springer.

Boogerd, F. C., et al. (Eds.). (2007). *Systems biology: Philosophical foundations.* Amsterdam: Elsevier.

Braendle, C., & Flatt, T. (2006). A role for genetic accommodation in evolution? *BioEssays: News and Reviews in Molecular, Cellular and Developmental Biology, 28*(9), 868–873.

Caetano-Anollés, G., Yafremava, L., & Mittenthal, J. E. (2010). Modularity and dissipation in evolution of macromolecular structures, functions, and networks. In *Evolutionary genomics and systems biology* (pp. 431–449). Hoboken: Wiley.

Carlson, J. M., & Doyle, J. (2002). Complexity and robustness. *Proceedings of the National Academy of Sciences of the United States of America, 99*(Suppl 1), 2538–2545.

Csete, M. E., & Doyle, J. (2002). Reverse engineering of biological complexity. *Science, 295*(5560), 1664–1669.

Csete, M., & Doyle, J. (2004). Bow ties, metabolism and disease. *Trends in Biotechnology, 22*(9), 446–450.

de Visser, J. A. G. M., et al. (2003). Perspective: Evolution and detection of genetic robustness. *Evolution; International Journal of Organic Evolution, 57*(9), 1959–1972.

Delattre, M., & Félix, M.-A. (2009). The evolutionary context of robust and redundant cell biological mechanisms. *BioEssays, 31*(5), 537–545.

Devert, A., Bredeche, N., & Schoenauer, M. (2011). Robustness and the halting problem for multicellular artificial ontogeny. *IEEE Transactions on Evolutionary Computation, 15*(3), 387–404.

Donzé, A., et al. (2011). Robustness analysis and behavior discrimination in enzymatic reaction networks J. Parkinson, ed. *PLoS ONE, 6*(9), e24246.

Duc-Hau Le, D. H., & Kwon, Y.-K. (2013). A coherent feedforward loop design principle to sustain robustness of biological networks. *Bioinformatics (Oxford, England), 29*(5), 630–637.

Edelman, G. M., & Gally, J. A. (2001). Degeneracy and complexity in biological systems. *Proceedings of the National Academy of Sciences of the United States of America, 98*(24), 13763–13768.

Edery, I. (2000). Circadian rhythms in a nutshell. *Physiological Genomics, 3*(2), 59–74.

Endy, D. (2005). Foundations for engineering biology. *Nature, 438*(7067), 449–453.

Ereshefsky, M. (2012). Homology thinking. *Biology and Philosophy, 27*(3), 381–400.

Feinerman, O., et al. (2008). Variability and robustness in T cell activation from regulated heterogeneity in protein levels. *Science (New York, N.Y.), 321*(5892), 1081–1084.

Félix, M.-A., & Wagner, A. (2008). Robustness and evolution: Concepts, insights and challenges from a developmental model system. *Heredity, 100*(2), 132–140.

Fernandez-Leon, J. A. (2011). Behavioral robustness and the distributed mechanisms hypothesis: Lessons from bio-inspired and theoretical biology. *Ciencia y Tecnología, 11*(2), 85–107.

Force, A., et al. (2005). The origin of subfunctions and modular gene regulation. *Genetics, 170*(1), 433–446.

Freeman, M. (2000). Feedback control of intercellular signalling in development. *Nature, 408*(6810), 313–319.

Fusco, G., Carrer, R., & Serrelli, E. (2014). The landscape metaphor in development. In A. Minelli & T. Pradeu (Eds.), *Towards a theory of development* (pp. 114–128). Oxford: New York.

Galison, P. (1997). *Image and logic: A material culture of microphysics.* Chicago: University of Chicago Press.

Gibson, G. (2002). Developmental evolution: Getting robust about robustness. *Current Biology, 12*(10), 347–349.

Giuliani, A., Filippi, S., & Bertolaso, M. (2014). Why network approach can promote a new way of thinking in biology. *Frontiers in Genetics, 5*(APR), 1–5.

Gonze, D., Halloy, J., & Goldbeter, A. (2002). Robustness of circadian rhythms with respect to molecular noise. *Proceedings of the National Academy of Sciences of the United States of America, 99*(2), 673–678.

Goodwin, B. C., Kauffman, S., & Murray, J. D. (1993). Is morphogenesis an intrinsically robust process? *Journal of Theoretical Biology, 163*(1), 135–144.

Gorban, A. N., & Radulescu, O. (2007). Dynamical robustness of biological networks with hierarchical distribution of time scales. *IET Systems Biology, 1*(4), 238–246.

Gorman, M. E. (Ed.). (2010). *Trading zones and interactional expertise. Creating new kinds of collaboration*. Cambridge, MA: MIT Press.

Gu, Z., et al. (2003). Role of duplicate genes in genetic robustness against null mutations. *Nature, 421*(6918), 63–66.

Guo, H. H., Choe, J., & Loeb, L. a. (2004). Protein tolerance to random amino acid change. *Proceedings of the National Academy of Sciences of the United States of America, 101*(25), 9205–9210.

Hagios, C., Lochter, A., & Bissell, M. J. (1998). Tissue architecture: The ultimate regulator of epithelial function? *Philosophical Transactions of the Royal Society of London. Series B, Biological Sciences, 353*(1370), 857–870.

Hammerstein, P., et al. (2006). Robustness: A key to evolutionary design. *Biological Theory, 1*(1), 90–93.

Hartman, J. L., Garvik, B., & Hartwell, L. (2001). Principles for the buffering of genetic variation. *Science (New York, N.Y.), 291*(5506), 1001–1004.

Hartwell, L. H., et al. (1999). From molecular to modular cell biology. *Nature, 402*(6761 Suppl), C47–C52.

Hejnol, A., & Lowe, C. J. (2015). Embracing the comparative approach: How robust phylogenies and broader developmental sampling impacts the understanding of nervous system evolution. *Philosophical Transactions of the Royal Society of London. Series B, Biological Sciences, 370*(1684).

Hintze, A., & Adami, C. (2008). Evolution of complex modular biological networks. *PLoS Computational Biology, 4*(2), e23.

Hunter, P. (2009). Robust yet flexible. In biological systems, resistance to change and innovation in the light of it go hand in hand. *EMBO Reports, 10*(9), 949–952.

Jen, E. (2003). Stable or robust? What's the difference? *Complexity, 8*(3), 12–18.

Jin, Y., & Meng, Y. (2011). Morphogenetic robotics: An emerging new field in developmental robotics. *IEEE Transactions on Systems, Man, and Cybernetics, Part C (Applications and Reviews), 41*(2), 145–160.

Jones, C. B. (2012). *Robustness, plasticity, and evolvability in mammals: A thermal niche approach*. New York/Heidelberg/Dordrecht/London: Springer.

Kacser, H., & Burns, J. A. (1973). The control of flux. *Symposia of the Society for Experimental Biology, 27*, 65–104.

Kitano, H. (2004). Biological robustness. *Nature Reviews. Genetics, 5*(11), 826–837.

Kitano, H., & Oda, K. (2006). Self-extending symbiosis: A mechanism for increasing robustness through evolution. *Biological Theory, 1*(1), 61–66.

Klopčič, M., et al. (2009). *Breeding for robustness in cattle*. Wageningen: EAAP publication/ Wageningen Academic Publishers.

Krakauer, D. C. (2005). Robustness in Biological Systems: A provisional taxonomy. In T. S. Deisboeck & J. Kresh (Eds.), *Complex systems science in biomedicine, Santa Fe Institute Working Papers* (pp. 185–207). New York: Plenum Press.

Krakauer, D. C., & Plotkin, J. B. (2001). Redundancy, antiredundancy, and the robustness of genomes. *Proceedings of the National Academy of Sciences of the United States of America, 99*(3), 1405–1409.

Leibler, S., & Barkai, N. (2000). Biological rhythms: Circadian clocks limited by noise. *Nature, 403*(6767), 267–268.

Lesne, A. (2008). Robustness: Confronting lessons from physics and biology. *Biological Reviews, 83*(4), 509–532.

Levin, M. (2012). Morphogenetic fields in embryogenesis, regeneration, and cancer: Non-local control of complex patterning. *Bio Systems, 109*(3), 243–261.

Levin, S. A., & Lubchenco, J. (2008). Resilience, robustness, and marine ecosystem-based management. *Bioscience, 58*(1), 27.

Levy, S. A., & Siegal, M. L. (2012). The robustness continuum. In O. S. Soyer (Ed.), *Evolutionary systems biology* (pp. 431–452). New York: Springer.

MacLeod, M., & Nersessian, N. J. (2013). Coupling simulation and experiment: The bimodal strategy in integrative systems biology. Studies in history and philosophy of biological and biomedical sciences. *Studies in History and Philosophy of Science Part C: Studies in History and Philosophy of Biological and Biomedical Sciences, 44*(4), 572–584.

MacNeil, L. T., & Walhout, A. J. M. (2011). Gene regulatory networks and the role of robustness and stochasticity in the control of gene expression. *Genome Research, 21*, 645–657.

Mayr, E. (1993). Proximate and ultimate causation. *Biology and Philosophy, 8*, 95–98.

Minelli, A., & Pradeu, T. e. (2014). *Towards a theory of development.* Oxford: Oxford University Press.

Mitchell, S. D. (2003). *Biological complexity and integrative pluralism* (Cambridge studies in philosophy and biology, p. 244). Cambridge University Press: Cambridge

Mitchell, S. D. (2009). Unsimple truths. In *Science, complexity, and policy.* Chicago: University Of Chicago Press.

Morohashi, M., et al. (2002). Robustness as a measure of plausibility in models of biochemical networks. *Journal of Theoretical Biology, 216*(1), 19–30.

Muir, W. M., Cheng, H. W., & Croney, C. (2014). Methods to address poultry robustness and welfare issues through breeding and associated ethical considerations. *Frontiers in Genetics, 5*(NOV), 1–11.

Nijhout, H. F. (2002). The nature of robustness in development. *BioEssays: News and Reviews in Molecular, Cellular and Developmental Biology, 24*(6), 553–563.

Noman, N., et al. (2015). Evolving robust gene regulatory networks. *PLoS One, 10*(1), 1–21.

Pattee, H. H. (1973). *Hierarchy theory: The challenge of complex systems.* New York: George Braziller.

Pichersky, E. (2005). Is the concept of regulation overused in molecular and cellular biology? *The Plant Cell, 17*(12), 3217–3218.

Pigliucci, M., Murren, C. J., & Schlichting, C. D. (2006). Phenotypic plasticity and evolution by genetic assimilation. *The Journal of Experimental Biology, 209*(Pt 12), 2362–2367.

Pow, H. (2013). Meet the toughest animal on the planet: The water bear that can survive being frozen or boiled, float around in space and live for 200 years (shame it isn't much to look at). Daily Mail Online, February 18, at http://www.dailymail.co.uk/news/article-2280286/Meet-toughest-animal-planet-The-water-bear-survive-frozen-boiled-float-space-live-200-years.html. Accessed 1 Mar 2017.

Pradeu, T. et al. (2016). Defining "Development." Current Topics in Developmental Biology

Pumain, D. (Ed.). (2006). *Hierarchy in natural and social sciences.* Berlin/Heidelberg: Springer.

Ramos-Jiliberto, R., et al. (2012). Topological plasticity increases robustness of mutualistic networks. *Journal of Animal Ecology, 81*(4), 896–904.

Rollins, L. (1999). Robust control theory. In P. Koopman (Ed.). *Topics in dependable embedded systems.* Carnegie Mellon University Electrical and Computer Engineering Department.

Rutherford, S. L. (2000). From genotype to phenotype: Buffering mechanisms and the storage of genetic information. *BioEssays: News and Reviews in Molecular, Cellular and Developmental Biology, 22*(12), 1095–1105.

Safonov, M. (2012). Origins of robust control: Early history and future speculations. *Annual Reviews in Control, 36*(2), 173–181.

Sastry, S., & Bodson, M. (1989). *Adaptative control. Stability, convergence, and robustness.* Englewood Cliffs: Prentice-Hall.

Savageau, M. A., et al. (2009). Phenotypes and tolerances in the design space of biochemical systems. *Proceedings of the National Academy of Sciences of the United States of America, 106*(16), 6435–6440.

Serrelli, E. (2016). Removing barriers in scientific research: Concepts, synthesis and catalysis. In *Understanding cultural traits* (pp. 403–410). Cham: Springer.

Shahbazi, Z., Kaminski, A., & Evans, L. (2015). Mechanical stress analysis of tree branches. *American Journal of Mechanical Engineering, 3*(2), 32–40.

Siegal, M. L., & Bergman, A. (2002). Waddington's canalization revisited: Developmental stability and evolution. *Proceedings of the National Academy of Sciences of the United States of America, 99*(16), 10528–10532.

Soler, L., et al. (Eds.). (2012). *Characterizing the robustness of science: After the practice turn in philosophy of science*. Dordrecht/Heidelberg/London/New York: Springer.

Stegenga, J. (2009). Robustness, discordance, and relevance. *Philosophy of Science, 76*(5), 650–661.

Tëmkin, I., & Eldredge, N. (2015). Networks and hierarchies: Approaching complexity in evolutionary theory. In E. Serrelli & N. Gontier (Eds.), *Macroevolution. Explanation, interpretation and evidence* (pp. 183–226). Berlin: Springer.

Tempesti, G., Mange, D., & Stauffer, A. (1997). A robust multiplexer-based FPGA inspired by biological systems. *Journal of Systems Architecture, 43*(10), 719–733.

Teng, S.-W., et al. (2013). Robust circadian oscillations in growing cyanobacteria require transcriptional feedback. *Science, 340*(6133), 737–740.

Thieffry, D., & Romero, D. (1999). The modularity of biological regulatory networks. *Biosystems, 50*(1), 49–59.

Treviño, S., et al. (2012). Robust detection of hierarchical communities from escherichia coli gene expression data. *PLoS Computational Biology, 8*(2), e1002391.

van der Krogt, M. M., et al. (2009). Robust passive dynamics of the musculoskeletal system compensate for unexpected surface changes during human hopping. *Journal of Applied Physiology (Bethesda, Md.: 1985), 107*(3), 801–808.

Waddington, C. H. (1940). *Organisers and genes*. Cambridge, MA: Cambridge University Press.

Wagner, A. (2005a). Distributed robustness versus redundancy as causes of mutational robustness. *BioEssays, 27*(2), 176–188.

Wagner, A. (2005b). *Robustness and evolvability in living systems*. Princeton: Princeton University Press.

Wagner, A. (2008). Robustness and evolvability: A paradox resolved. *Proceedings of the Royal Society of Biological Sciences, 275*(1630), 91–100.

Weisberg, M. (2006). Robustness analysis. *Philosophy of Science, 73*, 730–742.

West-Eberhard, M. J. (2005). Developmental plasticity and the origin of species differences. *Proceedings of the National Academy of Sciences of the United States of America, 102*(Suppl (2)), 6543–6549.

Whitacre, J. M. (2012). Biological robustness: Paradigms, mechanisms, systems principles. *Frontiers in Genetics, 3*(MAY), 1–15.

Wilke, C. O. (2001). Selection for fitness versus selection for robustness in RNA secondary structure folding. *Evolution, 55*(12), 2412–2420.

Wimsatt, W. C. (1980). Robustness, reliability and multiple determinism in science: The nature and variety of a powerful family of problem-solving heuristics. In M. Brewer & B. Collins (Eds.), *knowing and validating in the social sciences: A tribute to Donald T. Campbeii*. san francisco: Jossey-Bass.

Wimsatt, W. C. (1981). Robustness, reliability and overdetermination. In M. Brewer & B. Collins (Eds.), *Scientific inquiry and the social sciences* (pp. 124–163). San Francisco: Jossey-Bass.

Woodward, J. (2011). Scientific explanation. In E. N. Zalta (Ed.), *The Stanford encyclopedia of philosophy*.

Marta Bertolaso is Associate Professor of Philosophy of Science at University Campus Bio-Medico of Rome, Italy. After a degree in Biological Sciences and some years in the lab, she developed her academic career in the Philosophy of Life Sciences. Her expertise in philosophy of cancer, scientific practice and philosophy of complex organized systems has allowed her to collaborate with various journals and to publish monographs and papers sometimes in collaboration with established scientists. Supported by Visiting or Scholar Fellowships, she has worked in prestigious international research centres for Philosophy of Science including Pittsburgh (PA, USA), Exeter (UK), Paris (France) and the European Oncology Institute (IEO). She promotes and coordinates an international and interdisciplinary research network, participates in funded national and international research and educational projects. Lecturer for philosophy and epistemology in scientific practice in different courses at Campus Bio-Medico, she has been giving seminars and talks in different Universities and countries.

Emanuele Serrelli is a philosopher of science interested in interdisciplinarity within and across the natural and social sciences. He works with several Italian universities. He was visiting scholar at the University of Utah, and visiting fellow at the Sydney Centre for the Foundations of Science, at the Lisbon Applied Evolutionary Epistemology Lab, and at NESCent National Center for Evolutionary Synthesis, Durham, NC. As a member of the scientific board of CISEPS Center for Interdisciplinary Studies in Economics, Psychology and Social Sciences, University of Milano Bicocca, he works in the "Cultural Evolution" research program. As a philosopher of biology, he is trained in evolutionary theory, where he also studies interdisciplinarity and modeling.

Silvia Caianiello is Senior Researcher at the "Istituto per la Storia del Pensiero Filosofico e Scientifico Moderno" (Institute for the History of Science and Philosophy in Modern Age ISPF) of the Italian National Research Council (CNR) in Naples. She is co-editor of the Book Series "Filosofia e saperi. Crossing boundaries between humanities and life sciences" (CNR Edizioni), Vice-Director of the Italian inter-university Centre on epistemology and history of the life sciences "Res viva" (www.resviva.it), and, since 2017, associated to the Zoological Station "Anton Dohrn" in Naples, where she is member of the "Science and Society" Committee. Her research interests and experiences range from History of European Philosophy to History and Philosophy of Life Sciences. Her methodological focus is on conceptual interactions across different disciplines and epistemic approaches. She conducted extensive research on the correlation between representations of time and epistemologies of history from XVIII to XX Century, and on conceptual exchanges between Human and Life Sciences since the XIX Century. Her current research field is the history and philosophy of evolutionary theory and evo-devo (Evolutionary Developmental Biology), with particular focus on the breakthrough of systemic and hierarchical approaches to biological organization and evolutionary change.

Chapter 2
Prolegomena to a History of Robustness

Silvia Caianiello

Abstract The paper outlines a historical reconstruction of the spread of the concept of robustness across different disciplinary fields, and of the major significant shifts, which comprise the stratigraphy of the semantic expansion of this notion. Starting from the emergence of the modern notion in statistics, which inspired also its actual epistemic instantiations, the paper examines the historical relationship between dynamical systems theory and the notion of robustness, and analyzes the developments that prompted the shift from "modern" to "robust" control theory in engineering. It further deals with the first instantiations of the concept in biology in the 1990s, in order to highlight the turn impressed on the concept by Systems Biology, focusing particularly on its implications as to the relationship between robustness and complexity.

Keywords Robustness · Statistics · Dynamical systems theory · Systems biology · Engineering · Control theory · History of science · History of concepts · Organized complexity · Internal model principle · Historical epistemology

In the present intellectual climate, characterized by a "dialectic of dis/unity" in the sciences, particular relevance have acquired "concepts that have a way of expanding by cannibalizing other concepts" (Callebaut 2010, p. 448). This appears to be the case for robustness, if we look for a parsimonious explanation of this sketchy representation of the recent exponential increase of its frequency in scientific literature (Fig. 2.1), which registers an abrupt leap in the curve slope since the late 1990s, when its growth starts going exponential.

Wagner, in his influential book on biological robustness, deemed as hopeless any attempt to find a unifying thread in the actual extension of the term robustness

S. Caianiello (✉)
Institute for the History of Philosophy and Science in Modern Age (ISPF), National Research Council, Naples, Italy

Zoological Station Anton Dohrn, Naples, Italy
e-mail: caianiello@ispf.cnr.it

© Springer Nature Switzerland AG 2018
M. Bertolaso et al. (eds.), *Biological Robustness*, History, Philosophy and Theory of the Life Sciences 23, https://doi.org/10.1007/978-3-030-01198-7_2

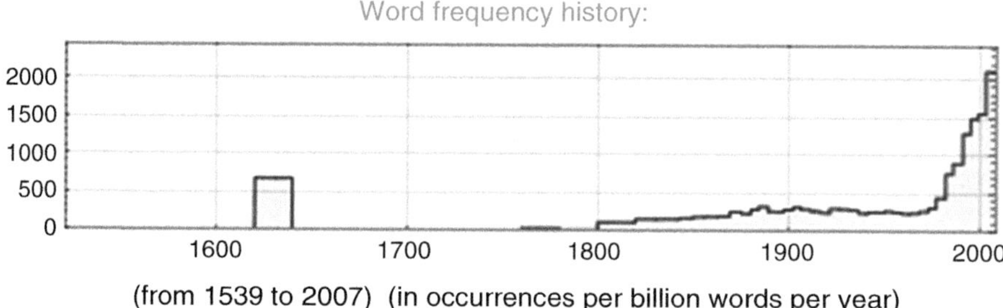

Fig. 2.1 Word frequency history of the term "robustness". (Data from Wolfram-Alpha (https://
www.wolframalpha.com))

(Wagner 2005, p. 297). My approach to this thorny issue will be a historical one,
aimed primarily at identifying the most significant shifts that accompanied the
spreading of the concept and its superposition with formerly related concepts.
However, the complexity of this stratification imposes several restrictions on the
ambitions of such an endeavor. First, it should be regarded as preliminary, as "prolegomena" just laying the ground for more detailed and encompassing research.
Moreover, I will focus mainly on the transitions, that is to the time in which the
conceptual transfer across different disciplines became established, renouncing to
an exhaustive history of its further evolution in the field of its origin as well as to an
accurate record of its disciplinary diversification in the subsequently "infected"
fields.

Even with these inescapable limitations, I consider such a historical approach
worth an attempt. In fact, such cannibalizing concepts (insofar as they devour their
kindred) are not just fashionable words, meant to attract funding or to advertise a
brand new view of phenomena. Irrespective of whether they arise from shifting or
expansion of meanings, or as neologisms, they appear rather to play a programmatic
role in opening channels of communication among disparate fields, as bridge-
building devices. A historical perspective, especially if it succeeds in identifying the
different layers superposed in the polysemy of a notion and relate them to their
original epistemic context, may provide tools for disentangling the tacit commitments involved in its uses, and promote a critical stance as to the effectiveness of the
unification they promise.

The role robustness has recently acquired in Systems Biology reflects fully this
kind of promise. The ability of a system to carry on with its essential functions
despite internal or external perturbations, unpredictable environment and unreliable
components (Cogan 2006) is hailed as the foremost "unifying design principle",
such as to unveil "the conceptual similarity of seemingly unrelated systems" (Alon
2006, p. 237). The stated aim is to develop a completely "formalized" (Kitano 2007,
p. 3), "a single… integrated theory of robustness" (Alderson and Doyle 2010,
p. 843; cf. Krakauer 2005, p. 202), as the system-level property that bridges natural
and artificial systems, and in particular the latest advancements of sophisticated
engineering to biology (Csete and Doyle 2002).

The path to this latest notion of robustness is far from linear, as it involves significant shifts and their respective embedding in different, as well intra- as interdisciplinary, conceptual frameworks. I will briefly retrace some of its main steps, starting with the emergence of the modern notion of robustness in statistics (Sect. 2.1); I will further examine the historical relationship between dynamical systems theory and robustness notions (Sects. 2.2 and 2.2.1); and analyze the developments that prompted the shift from "modern" to "robust" control theory, as the further layer which is necessary, although not sufficient, to bring into the picture for reconstructing the stratigraphy of the semantic expansion of the notion of robustness (Sect. 2.2.2). In order to appreciate the new twist imposed on the notion of robustness in Systems Biology since the early 2000s, I will further compare two early and conflicting instantiations of the notion in the life sciences before 2000 (Sect. 2.3). In the conclusive paragraph, I will highlight the turn impressed on the concept by Systems Biology, focusing particularly on its implications as to the relationship between robustness and complexity.

2.1 Origin of the Modern Meaning

Robustness may be less important to a system that has to deal with idealized abstractions, but it is vital for a system that has to deal with the contingencies of the real world (Cilliers 1998, p.70)

The history of the term robustness is marked by significant shifts. The original, and still commonsense, meaning draws from the Latin "robur" (oak) which conveys the basic sense of solidity and hardness, metaphorically extended to physical strength in general. A scientific acceptation of this original meaning took root in morphology, endowed with a measurable characterization, as "proportion of thickness to length".[1] In its first semantic layer, robustness has just a simple opposite: frailty, gracility.

However, the first instantiation of the *modern* meaning of robustness was a far cry from the original one, and occurred probably in the field of statistics, as a challenge to the extant methods for appraising uncertainty.[2] The new term, denoting the property of a statistical test to be "insensitive to outliers or more generally to possible variations in the underlying form of sampling distributions",[3] was introduced by the English statisticians George E.P. Box in 1953 (Box 1953, Dickson and Baird

---

[1] Such is the case of *Dinornis robustus* (Owen 1883, p. 260)

[2] In the 1940 the japanese engineer Taguchi laid the basis of a theory of robust design in quality engineering, aimed at making "systems of products or production processes less sensitive to sources of uncontrollable noise, or outside influences, through parameter design methods" (Arvidsson 2008, p. 25). However, Taguchi's writings were translated in English only in 1979, at a time when the term "robust" employed in the translation was already gaining ground in control engineering, see infra.

[3] Wilkinson 1979; cf. ASTM 2005, p. 519: "insensitivity of a statistical test to departures from underlying assumptions".

2011, Gigerenzer et al. 1989, p. 90), but its full theoretical development as a subdiscipline of statistics was primarily the work of John W. Tukey and Peter J. Huber in the 1960s (Stigler 2010).

In statistics, robustness represented a move beyond the extant "dogma of normality", that is the assumption of a "continuity and stability principle"[4]: the expectation that "a minor error in the mathematical model should cause only a small error in the final conclusions" (Huber 1981, chap. 1). As it became increasingly evident that classical estimation methods, based on the normal (Gaussian) distribution of data errors, were extremely sensitive to outliers, their unrealistic assumptions engendered a gap between theory and practice. "Theoretical statisticians", for whom "statistical variability was just a nuisance to get rid of (…) tended to shun the subject as inexact and 'dirty'", while practical statisticians devised alternative estimators more resistant to outliers. The shift to robustness gained the task of construing methods endowed with sufficient "insensitivity to small deviations from the assumptions of the mathematical model" (Huber 1972) a sit at the high table of statistics.

> The property of robustness I believe to be even more important *in practice* than that the test should have maximum power and that the statistics employed should be fully efficient. Where necessary I believe that the latter qualities should be sacrificed to ensure the former (Box 1953, p. 333-334, my italics).

Within this definition, robustness had no longer a *simple* opposite; furthermore, the opposition took on a dynamical characterization, in the form of necessary trade-offs: bias minimizing vs optimality, safety vs efficiency and accuracy (Huber 1981).

This first use of the term can be defined as "epistemic", in the specific sense that it did not deal with a property exhibited by complex systems in their evolution, even less with intervening in their dynamics (Woodward 2006, p. 236). Its scope was rather devising mathematical tools for *representing uncertainties* in the process of data interpretation: "the problem of planning for inference when the assumed model could neither be fully trusted nor adequately checked" (Stigler 2010, p. 278). Notwithstanding the conceptual continuity between "epistemic" and "ontic" robustness (cf. Savageau 1971; Cascante et al. 1995; Morohashi et al. 2002), the questions addressed by the epistemic and the ontic notions overlap only partially. Both are committed to ensure that the approximation due to measurement uncertainties captures enough of the modelled system as to bridge the gap between mathematical idealization and reality. But the uses of the model are different; while epistemic robustness is committed to establish the reliability of the inferences that can be derived from the model, ontic robustness is interested in modeling systems dynamics: in predicting their evolution in time, their capability to counteract perturbations, and/or in devising tools for intervening in their evolution to steer them toward a desired behavior.[5] The gap between these two commitments is indirectly corroborated

---

[4]As described by Wilkinson 1979, the "Continuity Principle" states that "small perturbations of the evidence should produce corresponding small changes in the inferences therefrom".

[5]As emphasized by Woodward 2006, "physically intervening to alter some feature of the world is a fundamentally different operation from adopting one or another specification in the process of estimation". However, I prefer "ontic" to his definition of "causal robustness", as this latter relies

by the chronological delay in the appearance of the epistemic and the "ontic" notions of robustness.

The robustness approach spread soon from statistics across several scientific fields (Johnson 1968; Hansen and Sargent 2008). While this is unsurprising, given the intrinsic "interfield" (Maull and Darden 1977) nature of statistical theory, robustness did not acquire yet a fully substantivized, "conceptual" status. The term remained rather "technical" (and adjectival) along the 1970s and 1980s, and surely not yet stabilized in a wider programmatic meaning, the less so a unifying one.

But the reliability of statistical models can be regarded as a "meta-theory of statistical inference for reasoning about scientific theories of the real world in the face of uncertainties engendered by observation of it" (Wilkinson 1979, p. 204). Thus, robustness was already settled to become more than just a statistical notion.

There might be, in fact, a more direct filiation than usually recognized throughout the epistemic trend inaugurated by the new statistical acceptation of the term. Box himself, in 1979, generalized the issue of robustness to a "strategy of scientific model building":

> Robustness may be defined as the property of a procedure which renders the answers it gives insensitive to departures, of a kind which occur in practice, from ideal assumptions (Box 1979, p. 201).

Thus, as Richard Levins in 1966 paved the ground for a theoretical reflection on the extent and limits of the validity of models in mathematical population biology, his celebrated catchphrase "our truth is but the intersection of independent lies" (Levins 1966, p. 423) might have more than a superficial resemblance with Box's equally successful catchphrase "all models are wrong, but some are useful" (Box 1979, p. 202).[6] They shared a concern for robustness, in the new meaning of the term they both endorsed, as the truly realistic epistemic requirement models should conform to. Furthermore, they both promoted the awareness that robustness inevitably involves a trade-off, that Levins similarly described as "between generality, realism and precision" (Levins 1993).

William Wimsatt reappraised this pragmatic approach under the same label of "robustness analysis" in the 1980s. Directly inspired by Levins' proposal of "intersecting lies" of different models for ensuring the robustness of the knowledge acquired, Wimsatt pioneered a new practice-oriented approach in philosophy of science, toward a view of model-building as a "tentative and exploratory activity with known false or oversimplified conceptual tools" (Wimsatt 2007; cf. Soler et al. 2012).

More generally, the onset of the new meaning of robustness in statistics sheds light on the process of establishing new conceptual tools for a changed intellectual

---

on a notion of invariance of causal relationships under perturbations and interventions which is at odds with definitions of robustness related to the system's dynamic properties, such as the system's ability to switch among multiple alternatives and strategic options (Kitano 2004; Jen 2005).

[6] Furthemore, Box himself supported in the 1980s an "ecumenical" approach to statistical modeling – frequentist and bayesian – which consisted in applying «different approaches to the same set of data» (Gigerenzer et al., p. 90; cf. Box 1983).

scenario, in which different, and more "ontologically" committed, instantiations of dogmas of stability across different scientific fields were to be irreversibly shuttered.

2.2 Robustness and Control Theory

The upturn of the slope in the Wolfram-Alpha curve (Fig. 2.1) correlates rather well with the breakthrough of "Robust Control Theory" in the field of systems engineering since the late 1980s (Bryson 1996; Zames 1996; Safonov and Fan 1997; Neculai 2005; Safonov 2012). The chronological gap between the statistical and epistemic acceptance and the spread of robustness in engineering is not the only reason to claim the relative independence of the two notions. Engineering robustness, in fact, entails a parallel narrative, whose roots trace back to dynamical systems theory and its merging with Control Theory after World War II. Thus, a historical approach to the "ontic" acceptance of robustness is due to move from the theories and mathematical representations of stability that Control Theory will incorporate for coping with nonlinear, time-varying and noisy complex systems (Lin 2007).

2.2.1 From Nonlinear Stability Analysis to Robustness

Although rooted in the long history of nonlinear dynamical systems theory, from Poincaré and Lyapunov to Smale, the awareness that "our physical world is …a world of instabilities and fluctuations" (Nicolis and Prigogine 1989, p. IX) took a longer time and the convergence of multiple, techno-scientific as well as sociocultural, factors to break through, and thus to shutter the multifarious instantiations of the "dogma of stability" across several disciplines.[7] This shift of perspective – usually traced back to the chaos "revolution" in the 1970s (Aubin and Dahan Dalmedico 2002) – overturned the traditional relationship between instability and stability, whereby the latter, instead of the former, became the veritable *explanandum* (Schmidt 2008, p. 216).

The mathematics for dealing with instabilities was developed within the framework of Poincaré's qualitative analysis of differential equations, as a tool for representing the evolution of nonlinear dynamical systems. When differential equations

---

[7]Diverse declinations of the assumption that stability and the recovery of equilibrium after disturbances were the natural course of systems evolution can be found in several disciplines; see for instance Nicolis and Prigogine 1989, p. 18: "For a long time chemists thought that a homogeneous, time-independent state similar to equilibrium should eventually emerge from any chemical transformation". In ecology, an analogous "stability principle" (Hagen 1992, p. 106) underlying the evolution of ecosystems was shuttered by R. May in the 1970s. In Bailey's 1994 periodization of social systems theory, the "age of equilibrium", borne out of thermodynamics, extends from 1850 to 1950.

are not analytically solvable, a "qualitative" approach, describing the geometry of the set of trajectories characterizing the system, allows the study of its behavior in time, or phase portrait within the phase space. The topological characterization of stability corresponds to the behavior of returning, after perturbation, "in the vicinity of a given state of equilibrium for an arbitrarily long time" (Andronov et al. 1966, p. xxvii).

The qualitative stability of a system can be described with reference to two major kinds of sensitivity (Shivamoggi 2014, p. 13). The first, dynamic stability, focuses mainly on the stability (insensitivity) of the system to variations (uncertainty or perturbation) in the initial conditions, such as the convergence of dynamics to fixed points or limit cycles. The second, firstly investigated by Andronov and Pontryagin in the 1930s, focuses on the stability to small variations in a parameter or "in the mathematical model itself", that is the functions defining the dynamical system.[8] This "structural" stability may, therefore, take into accounts also perturbations in the system's parameters occurring along the evolution of the system (Lee 1992), and has the potential "to describe not only structurally stable systems, but also degenerate or bifurcation points at which arbitrarily small perturbations can produce qualitatively different behaviors" (Holmes 2005, p. 2705).

Thus, in its wider, double extension, nonlinear stability stands for insensitivity both to *initial conditions* and *parameter variations*. Historically, in fact, robustness is more tightly connected with the notion of "coarseness", introduced by Andronov in 1937 and translated by Lefschetz in 1949 as "structural stability" (cf. Schmidt 2008; Aubin and Dahan Dalmedico 2002; Pugh and Peixoto 2008). "Grubye systemi" ("systemes grossieres", Andronov and Pontryagin 1937) designate in fact a class of nonlinear, dissipative systems whose oscillations are "self-sustained": their period and amplitude are not dependent from initial conditions, but only from the system's property. "Coarseness" describes their property that a small variation in the equations of the system does not alter its qualitative topology, i.e. phase portrait.

Moreover, the characteristic technoscientific framework of Soviet Russia in which Andronov and his Gorki School[9] were embedded, driven by the demands of military research, promoted a tight interdisciplinary interaction between mathematicians, physicists, and engineers, intermingling fundamental and applied research. Unsurprisingly, Andronov turned since the 1930s to control theory and automatic regulation, adapting the qualitative approach to cope with self-oscillatory behaviors and parasitic noise arising in a wide range of technical nonlinear dynamical systems, from the stabilization of oscillations in high-speed train wagons to the new discipline he baptized radiophysics (Dahan Dalmedico, 2004). To this aim,

---

[8] Variations in parameters can be interpreted as well as measure of the degree of ignorance about the behavior of the parts of the system, that is of the uncertainty about the matching between the system and its model.

[9] Aubin and Dahan Dalmedico 2002. According to Dahan Dalmedico 2004's periodization, the Gorki (later Nizhny Novgorod) school proper ended soon after the Gorelik affair in 1950s. See also Bissell 1998.

Andronov and collaborators developed in 1944 the "point transformation method" for representing the states of the systems as points in the phase space. This method, inspired by Poincaré's first return map, but adapted for discrete state nonlinear dynamical systems,[10] will prove crucial in the shift from the "classical" to the Modern Control Theory.[11]

On the other side, the notion of structural stability came laden with inherent metaphysical implications, as the dogma of stability impinged directly on the realism of models (Schmidt 2008, pp. 207–210, 231). Structural stability was in fact long imposed as a prior restriction on "good" models of physical phenomena (Guckenheimer and Holmes 1983, p. 259; Abraham and Marsden 1978, p. IX), a proof that the model was sufficiently reliable as to identify physically verifiable properties (Lee 1992, p. 242). Therefore, encompassing "coarseness" ensured the matching between mathematical idealization and reality. Thus, the ontological tension appears constitutive of the notion of structural stability, as the ridgeline at the same time fostering and challenging the possibility of scientific knowledge.

In the 1970s René Thom and Robert May applied notions of structural stability to biological phenomena (respectively, to morphogenesis and ecosystems), embracing the same basic commitment to a more "realistic" approach to modeling complexity. In the meanwhile, the Andronov-Pontryagin criterion for structural stability had proved consistent only with low dimensional systems (Lee 1992, p. 314; Schmidt 2008, p. 230–231). Smale's attempts at extending the property of structural stability to higher dimensions had in fact failed, resulting in the chaotic dynamics of his horseshoe map (Smale 1966). What came to the fore was that, in complex multidimensional systems, regions of structurally stable behavior in the parametric space could coexist with whole areas which exhibited structural instability (Nikolov et al. 2007), making of the former even more of an *explanandum*.

Structural stability ceased to be a "dogma". As Thom stated in 1973, "whatever is the ultimate nature of reality (…) it is indisputable that our universe is not chaos" but inhabited by "forms and structures endowed with a *degree* of stability", it was with the important caveat that such stability was to be cast into the model as an additional hypothesis (Thom 1975, p. 16, my italics; see also p. 1, 39; Lee 1992, p. 22).

Thom's "dynamical structuralism" (Thom 1974), directly inspired by Waddington's concept of epigenetic landscape, aimed at modeling, with the tools of qualitative dynamics and differential topology, morphogenetic processes independently from the "special properties of the substrate of forms, or the nature of the forces that create them" (Thom 1975, p. 8). In particular, he ascribed self-reproducing systems to a class of systems characterized by a finite set of structurally stable

---

[10] This work was known in the West only after Wold War II, due to Lefschetz translation (Aubin and Dahan Dalmedico 2002).

[11] In fact, the major innovation of the post-World War II phase of Control Theory was the adoption of the time-domain state space representation – fostered by the increased computational power of digital computers – as more suitable to deal with the multivariable problems of real-world nonlinear complex systems, with respect to the frequency domain analysis, characteristic of the "classical" phase of Control Theory (1935–1950) (Bennett 1996; Neculai 2005). See below, § 2.2.

attractors, in which "even bifurcations and catastrophes may occur in a structurally stable way, according to a fixed algebraic model" (Thom 1969, p. 323). Only quite special topological properties, "generic" to a carefully restricted class of – living as well as nonliving – systems could account for the structural stability not only of attractors ("islands of determinisms" corresponding to Waddington's chreods) but also of bifurcations (zones of instabilities).

Thom's antireductionist stance fostered a dynamic interpretation of genetic constraints (Thom 1975, p. 303; Thom 1983), as well as the merging of dynamic and informational approaches (Thom 1975, p. 126ff). However, it was probably the identification of a "generic" class of self-reproducing systems endowed with structural stability that exerted the foremost influence on later "generic" interpretations of the origin of order in biosystems (see Sect. 2.3).

As, one year after the publication of Thom's *Structural stability and morphogenesis*, Robert May speculated on the structural stability of ecosystems, he followed a different pathway, moving from his scientific practice of mathematical modeling of ecosystems. If ecosystems escape the challenge of chaotic behavior, it must be because of the "intricate evolutionary processes" (related to the "long history of coevolution of their constituent plants and animals"), that ought to have pushed them into the highly improbable, "tiny and mathematically atypical regions of parameter space which endow them with long-term stability". To investigate how evolution sought out precisely this tiny region, a more realistic mathematical modeling is needed, such as to carry "built into the fabric of the model" the "incessant sequence of (…) perturbations" of the real world.[12] A similar pathway toward a more realistic modeling of "systems with input" appears to have driven the history of control theory, the next layer that must be brought into the picture in the stratigraphy of the semantic expansion of the notion of robustness.

As already highlighted, it is mostly structural stability – as insensitivity to changes in (specific) parameters (Umulis et al. 2008, p. 70) – that has been increasingly associated to robustness,[13] even if the actual extension of the term often comes to cover both aspects of nonlinear stability.[14] Actually, the insensitivity of some

---

[12] May 1973, p. 110. Interestingly, these considerations lead May to the speculation that structural complexity is intrinsically accompanied by fragility: "complex and stable natural systems are likely to be *fragile,* tending to crumple and simplify when confronted with disturbances beyond their normal experience (that is, tending to instability when carried out of their small and particular stable region of parameter hyperspace)" (cf. Huneman 2010). In May's argument there is no occurrence of robustness as a term. However, robustness in a cognate meaning and context did appear in Roberts and Tregonnin 1980, where a reference to control engineering is implicit in the reference to Dragoslav D. Šiljak, who was at the time already active in the field of Robust Control Theory. Roberts and Tregonnin's "developmental" model "contributed the insight that, although stable complex systems may be rare in the universe of permitted parameter combinations, they can be readily constructed over time by elimination from an initially large but unstable system" (Taylor 1988).

[13] See for instance Nicolis and Prigogine 1989, p. 96, where robustness is synonymous to structural stability.

[14] Cf. von Dassow et al. 2000, where the robustness of the segment polarity network in Drosophila is relative both to parameter values and initial conditions.

property of the system to changes in parameters relevant for a specific function is intuitively related to the characteristic way biosystems cope with environmental changes and with internal sources of noise. Rosen characterized in 1970 the unique adaptive property of living systems as "parametric control", the ability to modify the very transfer function characterizing the system as to change its response to a given input (Rosen 1970, p. 21). At least in biosystems, robustness captures a crucial aspect of autonomy, the fact that "system's parameters are controlled by the system itself" (Nikolov et al. 2007, p. 428; see Chap. 5 in this volume).

However, the quasi-synonymy between robustness and structural stability is not original of dynamical systems theory, and it is more probably retrospectively derived from the breakthrough of the term in the "post-modern" phase of control theory. From a historical perspective, this quasi-synonymy might be considered a phase in the evolution of the term, were such stance not too irenic. In fact, the early "substantivizations" of the term in biology in the 1990s, before the programmatic launch of the notion in Systems Biology, reflect rather the emerging contrast between two different systemic stances, concerning not only the conception of robustness, but also the origin of complexity and order in living systems (cf. below, Sect. 2.3).

2.2.2   *From Classical to Robust Control Theory*

As our technologies become more complex and intertwined, it's clear that they resemble biological systems more than those of physics (Abersman 2016).

"Robust" control broke through in Control and Systems Engineering since 1976, soon to become so intrinsic to the field as to make the attribute unnecessary (Safonov 2012).[15]

In a sense, the new label conveyed but a new approach to the constitutive goal of feedback control. In fact, the closed loop or feedback's corrective action on the input, by means of the information on the system's state provided by the output, is aimed precisely at making the system insensitive both to external disturbances and to variations in its individual elements (Åström and Murray 2008, p. 3). "The fundamental role of feedback control is to remove uncertainty from a system or to move it from one part of the system … to another part", which is considered less critical with respect to the desired behavior (Bennett 2004, p. 103).

However, it was known since the 1840s that feedback or closed loop control "is a double-edged sword: it can render a system insensitive to uncertainty (…), but it can also increase sensitivity to it" (Dullerud and Paganini 2000, p. 15); the tradeoff it involves makes of it both a powerful and a dangerous strategy for creating robustness (Csete and Doyle 2002, p. 1667).

---

[15]The author is indebted to Gabriele Oliva for many helpful discussions concerning this paragraph.

Thus, the history of Control Engineering can be framed in terms of a coevolution of mathematical techniques and technological advancements for coping with the "robustness" problem (Michel 1996) in increasingly realistic models of system behavior (as well as in the increasingly complex "plants" arising in technological evolution, cf. Bennett 1996), up to the challenge of Robust Control Theory of controlling "unknown plants with unknown dynamics subject to unknown disturbances" (Chandraseken 1998).

The first steps of Control Engineering at the time of its official birth in the 1920s (Neculai 2005), in the phase retrospectively labeled Classical Control Theory, were devoted to the development of sensitivity analysis for predicting stability and performance of closed-loop control systems. Until the 1950s, the prevalent approach to feedback stability was framed in the frequency domain paradigm, based on the representation of a signal by frequency components, such as Fourier transforms, rather than time, and originally limited to linear time-invariant systems with single input and single output (SISO). The frequency response is the relationship between the input and of the output signals measured as a function of frequency, and on some form of gain and phase plot, which could be represented by the powerful graphical technique of block diagrams.

The resulting model of the behavior of the system is derived from the experimental measure of the frequency response. It is, therefore, an "external" model, in which any explicit knowledge of the physical structure of the system or of the mechanisms by which inputs are transformed into outputs is black boxed (Bennett 1996).

The application of the frequency domain approach for devising stability criteria for feedback controllers was mostly the work of engineers. At the Bell Laboratories, telecommunication engineers Harry Nyquist, Harold Black and Hendrik Bode laid the basis of negative feedback control and developed the techniques for building negative feedback amplifiers, in order to correct distortion of input signals over long distances. Sensitivity analysis of feedback systems led to the Nyquist stability criterion that, by looking at the properties of the open loop transfer function, enabled the choice of a controller such that "an impressed small disturbance, which itself dies out, results in a response which dies out" (Nyquist 1932; Åström and Murray 2008, p. 269). Thus, the leading assumption was that, in a linear well-behaved system, bounded inputs will result in bounded outputs, and consequently small changes in the former will engender equally small changes in the latter. Bode contributed to an early insight into the controller robustness issue, by developing notions of phase and gain margin as the quantitative measure of the closed loop insensitivity to changes in the system's parameters (and/or to modeling errors), and he was the first to highlight the *tradeoff* between insensitivity to parameter variations and logarithmic gain (Bennett 1993; Bissell 2009; Dorato 1987).

This early approach to feedback control can be considered "hard" (Fig. 2.2), in the sense that it aimed at minimizing the system's response to expected perturbations by reducing its degrees of freedom (Bennett 2004, p. 122).

However, the frequency domain paradigm, which dominated the design of linear servomechanisms along the 1940s, proved increasingly inadequate for dealing with nonlinear and nondeterministic, noisy systems, such as those relevant for aerospace

Fig. 2.2 From Zhou, Doyle and Glover 1995

and ballistic control in World War II. Even the early attempts at coping with nondeterministic inputs by means of statistical methods by Wiener in 1942 did not at the time fare better than Bode's geometrical fire-control predictor (Galison 1994).

Since the 1950s, the time at which Thomas Hughes has located the post-modern turn in engineering,[16] the focus of Control Theory shifted "beyond feedback amplifiers and single-loop servomechanisms (…) to large-scale, complex systems" (Bennett 1993). The problems related to the design of controllers for multivariable, nonlinear or time-varying complex systems demanded a more sophisticated representation of the system's dynamics. In the period from 1959 to 1975, the urgency to tackle "nonlinear stability" issues (Zames 1996) led to the adoption of the statespace approach, that Andronov had already adapted to serve Control Theory.

If the frequency domain paradigm was dominated by engineers, directly involved in the practice of building real controller devices, in the post-World War II phase of Control Theory – the so-called Modern or "Optimal" Control Theory – both science and technology shifted the discourse to a higher level of abstraction. The increased

---

[16] Hughes 1993, 1998. Implicit in this shift is a notion of control not "merely in the narrow sense of the control of mechanisms but in the wider sense of the control of any dynamic system (…), in general stochastic and imperfectly observed" (Whittle 1996). See Vajk et al. 2009, who name "postmodern control theory" the one that "places special emphasis on modeling of uncertainties" (p. 187); cf. also Doyle et al. 1991, p. 1232.

computational power of digital computers started making tractable more complex, time-varying and multivariate MIMO (multiple-input and multiple output) systems (Bryson 1996); and the pioneering work of R.E. Bellman, L. Pontryagin and particularly of Rudolph E. Kalman ensured the new mathematical framework for the time domain approach in Control Theory.

Modern Control Theory laid new emphasis on the search for optimal control, consistently with the optimality principle as formulated by Bellman in the early 1950s and later developed in his Dynamic Programming algorithm. Optimality was a stringent requirement for effectively managing the increased complexity of industrial and military plants. In Control Theory terms, it deals with the search for a feedback law capable of steering the system into the trajectory that better minimizes a cost function for the intended performance.

The strategy outlined by Bellman moved from a control law assumed as optimal for a given initial state, followed by the computation at each next stage of the optimal value for the control to remain effective for that state. In this framework, control actions are chosen based on local information, without looking ahead to future scenarios, nor back to the previous history of the system (Ogata 2010). But, although it exploited – as first, see Bennett 2004, p. 120 – the digital computer, the Dynamic Programming algorithm remained easily entrapped into the curse of dimensionality: systems with more than 3 variables exceeded the storage capacity of those early machines (Bryson 1996).

Kalman enriched the state-space approach with the pivotal concepts of controllability and observability of a dynamical system. Controllability is the condition that an input exists which can transfer the system from any initial state to any other state in a finite time interval. Observability is the ability to determine the internal state of a system from the observation of the output over a finite time interval. Only if both these conditions are given it is possible to design a controller. Among several possibilities, the Kalman Filter proved its effectiveness in noisy systems.

Within such a technique, starting from a reasonably accurate knowledge of the initial state of the system, predictions about its trajectory are recursively updated by consecutive measurements. At each next state, the merging of the probability distribution stemming from the two distinct sources of knowledge about the system's behavior (a model of the system, used for predicting the next state, and the measurement, which is used to correct the current estimate via real output measurements) allows such an observer to reconstruct the state of the system in a probabilistic sense (i.e., in terms of expected value and covariance).

Although the Kalman filter is proved optimal only for linear systems and assuming Gaussian noises, similar techniques could be devised also for nonlinear systems, although with decreasing reliability (e.g., Extended Kalman Filter, Particle Filter, etc.). Nonetheless, the basic philosophy of optimal control relied on the assumption that the knowledge the observer gathers provides a sufficient statistic for the controller to circumscribe the uncertainty of the system.

However, as the term "robust" reappeared in 1976 in control engineering, it conveyed again the need to fill the gap between mathematical modeling and real-world challenges, which had been made dramatically evident by the catastrophic failures in the performance of the F8c aircraft and of the military submarine Trident in the early 1970s (Safonov 2012). The new paradigmatic shift which took place in Modern Control Theory in response to these failures was from "optimality to robustness", that is from the reduction of feedback design to a mere problem of mathematical optimization, to the concern for optimizing its performance "in face of uncertainty" (Doyle et al. 1982). Furthermore, it emphasized that optimality for a given function should not be simply modeled "one loop at the time", but that the nonlinearity arising from simultaneous loop gain variations and interconnected MIMO systems had to be encompassed in the establishment of stability margins (Doyle 1979).

The expression "robust control" was introduced by Davison and Goldenberg (1975), for the control law that would "allow parameter variations". The general concern which inspired Robust Control Theory was to encompass uncertainty within the process of synthetizing a "controller structure" which would ensure "structural stability" (Francis and Wonham 1976), that is closed loop stability in the face of variations in the system parameters. In fact, Optimal control theory was based on the assumption that uncertainty can be effectively measured. The knowledge about the system gathered by iteratively observing state variables and feeding them back into the control law would be sufficient to select the best controller design and thus build the controller on the estimate of this bounded uncertainty. No room was left for stability margins.

Robust control theory challenged this assumption and proposed an alternative strategy, in which the concern for optimality had to be complemented with an estimate of the modeling error and of its effect on the performance of the controller. This was accomplished by means of the H-infinity technique introduced by Zames in 1981 – originally within the frequency domain framework. Instead of moving from a bounded uncertainty assumption, it encompasses the maximum range of uncertainty tolerance by optimizing stability margins to the worst possible performance for many classes of input signals. In this worst-case optimization strategy, "some performance may be sacrificed in order to guarantee that the system meets certain requirements" (Rollins 1999). Only a negotiation among conflicting needs could ensure "proper control (…) even if the model deviates from the real system" (Kitano 2007). Nonetheless, what characterizes the new, "postmodern" epistemic stance (see Fig. 2.2) is the awareness that controller's robustness is effective only if the prediction of the uncertainty range proves correct.

The adaptation of extant mathematical optimization methods to the new scope of Robust Control Theory was long and complex. The recasting of the H-infinity technique into the state space approach, more suitable to MIMO systems, while keeping it sufficiently simple and tractable was achieved only in 1989 (Doyle et al. 1989; Zhou et al. 1995).

However, already in 1976, a crucial design principle for robust control was firstly formulated by Francis and Wonham: "a structurally stable synthesis must incorporate in the feedback path a suitably reduplicated model of the dynamic structure of the disturbance and reference signals". The "internal model principle" (originally formulated for "linear and weakly nonlinear systems") states that the control must contain a dynamical model of the process being controlled, encompassing the disturbances the system is to reject and the reference signals it is to track. The "internal model" strategy, brought to the fore by Modern Control Theory, was thus complemented with a representation not only of the states of the system but also of the possible perturbations the system was liable to encounter.

The necessary condition for a system to remain insensitive to a given class of inputs is that it embeds a subsystem that can generate that class (Umulis et al. 2008; see Chap. 9 in this volume). In other terms, the internal model principle operationalizes Ashby's notion of requisite variety (Ashby 1958; cf. Whitacre 2012), and thus entails that internal models must contain (at least) enough variety as to cope with the variations in the external environment. The novelty with respect to former feedback design is in fact that the model "is generally richer in structure than the exogenous model adopted apriori" (Francis et al. 1974). It is this embedded redundancy that provides robust systems with an anticipatory capacity in front of upcoming perturbations (Bryson 1996), and that limits the parsimony prescription of the optimality approach.

Such new, "sophisticated" engineering (Csete and Doyle 2002) method endorses a softer notion of control. Instead of forcing the system directly to the desired state, it exploits fluctuations and instabilities to drive it to that state without applying intense forces; in this sense, control needs fluctuations (Neculai 2005). The other side of the coin is, however, the "spiraling complexity" of the control systems, which are driven to "almost biological levels of complexity" (Csete and Doyle 2002). The intrinsic complexity of Robust Control narrows the gap between natural and artificial systems, coming closer to the "baroque" richness of biological network design, characterized by "multiple feedback signals, non-linear component dynamics, numerous uncertain parameters, stochastic noise, parasitic dynamics", and other forms of uncertainty (Csete and Doyle 2002).

Unsurprisingly, the way to robustness by means of an internal model principle underlays the very first example of cross-fertilization between Robust Control Theory and biological modeling highlighted in the literature under the explicit heading of robustness, the subject of the next paragraph.

2.3 Early Inceptions of Robustness in Biology: Organizing Vs Design Principles

Robustness, in a substantivized and therefore already "conceptual" form, appears in biology since the early 1990s. Without claiming a complete coverage of its occurrence, I will briefly turn to two of the earliest appearances of the term in biology, dated respectively 1993 and 1997. The two notions are largely independent of one another, albeit they share several features and goals. In both cases, robustness relates to parameter insensitivity, and is meant to challenge notions of close genetic control fine-tuning parameter values throughout the biological process under study. Both rely heavily on computational methods and simulations,[17] and both are representative of systemic approaches. However, the very way they conceive the task of a systemic approach in biology is different.[18] The difference they exhibit matches quite well the historical tension between *organizing* principles vs *design* principles (Green and Wolkenhauer 2013).

The first paper, "Is morphogenesis an intrinsically robust process?", (Goodwin et al. 1993), appeared on the wake of Goodwin's structuralist manifesto (Goodwin 1990), focused on the search for the *organizing* principles by which biological morphologies become established as attractors in the space of morphogenetic field solutions. Robustness in the new sense, although inspired by Thom, takes a step further with respect to his modeling approach to structural stability, claiming that "to determine the properties that make systems stable to a range of disturbances (…) provides no insight into how they come into existence in the first place" (Goodwin 1990, p. 298).

In the 1993 paper, the claim that morphogenesis is "intrinsically" robust raises a challenge to the notion of a "genetic program", as the implausible hypothesis that natural selection could exert such a close control on parameter values as to drive development throughout the whole set of "remarkably coherent transformations that generate the basic body plans of different phyla". The endorsed alternative hypothesis is that the stability and repeatability of developmental outcomes arise from the nonlinear dynamic coupling of different "mechanisms", understood here as "self-organizing" mechanisms (Lander 2011), such as Turing pattern and cell sorting. Robustness is the dynamical result of the fact that the nonlinear coupling of many such mechanisms on one side enlarges the state and parameter space for possible forms, while at the same time restricting the set of reachable morphologies. The emergent result is that "the *non-linearity* and the *coupling* … enhance the strength of the basin of attraction of specific patterns", so that "robust patterns correspond to large domains in parameter space" (Goodwin et al., 1993, p. 141–142).

It is beyond the scope of this paper to provide an updated evaluation of this early attempt at a "free lunch view" (Goodwin 1990) of morphogenesis, which has been labeled as "innocent of genes" (Gilbert et al. 1996), as it ignored the contemporary evidence of boundary-driven organization of morphogenetic fields (De Robertis et al. 1991); or reappraised for encompassing the many non-genetic mechanisms occurring at different levels of the developmental process.[19] However, a remarkable consequence of this kind of dynamical system approach to robustness, increasingly associated to the "physicist" point of view in the literature (Keller 2002; Carlson

---

[17] Albeit at very different scale: computational modeling of morphodynamics in the former and of biochemical networks in the latter.

[18] One might speak of an increasing divergence since different "systemic stances" appear to have coexisted and cross-talked under the banner theoretical biology along the line traced by the Waddington's Serbelloni symposia up to the late 1980s; see for instance the collected work edited by Goodwin and Saunders 1989.

[19] Minelli 2003, p. 9ff; for further developments of this approach to robustness see Aldana et al. 2007 and Igamberdiev et al. 2012.

and Doyle 2002), regards the relation between complexity and robustness. In the specific class of complex systems characterized by "many coupled elements", operating at the edge of chaos, such as developmental systems or biological networks (Kauffman 1990, 1993), robustness is hailed as a statistically *generic* property, assumed to "flow from complexity itself" (i.e. the very existence of many coupled elements; Goodwin et al. 1993; Kauffman 1993, p. 637). This is the opposite of the view actually held by prominent systems biologists. As we will see (Sect. 2.4), they move from a functional and utterly nongeneric definition of robustness, to conclude that the specificity of biological and engineered complexity is that it is "driven by robustness".

As Barkai and Leibler employed in 1997 the term robustness for describing the relative insensitivity of "key properties of biochemical networks (…) to the precise values of biochemical parameters", Hartwell, in the same journal, did not hesitate to salute the birth of "a new concept".[20]

Hartwell's comment was nonetheless accurate. The theoretical and technical background for this notion of robustness lays elsewhere, in the systemic approach to metabolic regulation, as developed since the early 1970s by Sagaveau's Biochemical Systems Analysis and Metabolic Control Analysis (Fell 1992; Papin et al. 2003; Voit 2013), acknowledged sources of actual Systems Biology (Westerhoff and Palsson 2004; Cornish-Bowden et al. 2004). The focus in this tradition is the functional organization of biological networks, by means of a notion of function which could be seen as a pragmatic synthesis of the actual opposition between evolutionary ("etiological") and "systemic" ("dispositional") approaches (cf. Bertolaso 2017). "Network function, design and evolution" can be addressed within a unitary framework by looking at the *design principles* that are "deeply rooted in the system structure", and that ultimately determine "which type of dynamic behavior the system will exhibit" (Savageau 1991). In fact, such *design principles*, shaped by the "pressure which selection exerts on nearly every aspect of the structure, function, and performance of biological individuals" (Rosen 1967) also account for the present degree of optimality of the system's functioning. This notion of design principles traces back to Rosen's Principle of Optimal Design (Rosen 1967, p. 7[21]), which posited "optimality as the basis of selection" (Savageau 1974). Interestingly, Rosen formulated this principle in the context of his endeavor to update the engineering approach to biological systems to the program of Modern Control Theory, which emphasized optimality issues and state space representation (Rosen 1967, chap. 10). However, it must be reminded that this notion of design is distinctly cybernetic,

---

[20] Barkai and Leibler 1997; Hartwell 1997. About 1999 also the term "mutational robustness" apparently started to spread (Nimwegen et al. 1999; Wagner and Stadler 1999) pursuing the same theoretical pathway and computational approach.

[21] See Rosen 1967, p.7: "the fundamental hypothesis that biological structures, which are optimal in the context of natural selection, are also optimal in the sense that they minimize some cost functional derived from the engineering characteristics of the situation". Some reservations about this conflation of different meanings of optimization are expressed by Kacser and Burns 1979.

very different from the idea of a plan "imposed upon matter", but rather evolutionary and intrinsically dynamic (Pickering 2010, p. 32).

This cybernetic legacy is substantial in corroborating, from a historical perspective, a fundamental disjunction between design principles and reductionism (cf. Green 2015). In fact, the application of engineering methods to molecular biology, pioneered by Savageau (Yi et al. 2000), was meant to complement the powerful reductionist approach of mainstream molecular biology with an integrative, "reconstructionist" program, capable of facing the issues of complexity and dynamical organization which had been emerging in the quantitative analysis of biochemical networks.

Actually, the computational and experimental analysis of sensitivity coefficients, which allowed measuring the importance of a single enzyme on the control of flux, unveiled what Kacser and Burns labeled "molecular democracy". "Control is shared among all enzymes": the role of the single enzyme was shown to depend not on its nature or quantity but on "its relation to a specific set of fellow enzymes", so that it could be altered by a change in another point of the pathway (Kacser and Burns 1979). At the same time, the discovery that the "measurement of 'local' properties of the enzymes (…) yields 'global properties' such as the system sensitivities of fluxes and metabolites" made it feasible to relate "systemic behavior to the properties of the underlying molecular determinants" through a quantitative method (Savageau 1991; Westerhoff and Palsson 2004).

In this framework, therefore, as it will become even more apparent in the postgenomic era, different notions of causality and control come to the fore than those entailed in the metaphor of a software "program".[22] Assuming that natural selection does not target the properties of single components – such as the kinetic property of an enzyme – but rather control mechanisms, shifts the focus from the part to the dynamic organization of the whole. Control mechanisms can no longer be thought of as isolated controllers (cf. Whitacre 2012), but rather as implementations of evolved design principles, whose understanding can only "emerge at the system level" (Savageau 1991). The complexity of this "systemic" shift is reverberated in the actual debate on the kind of selection at work in the explanations of robustness (Bertolaso and Caianiello 2016), giving way to less straightforward notions of selection, such as "dynamic selection" (Lesne 2008) or "second order" selection (Wagner 2005, p. 248): a selection that no longer affects directly a state of the biological system (such as a character state), but touches on organizational and variational properties. Consistently, Csete and Doyle emphasize selection pressure on

---

[22] On the ambiguity of the notion of "molecular control of cellular activity" see Weiss 1963, pp. 322–323: "'controlling' molecules have themselves acquired their specific configurations, which are the key to their power of control, by virtue of their membership in the population of an organized cell, hence under 'cellular control' (…) the distinction between molecular control of cellular activity and cellular control of molecular activity is based on the semantic ambiguity of the term 'control'". As examples of the huge debate about causality in nonlinear systems, I will just mention Wagner 1999 on the unfeasibility of the "regularity notion of causality" for nonlinear systems; Schmidt's outline of the notion of "weak causality" in postmodern physics (Schmidt 2008, pp.322–323); Keller 2002, Chap. 3.

"protocols", the rules for the configuration and/or interaction of system components (Alderson and Doyle 2010), as fostering the evolutionary conservation of those that "both facilitate evolution and are difficult to change" (Csete and Doyle 2002; see below § 4).

The notion of inbuilt design principles contrasted, in Savageau's intention, earlier assumptions that control mechanism may simply arise by "accidents of history", with the claim that specific modes of control are positively selected for the advantage they provide in specific ecological conditions (Savageau 1991; cf. Savageau 2001; Alon 2006, chap. 11). Thus, the tradition of robustness pioneered by Savageau's studies on parameter sensitivity and stability analysis of biochemical networks (Savageau 1971,[23] 1976) shares a further common goal with the structuralist approach mentioned earlier, that is to correct unduly emphasis on the historical contingency view of biological constraints (Kauffman 1993, p. 13).

The rethorics of the approach in Barkai and Leibler 1997 paper is quite similar to the earlier robustness paper, even if the challenge is not to the notion of genetic control but on the way such control is to be conceived: a way that has to take into account "the high-dimensional dynamics of complex nonlinear systems" (Zhou et al. 2016).

In their analysis of the bacterial chemotaxis network – the network which controls the movement of bacteria in response to chemical stimuli – Barkai and Leibler investigate whether the stability of such biochemical networks is due to a close control on their parameter values (by which rate constant and enzymatic concentrations "need to be chosen in a very precise manner", so that "any deviation from the fine-tuned values will ruin the network's performance"), or whether some key properties of the network itself might be robust or "insensitive to key values of biochemical parameters". The earlier case would imply that natural selection "has fine-tuned the kinetic parameters and the amount of proteins to achieve the desired behavior" (Hartwell 1997). The proposed alternative, on the contrary, implies that the selective pressure has rather favored the instantiation of a specific design principle that makes the chemotaxis pathway robust with respect to the natural genetic polymorphism that is known to affect kinetic parameters or amounts of proteins (Hartwell 1997). The evolutionary advantage, in this case, would be "having a larger parameter space in which to evolve and adjust to environmental change".

In the case of the bacterial chemotaxis pathway, such a property is exact adaptation, its ability to reset precisely the tumble frequency to the prestimulus value of no signal after perturbation, i.e. an addition or a change in concentration of a chemical stimulant. Robust adaptation refers to the fact that the output of the pathway asymptotically approaches a constant value independently of the actual attractant concentration. Such a desensitization to a continued stimulation allows the organism to restore its sensitivity to chemical gradients in order to respond readily to further stimuli.

The computational simulations performed by Barkai and Leibler on the basis of a two-state (active/inactive) model of the chemotactic network showed that the

---

[23] This paper is often quoted in Systems Biology's genealogies of the notion of robustness, although the term does not yet appear.

adaptation property was robust to even dramatic variations in any of the biochemical parameters of the network.

The explanation for the adaptation process lays for them in the "connectivity" of the network itself. In the complex wiring of the chemoreceptor complex, adaptation is achieved by changes in the level of methylation of the chemoreceptors. By means of the activity of two proteins, CheR and CheA, the levels of methylation are adjusted in order to compensate the action of the attractants on the receptors. As these latter drive the receptors toward the kinase-off state, the methylation level is raised in order to overcome the suppression of receptor activity caused by attractant binding. Thus, the negative feedback mechanism depends solely on the "system activity" and reacts to environmental changes insofar as they affect this activity (Barkai and Leibler 1997).

Barkai and Leibler's computational model was soon experimentally confirmed (Alon et al. 1999), and then recast into a thorough engineering description involving an integral feedback control mechanism (Yi et al. 2000): a mechanism by which the system's activity is constantly compared to a reference steady-state value, and the time integral of the system error – the difference between the actual and the desired output – is fed back into the system.

The peculiarity detected in the chemotaxis integral feedback control is that the mechanism is inbuilt into the system as a "structural property" (i.e. an intrinsic property of the connectivity of the network, as claimed by Barkai and Leibler). This topological configuration embodies a "special case of the internal model principle" (Yi et al. 2000), by which the integrator is implemented inside the loop (Zhou et al. 1995, p. 450). The level of methylation of the receptors acts therefore as the integrator, the mechanism that generates a model of the external signal in order to counteract its effect.

It is however not to the chemotaxis network as a whole, but to the specific functional configuration by which it attains adaptation that robustness is ascribed. Other functions, such as adaptation time, resulted fine-tuned so that they could be varied according to the stochastic variation to which they are known to be prone.

This qualification entails two important tenets of the "new concept" of robustness: firstly, that the isolation of a function – the functional "system identification" – logically precedes the assessment of robustness; secondly, that robustness is a highly conditional property, which must be expressed in an algebraic form as "A [property] of a [system] is robust if it is [invariant] with respect to a [set of perturbations]" (Alderson and Doyle 2010).

The chemotaxis case instantiated the first template of robust control in a biological network, fulfilling the property to "produce a desired level of output in an uncertain environment (e.g., varying levels of stimulant) with uncertain components (e.g., varying concentrations of protein)" (Yi et al. 2000). It provided an accurate example of the possibility to reverse engineering biological complexity "from function to mechanisms" (Csete and Doyle 2002). At the same time, the relative independence of control mechanism from biochemical parameters endorsed the possibility to extract "some of the principles underlying cell function without a full knowledge of the molecular detail" (Barkai and Leibler 1997; cf. Hartwell 1997).

2.4 Robustness and Complexity

"It is in the nature of their robustness and complexity that biology and advanced engineering are most alike". This convergence, boosted by the latest developments of sophisticated engineering, prompts a renewal of the alliance between the fields "in which function appears naturally", with robustness as the major "conceptual and theoretical bridge" (Hartwell et al. 1999, p.C49; Csete and Doyle 2002, p. 1664–1665). It is patent that the new twist superimposed on the notion of robustness by a mainstream trend in Systems Biology since the early 2000s is that of a demarcation. Notably, the demarcation aims to divide robustness as a property of functional organization, common to living and engineered systems, and the form of stability spontaneously attained by complex, nonlinear inanimate systems.

This demarcation is the consequence of referring robust stability to *functions*,[24] and not simply to *states* of equilibrium to which the system approximates in time (Kitano 2007, p. 1).

The difference from notions of nonlinear stability in inanimate systems is substantial, as it impinges dramatically on the system's identification, and therefore on the range of changes a system can endure while preserving the same "qualitative" behavior. In so far as a system is circumscribed according to a definite function it performs, robustness is its property of maintaining its functional identity even in front of changes in its "structure and components", or to "change its mode of operation in a flexible way" (Kitano 2004, p. 827). It may do so by exploiting noise and other kinds of instabilities, "the very variability of the elements", or even by increasing instability in a part of the system for coping with internal and external perturbations (cf. Kitano 2007; Lesne 2008). The notion of robust stability becomes therefore conceptually different from all notions that describe a return of the system to a previous state (resilience), or in the vicinity of a state of equilibrium (dynamical and structural stability) after perturbation.

This flexibility in the determinants of the phenomenology of robust behaviors involves a coupling of topology and dynamics, in which the system's ability to switch among multiple alternatives and strategic options (Kitano 2004; Jen 2005) plays a pivotal role. Thus, the organizational requirements for robustness – modularity, regulatory complexity, redundancy, degeneracy, distributed control, multistability, layered architecture, or specific designs of positive and negative feedback loops (Bertolaso and Caianiello 2016) – all represent "alternative mechanisms" which need to be coupled with "system control" to effect robust behaviors. Therefore, all such properties, many of which independently emphasized in the former literature, are restated within a unified conceptual framework as instantiations

---

[24] Obviously, "various meanings and measures of function pertain to different types of networks and so conclusions drawn from one type of network, or from abstract representations of idealized network, might have little relevance to other types of network" (Siegal et al. 2007, p. 84). I will not delve into the complex philosophical debate on the definition of function (see for instance Mossio et al. 2009; Huneman 2013), because the demarcation I wish to emphasize may apply in principle to all definitions, at least in so far as they can be operationalized in a suitable representation.

Table 2.1 Characteristics of SOC HOT, and data

	Property	SOC	HOT and Data
1	Internal configuration	Generic, homogeneous, self-similar	Structured, heterogeneous, self-dissimilar
2	Robustness	Generic	Robust, yet fragile
3	Density and yield	Low	High
4	Max event size	Infinitesimal	Large
5	Large event shape	Fractal	Compact
6	Mechanism for power laws	Critical internal fluctuations	Robust performance
7	Exponent α	Small	Large
8	α vs. dimension d	$\alpha \approx (d - 1)/10$	$\alpha \approx 1/d$
9	DDOFs	Small (1)	Large (∞)
10	Increase model resolution	No change	New structures, new sensitivities
11	Response to forcing	Homogeneous	Variable

Differences between the "new sciences of complex networks" and organized complexity, after Carlson and Doyle (2002). Copyright (2002) National Academy of Sciences, U.S.A

of robustness, the crucial "system-level" property that drives the evolution of "evolving, complex dynamic systems" (Kitano 2004, p. 828).[25]

The focus on functionality imposes stricter requisites on what is to be understood as "organized" and what as "unorganized" complexity (Alderson and Doyle 2010; see Weaver 1948). "Organisms" and (sophisticated) "machines", on one side, and "thunderstorms", on the other (Keller 2008, 2009) do not only differ in the *kind of complexity* they exhibit, but in the *way they achieve it*, which is driven by the ratchet engendered by the requirements of functional robustness with its inherent tradeoffs.

The basic assumption is that the complexity found in robust living and manmade networks is comprised of "highly specific internal structures", whose details "matter enormously".[26] In virtue of this highly organized and specialized structure, the "design space of life" is characterized by nongeneric, self-dissimilar and rare configurations (Carlson and Doyle 2002, see Table 2.1). Most importantly, organized complexity grows "primarily to provide mechanisms to create robustness", so that the increase in complexity is concentrated "in control processes that regulate the internal state and respond to external changes" (Alderson and Doyle 2010, pp. 839, 842).

---

[25] For Alderson and Doyle 2010, a further argument for the demarcation between biology and technology and the other types of complex systems is that they exhibit system-level constraints which are distinct from those on their components.

[26] Carlson and Doyle, 2002. Cf. Alderson and Willinger 2005, p. 96: "This approach requires incorporating knowledge of the system's functional objectives, the details of its component parts, and the specifics of its operating environment to yield descriptions that explain the observed structure or behavior but are also fully consistent with engineering reality and available measurements".

Consequently, robustness is a property that can be predicated only of highly organized systems, whose specific structure is as "a consequence of specific *constraints* (...) on their functionality and/or behavior", independently "of the process by which this organization arises, whether by design or evolution". The gap between biological and artificial systems is further narrowed by the observation that even in engineering design constraints may be historical in their origin, or that "mix of the ad hoc, trial and error, accident, and history" to which also technological evolution is prone (Alderson and Doyle 2010, p. 840). Thus, it may be not entirely true that the contingency of evolutionary histories is a sufficient argument for a clear-cut distinction between engineering and biological optimization (Siegal et al. 2007; one might be reminded of the QWERTY keyboard standardization case so effectively presented by Gould 1992, chap. 4).

Leaving aside the open issues about the relative roles of self-organization and natural selection in optimizing complex biological networks (Wagner 2005, 2014; Kitano 2004; Whitacre 2012; Bertolaso 2017), it is important to emphasize that, in the robustness perspective, optimization is no longer synonymous with simplification or cost reduction. Robust optimization is more akin to what Herbert Simon labeled satisficing, or optimization under constraints (Callebaut 2007). As exemplified by the internal model principle (Sect. 2.2.2), robust control by its very definition cannot aim at "minimal function" (Csete and Doyle 2002, p. 1666; Alderson and Doyle 2010, p. 844), because it rather demands and exploits internal complexity to cope with the range of disturbances it is evolved or designed to withstand. The accumulation of cryptic genetic variation – the classical argument for Conrad H. Waddington's precursor notion of canalization (de Visser et al. 2003) –, as well as the conservation of gene duplicates in eukaryote metabolic networks with respect to bacterial (Papp et al. 2011) are concrete examples of how increase of complexity needs robustness.

The costs of complexity are also part of the tradeoffs that drive the system's evolution. They drove, according to Mattick and Gagen (2005, pp. 857ff), the evolution of aircraft control from mechanical to computational, as well as the shift from "analog protein-based regulatory systems in proteins to a digital RNA-based control architecture". Transitions that alter "the physical basis of the control architecture" in order to cope with the increased connection and organizational costs would be in fact compensated by increased efficiency and evolvability (Mattick and Gagen 2005. On the resolution of the apparent "paradox" between robustness and evolvability see Wagner 2008). The evolution of the control strategies is much more important than the addition of parts, as it is the control strategies – mediated by increasingly complex hierarchies of "protocols" – that allow in the first place the increasing complexity of the system, enabling its modular organization. Better protocols[27] are those that "supply both robustness and evolvability" (Csete and

---

[27] The informatic notion of protocol is proposed as an abstraction for a wide variety of biological intra- and interlevel communication and regulatory mechanisms (Csete and Doyle 2002). Its boundaries are not always very clear, especially with respect to recurrent feedback strategies and network motifs, as well as to design principles.

Doyle 2002), and are fixed by selection because of their efficiency and parsimony. They tend therefore to give rise to a "universal" code of control, shared by all interacting components – one might be reminded of the well-known genetic toolkit for eukaryote regulation (Carroll et al. 2001). However, it is in the very universality of this code that fragility lurks.

A "robust yet fragile" feature characterizes the extremely specialized architecture labeled "highly optimized tolerance" (HOT). The architectural model that best fits the requirements of *optimizing tolerance* to specific perturbations has been identified in the bow-tie network (Csete and Doyle 2004; Jones 2014), whose recurrence in biological systems is being increasingly proved, starting from metabolic networks (Ma and Zeng 2003; Friedlander et al. 2015).

Bow-tie architectures possess a tightly connected and highly conserved core ("giant strong component subnetwork"), shielded from perturbation in virtue of its high degree of degeneracy. The core, which is the site of basic and highly conserved biological functions (Ma and Zeng 2003, Csete and Doyle 2004), is connected to less constrained peripheral modules, characterized by higher variability and evolvability (Csermely et al. 2013). This compromise between conservation and variability makes it possible to counter the effects of noise and of perturbations, while keeping in place the "extreme heterogeneity that allows for robust regulation" (Csete and Doyle 2004, p. 447). At the same time, bow-ties architectures have inherent fagilities, and even extreme ones, as when parasites or cancerous processes hijack crucial control systems – where it is exactly the protocols, "the universal common currencies responsible for robustness", that allow hijacking (Csete and Doyle 2004, p. 447).

A remarkable novelty emerging in the new robustness perspective is that the magnitude of the perturbation is no longer the primary challenge. Robust systems can tolerate much larger, wider in range and multiple simultaneous perturbations than structural stability (Lesne 2008; Zhou et al. 2002). At the same time, they may be fragile to even infinitesimally small but unexpected or unexperienced perturbations, such as rare, non-recurrent environmental changes. Moreover, the ratchet of complexity which affects control processes managing "the interaction among components" (Alderson and Doyle 2010, p. 845) make robust systems extremely fragile also to "rearrangements of the interconnection of internal parts" (Carlson and Doyle 2002, p. 2539) and to "design flaws" (Zhou and Carlson 2000, p. 62).

Interestingly, the target of this renewed notion of robustness is the disparate family of "physics-based" theories supporting "order-for-free" views of biological systems, to which also one of the earliest papers on robustness belonged (Sect. 2.3). But the true battlefield on which to test notions of "self-organized-criticality" (SOC, Bak 1996), edge of chaos theories and their latest development, the "new sciences of complex networks" (Alderson and Doyle 2010) inspired by Barabasi and his school, is now computer simulation (Zhou et al. 2002; cf. Albert and Barabasi 2002). Here, anew, the claim is that of a gap between "abstractions" (Carlson and Doyle 2002, p. 2538) and more realistic, biomimetic modeling, and the major point of contention is whether the "design space of life" is inhabited by generic vs nongeneric configurations (Table 2.1).

Blinded by the deceiving ubiquity of power law distribution in all kinds (from physical to biological, technological and social) of complex systems,[28] the SOC modeling strategy upholds the scale-free network model as the universal pathway to (generic) forms of order. The assumption of SOC modeling is randomness without heredity; networks evolve by random growth via preferential attachment, "rich get richer" dynamics, but the order they spontaneously reach has none of the realistic features of life nor of engineering, and, more specifically, does not account for their specific form of robustness. In fact, SOC and HOT computational modeling "predict not just different but exactly opposite features of complex systems" (Carlson and Doyle 2002, p. 2538). While scale-free topology is robust to random rewiring, HOT modeling is not. By taking "into account the heterogeneity and structured nature of biological systems", and particularly including in the simulation the role of abiotic forcing and Darwinian mechanisms of evolution (Zhou et al. 2002; Carlson and Doyle 2002 p. 2542), HOT evolve quickly to a highly structured configuration, whose power law distribution is not only quantitatively different, – exhibiting much steeper power laws, which "extend to larger event sizes than the critical power laws" (Zhou et al. 2002, p. 2054) – but most of all qualitatively, as power laws arise "from tuning and optimization" of many internal variables (Carlson and Doyle 2002 p. 2540).

The proponents of the quite strong thesis that robustness is a property that demarcates biological and sophisticated engineering systems make no attempt to hide the gap between evolved "organizing" and engineered "design" principles, and specifically between the nature and degree of autonomy which characterizes robustness in living systems and the one actually attainable in even the most sophisticated engineering products (Nikolov et al. 2007, p. 428; Kitano 2004; Whitacre 2012). Yet, differences between these systems lay not only in the degree of their extant complexity.

As Lewontin observed in 1996, organisms "do not solve problems (…). Problem solving in the usual sense is a goal-seeking process carried on by a conscious actor who knows both the final state to be achieved and the repertoire of possible starting conditions (…) There is no natural analog to a knowledge of the final state" (Lewontin 1996).

Although not all of Lewontin's criticism applies to sophisticated engineering devices and to the tenets of "robust control", the issue raised by Rosen with the still impressionistic notion of "parametric control" still holds. As Kitano emphasizes, in the control engineering practice, "there is a certain set point, determined by the designer, that the system's state will approach, even when perturbed … in biological

---

[28] The Pareto (power-law or scale-free) probability distribution is characterized by heavy tails, which, when modelled with Graph Theory, correspond to a "small-world" network topology, that is one in which few nodes are highly connected ("hubs" which interact with many other nodes). As this distribution is not random, it has been taken as a "signature" of some ordering process at work, such as a "rich-gets-richer" multiplicative process, and in particular in SOC as the indication of critical phase transitions. This inference is challenged by Alderson and Doyle 2010 by means of technical arguments on the methods employed for devising power law topologies. See also Keller 2005; Lima-Mendez and van Helden 2009; Broido and Clauset 2018.

systems ... set point is implicit in the equilibrium state of the system, which often changes dynamically" (Kitano 2004, p. 835). Furthermore, "robust performance requires the precise specification of both a performance metric and the type/size of uncertainty", but "the performance metric is often difficult to be defined precisely in biology, as it is an implicit element of an evolved entity" (Doyle and Stelling 2006, p. 610).

Actually, the extant gap the two kinds of robustness is felt at once as the challenge and the unprecedented opportunity for building a new "interface" between biology and engineering (Doyle and Stelling 2006, Sontag 2004). However, for the time being, the promise of a unifying theory of robustness might be best fulfilled by stressing its nature as an Interfield Theory: that is a theory that highlights convergences and similarities, the whole "complex network of relationships between fields", without occulting differences, so that "even though new lines of research closely coordinate the fields after the establishment of the interfield theory... the fields retain their separate identities" (Maull and Darden 1977).

References

Abersman, S. (2016). *Overcomplicated. Technology at the limits of comprehension*. New York: Penguin Random House.

Abraham, R., & Marsden, J. E. (1978). *Foundations of mechanics* (II ed.). Reading: Addison-Wesley.

Albert, R., & Barabasi, A.-L. (2002). Statistical mechanics of complex networks. *Reviews of Modern Physics, 74*, 47–97.

Aldana, M., Balleza, E., Kauffman, S. A., & Resendiz, O. (2007). Robustness and evolvability in genetic regulatory networks. *Journal of Theoretical Biology, 245*, 433–448.

Alderson, D. L., & Doyle, J. C. (2010). Contrasting views of complexity and their implications for network-centric infrastructures. *IEEE Transactions on Systems, Man, and Cybernetics—Part A: Systems and Humans, 40*(4), 839–852.

Alderson, D. L., & Willinger, W. (2005). A contrasting look at self-organization in the internet and next-generation communication networks. *IEEE Communications Magazine, 43*(7), 94–100.

Alon, U. (2006). *An introduction to systems biology. Design principles of biological circuits*. Boca Raton: Chapman & Hall/CRC.

Alon, U., Surette, M. G., Barkai, N., & Leibler, S. (1999). Robustness in bacterial chemotaxis. *Nature, 397*, 168–171.

Andronov, A. A., & Pontryagin, L. (1937). Systèmes grossiers. *Doklady Akademi Nauk SSSR, 14*, 247–250.

Andronov, A. A., Vitt, A. A., & Khaikin, S. E. (1966). *Theory of oscillators (1937)*. London: Pergamon.

Arvidsson, M. (2008). Principles of robust design methodology. *Quality and Reliability Engineering International, 24*(1), 23–35.

Ashby, C. R. (1958). Requisite variety and its implications for the control of complex systems. *Cybernetica, 1*, 1–17.

ASTM dictionary of engineering science & technology. 2005. 10th ed. West Conshohocken: ASTM International.

Åström, K.-J., & Murray, R. M. (2008). *Feedback systems: An introduction for scientists and engineers*. Princeton: Princeton University Press.

Aubin, D., & Dahan Dalmedico, A. (2002). Writing the history of dynamical systems and chaos: Longue durée and revolution, disciplines and cultures. *Historia Mathematica, 29*(3), 273–339.

Bailey, K. D. (1994). *Sociology and the new systems theory. Toward a theoretical synthesis.* Albany: State University of New York Press.

Bak, P. (1996). *How nature works: The science of self-organized criticality.* New York: Springer.

Barkai, N., & Leibler, S. (1997). Robustness in simple biochemical networks. *Nature, 387,* 913–917.

Bennett, S. (1993). *A history of control engineering* (pp. 1930–1955). Stevenage: Peregrinus – IEET.

Bennett, S. (1996). *A brief history of automatic control. IEEE Control Systems 16* (3), 17–25

Bennett, S. (2004). Technological concepts and mathematical models in the evolution of control engineering. In M. Lucertini, A. Millàn Gasca, & F. Nicolò (Eds.), *Technological concepts and mathematical models in the evolution of modern engineering systems: Controlling managing organizing* (pp. 103–128). Basel: Birkhäuser Verlag.

Bertolaso, M. (2017). Robustez biológica. In C. E. Vanney, I. Silva, & J. F. Franck (Eds.), *Diccionario Interdisciplinar Austral.* http://dia.austral.edu.ar

Bertolaso, M., & Caianiello, S. (2016). Robustness as organized heterogeneity. *Rivista di Filosofia Neoscolastica, CVIII*(2), 293–303.

Bissell, C. (1998). A.A. Andronov and the development of Soviet control engineering. *IEEE Control Systems, 18*(1), 56–62.

Bissell, C. (2009). In Nof (Ed.), *A history of automatic control* (pp. 53–69). Heidelberg: Springer.

Box, G. E. P. (1953). Non-normality and tests on variances. *Biometrika, 40*(3/4), 318–335.

Box, G. E. P. (1979). Robustness in the strategy of scientific model building. In Launer & Wilkinson (Eds.), *Robustness in Statistics* (pp. 201–236). Madison: University of Wisconsin.

Box, G. E. P. (1983). An apology for Ecumenism in statistics. In G. E. P. Box, T. Leonard, & D. F. J. Wu (Eds.), *Scientific inference, data analysis, and robustness* (pp. 51–84). New York: Academic.

Broido, A. D., & Clauset, A. (2018). *Scale-free networks are rare.* Preprint https://arxiv.org/abs/1801.03400. Accessed 24 June 2018.

Bryson, A. E. J. (1996). Optimal control – 1950 to 1985. *IEEE Control Systems, 16*(3), 26–33.

Callebaut, W. (2007). Herbert Simon's silent revolution. *Biological Theory, 2*(1), 76–86.

Callebaut, W. (2010). The dialectic of dis/unity in the evolutionary synthesis and its extensions. In M. Pigliucci & G. B. Müller (Eds.), *Evolution. The extended synthesis* (pp. 443–481). Cambridge: The MIT Press.

Carlson, J. M., & Doyle, J. C. (2002). Complexity and robustness. *PNAS, 99*(1), 2538–2545.

Carroll, S. B., Grenier, J. K., & Weatherbee, S. D. (2001). *From DNA to diversity.* London: Blackwell.

Cascante, M., Curto, R., & Sorribas, A. (1995). Testing the robustness of the steady-state characteristics of a metabolic pathway: Parameter sensitivity as a basic feature for model validation. *Journal of Biological Systems, 3,* 105–113.

Cilliers, P. (1998). *Complexity and postmodernism.* London: Routledge.

Cogan, B. (2006). Computing robustness in biology. *Scientific Computing World, 2005/2006.* https://www.scientific-computing.com/issue/december-2005january-2006

Cornish-Bowden, A., Cárdenas, M. L., Letelier, J.-C., Soto-Andrade, J., & Guíñez Abarzúa, F. (2004). Understanding the parts in terms of the whole. *Biology of the Cell, 96,* 713–717.

Csermely, P., London, A., & Wu, L.-Y. (2013). Structure and dynamics of core/periphery networks. *Journal of Complex Networks, 1,* 93–123.

Csete, M., & Doyle, J. C. (2002). Reverse engineering of biological complexity. *Science, 295,* 1664–1669.

Csete, M., & Doyle, J. C. (2004). Bow ties, metabolism and disease. *Trends in Biotechnology, 22*(9), 446–450.

Dahan Dalmedico, A. (2004). Early developments of nonlinear science in Soviet Russia: The Andronov School at Gor'kiy (in collaboration with I. Gouzévitch). *Science in Context, 17,* 235–265.

Davison, E. J., & Goldenberg, A. (1975). Robust control of a general servomechanism problem: The servo compensator. *Automatica, 11,* 461–471.

De Robertis, E. A., Morita, E. M., & Cho, K. W. Y. (1991). Gradient fields and homeobox genes. *Development, 112*, 669–678.

Dickson, M., & Baird, D. (2011). Significance testing. In P. S. Bandyopadhyay & M. R. Forster (Eds.), *Philosophy of statistics* (pp. 199–229). Oxford: North Holland.

Dorato, P. (1987). A historical review of robust control. *Control Systems Magazine IEEE, 7*(2), 44–47.

Doyle, J. C. (1979). *Robustness of multiloop linear feedback systems.* In Proceedings of the 1978 IEEE conference on decision and control (pp. 12–18). New York: IEEE Press.

Doyle, F. J., III, & Stelling, J. (2006). Systems interface biology. *Journal of the Royal Society Interface, 3*(10), 603–616.

Doyle, J. C., Wall, J. E., & Stein, G. (1982). *Performance and robustness analysis for structured uncertainty.* 21st IEEE conference on decision and control (pp. 629–636). New York: IEEE Press.

Doyle, J. C., Glover, K., Khargonekar, P., & Francis, B. A. (1989). State-space solutions to standard H2 and H1 control problems. *IEEE Transactions on Automatic Control, 34*, 831–847.

Doyle, J. C., Packard, A., & Zhou, K. (1991). *Review of LFTs, LMIs, and μ.* In Proceedings of the 30th IEEE conference on decision and control. New York: IEEE Press.

Dullerud, G. E., & Paganini, F. G. (2000). *A course in robust control theory – A convex approach.* New York: Springer.

Fell, D. A. (1992). Metabolic control analysis: A survey of its theoretical and experimental development. *Biochemical Journal, 286*, 313–330.

Fox Keller, E. (2002). *The century of the gene.* Cambridge, MA: Harvard University Press.

Francis, B. A., & Wonham, W. M. (1976). The internal model principle of control theory. *Automatica, 12*, 457–465.

Francis, B. A., Sebakhy, O. A., & Wonham, W. M. (1974). Synthesis of multivariable regulators: The internal model principle. *Applied Mathematics & Optimization, 1*, 64–86.

Friedlander, T., Mayo, A. E., Tlusty, T., & Alon, U. (2015). Evolution of bow-tie architectures in biology. *PLoS Computational Biology, 11*(3), e1004055.

Galison, P. (1994). The ontology of the enemy: Norbert Wiener and the cybernetic vision. *Inquiry, 21*(1), 228–266.

Gigerenzer, G., et al. (1989). *The empire of chance.* Cambridge: Cambridge University Press.

Gilbert, S. F., Opitz, J., & Raff, R. A. (1996). Resynthesizing evolutionary and developmental biology. *Developmental Biology, 173*, 357–372.

Goodwin, B. C. (1990). Structuralism in biology. *Science Progress, 74*(2), 227–243.

Goodwin, B. C., & Saunders, P. (1989). *Theoretical biology: Epigenetic and evolutionary order from complex systems.* Edinburgh: Edinburgh University Press.

Goodwin, B. C., Kauffman, S. A., & Murray, J. D. (1993). Is morphogenesis an intrinsically robust process? *Journal of Theoretical Biology, 163*(1), 135–144.

Gould, S. J. (1992). *Bully for brontosaurus: Reflections in natural history.* New York: W. W. Norton & Company.

Green, S. (2015). Can biological complexity be reverse engineered? *Studies in History and Philosophy of Biological and Biomedical Sciences, 53*, 73–83.

Green, S., & Wolkenhauer, O. (2013). Tracing organizing principles: Learning from the history of systems biology. *History and Philosophy of the Life Sciences, 35*(4), 553–576.

Guckenheimer, J., & Holmes, P. (1983). *Nonlinear oscillations, dynamical systems, and bifurcations of vector fields.* New York: Springer.

Hagen, J. B. (1992). *An entangled bank: The origins of ecosystem ecology.* New Brunswick: Rutgers University Press.

Hansen, L. P., & Sargent, T. J. (2008). *Robustness.* Princeton: Princeton University Press.

Hartwell, L. (1997). Theoretical biology: A robust view of biochemical pathways. *Nature, 387*, 855–857.

Hartwell, L. H., Hopfield, J. J., Leibler, S., & Murray, A. W. (1999). From molecular to modular cell biology. *Nature, 402*, 47–52.

Holmes, P. (2005). Ninety plus thirty years of nonlinear dynamics: Less is more and more is different. *International Journal of Bifurcation and Chaos, 15*(9), 2703–2716.

Huber, P. J. (1972). Robust statistics: A review. *The Annals of Mathematical Statistics, 43*(4), 1041–1067.

Huber, P. J. (1981). *Robust statistics*. New York: Wiley.

Hughes, T. P. (1993). *Modern and postmodern engineering*. Paper presented at Seventh Annual Arthur Miller Lecture on Science and Ethics, MIT, April 8.

Hughes, T. P. (1998). *Rescuing Prometheus*. New York: Vintage Books.

Huneman, P. (2010). Topological explanations and robustness in biological sciences. *Synthese, 177*(2), 213–245.

Huneman, P. (Ed.). (2013). *Functions: Selection and mechanism*. Dordrecht: Springer.

Igamberdiev, A. U., Beloussov, L. V., & Gordon, R. (2012). Editorial to biological morphogenesis: Theory and computation. *Biosystems, 109*(3), 241–242.

Jen, E. (2005). Stable or robust? What's the difference? In E. Jen (Ed.), *Robust design. A repertoire of biological, ecological, and engineering case studies* (pp. 7–20). Oxford: Oxford University Press.

Johnson, H. G. (1968). The economic approach to social questions. *Economica N.S., 35*(137), 1–21.

Kacser, H., & Burns, J. A. (1979). Molecular democracy: Who shares the controls? *Biochemical Society Transactions, 7*(5), 1149–1160.

Kauffman, S. A. (1990). Requirements for evolvability in complex systems: Orderly components and frozen dynamics. *Physica, D, 42*, 135–152.

Kauffman, S. A. (1993). *The origins of order*. New York: Oxford University Press.

Keller, E. F. (2002). Developmental robustness. *Annals of the New York Academy of Sciences, 981*, 189–201.

Keller, E. F. (2005). Revisiting scale-free networks. *BioEssays, 27*(10), 1060–1068.

Keller, E. F. (2008). Organisms, machines, and thunderstorms: A history of self-organization, part one. *Historical Studies in the Natural Sciences, 38*(1), 45–75.

Keller, E. F. (2009). Organisms, machines, and thunderstorms: A history of self-organization, part two. *Historical Studies in the Natural Sciences, 39*(1), 1–31.

Kitano, H. (2004). Biological robustness. *Nature Reviews Genetics, 5*, 826–837.

Kitano, H. (2007). Towards a theory of biological robustness. *Molecular Systems Biology, 3*, 137.

Krakauer, D. C. (2005). Robustness in biological systems: A provisional taxonomy. In T. S. Deisboeck & J. Yasha Kresh (Eds.), *Complex systems science in biomedicine* (pp. 185–207). New York: Plenum Press.

Lander, A. D. (2011). Pattern, growth, and control. *Cell, 144*(6), 955–969.

Launer, R. L., & Wilkinson, G. N. (Eds.). (1979). *Robustness in statistics*. New York: Academic.

Lee, K. K. (1992). *Lectures on dynamical systems, structural stability, and their applications*. Singapore: World Scientific.

Lesne, A. (2008). Robustness: Confronting lessons from physics and biology. *Biological Reviews, 83*, 509–532.

Levins, R. (1966). The strategy of models building in population biology. *American Scientist, 54*, 421–431.

Levins, R. (1993). A response to Orzack and Sober: Formal analysis and the fluidity of science. *Quarterly Review of Biology, 68*, 547–555.

Lima-Mendez, G., & van Helden, J. (2009). The powerful law of the power law and other myths in network biology. *Molecular BioSystems, 5*, 1482–1493.

Lin, F. (2007). *Robust control design: An optimal control approach*. Chichester: John Wiley/RSP.

Ma, H. W., & Zeng, A. P. (2003). The connectivity structure, giant strong component and centrality of metabolic networks. *Bioinformatics, 19*, 1423–1430.

Mattick, J. S., & Gagen, M. J. (2005). Accelerating networks. *Science, 307*(5711), 856–858.

Maull, N., & Darden, L. (1977). Interfield theories. *Philosophy of Science, 44*(1), 43–64.

May, R. M. (1973). *Stability and complexity in model ecosystems.* Princeton: Princeton University Press.

Michel, A. N. (1996). Stability: The common thread in the evolution of feedback control. *IEEE Control Systems, 16*(3), 50–60.

Minelli, A. (2003). *The development of animal form: Ontogeny, morphology, and evolution.* Cambridge, MA: Cambridge University Press.

Morohashi, M., Winn, A. E., Borisuk, M. T., Bolouri, H., Doyle, J. C., & Kitano, H. (2002). Robustness as a measure of plausibility in models of biochemical networks. *Journal of Theoretical Biology, 216,* 19–30.

Mossio, M., Saborido, C., & Moreno, A. (2009). An organizational account of biological function. *The British Journal for the Philosophy of Science, 60*(4), 813841.

Neculai, A. (2005). Modern control theory. A historical perspective. *Studies in Informatics and Control, 10*(1), 51–62. https://camo.ici.ro/neculai/history.pdf.

Nicolis, G., & Prigogine, I. (1989). *Exploring complexity. An introduction.* New York: W.H. Freeman & Co.

Nikolov, S., Yankulova, E., Wolkenhauer, O., & Petrov, V. (2007). Principal difference between stability and structural stability (robustness) as used in systems biology. *Nonlinear Dynamics, Psychology, and Life Sciences, 11*(4), 413–433.

Nyquist, H. (1932). Regeneration theory. *Bell Systems Technical Journal, 2,* 126–147.

Ogata, K. (2010). *Modern control engineering* (V ed.). London: Prentice Hall.

Owen, R. (1883). On Dinornis (Part XXIV): Containing a description of the head and feet, with the dried integuments, of an individual of the species *Dinornis didinus. Transactions of the Zoological Society of London, 11*(8), 257–261.

Papin, J. A., Price, N. D., Wiback, S. J., Fell, D. A., & Palsson, B. O. (2003). Metabolic pathways in the post-genome era. *Trends in Biochemical Sciences, 28*(5), 250–258.

Papp, B., Notebaart, R. A., & Pál, C. (2011). Systems-biology approaches for predicting genomic evolution. *Nature Reviews Genetics, 12*(9), 591–602.

Pickering, A. (2010). *The cybernetic brain.* Chicago: The University of Chicago Press.

Pugh, C., & Peixoto, M. P. (2008). Structural stability. *Scholarpedia, 3*(9), 4008.

Roberts, A., & Tregonnin, K. (1980). The robustness of natural systems. *Nature, 288,* 265–266.

Rollins, L. (1999). *Robust control theory.* Carnagie Mellon University White Paper. https://users.ece.cmu.edu/~koopman/des_s99/control_theory/

Rosen, R. (1967). *Optimality principles in biology.* New York: Springer Science+Business Media.

Rosen, R. (1970). *Dynamical system theory in biology: Stability theory and its applications.* New York: Wiley.

Safonov, M. G. (2012). Origins of robust control: Early history and future speculations. *Annual Reviews in Control, 36*(2), 173–181.

Safonov, M. G., & Fan, M. K. H. (1997). Editorial (Special issue on multivariable stability margin). *International Journal of Robust and Nonlinear Control, 7,* 97–103.

Savageau, M. A. (1971). Parameter sensitivity as a criterion for evaluating and comparing the performance of biochemical systems. *Nature, 229,* 542–544.

Savageau, M. A. (1974). Optimal design of feedback control by inhibition. *Journal of Molecular Evolution, 4,* 139–156.

Savageau, M. A. (1976). *Biochemical systems analysis: A study of function and design in molecular biology.* Reading: Addison-Wesley.

Savageau, M. A. (1991). Reconstructionist molecular biology. *The New Biologist, 3*(2), 190–197.

Savageau, M. A. (2001). Design principles for elementary gene circuits: Elements, methods, and examples. *Chaos, 11*(1), 142–159.

Schmidt, J. C. (2008). *Instabilität in Natur und Wissenschaft: Eine Wissenschaftsphilosophie der nachmodernen Physik.* New York: Walter de Gruyter.

Shivamoggi, B. K. (2014). *Nonlinear dynamics and chaotic phenomena: An introduction* (II ed.). Dordrecht: Springer.

Siegal, M. L., Promislow, D. E. L., & Bergman, A. (2007). Functional and evolutionary inference in gene networks: Does topology matter? *Genetica, 129*, 83–103.

Smale, S. (1966). Structurally stable systems are not dense. *American Journal of Mathematics, 87*, 491–496.

Soler, L., Trizio, E., Nickles, T., & Wimsatt, W. C. (Eds.). (2012). *Characterizing the robustness of science: After the practice turn in philosophy of science*. Dordrecht: Springer.

Sontag, E. D. (2004). Some new directions in control theory inspired by systems biology. *Systems Biology, 1*(1), 9–18.

Stigler, S. M. (2010). The changing history of robustness. *The American Statistician, 64*(4), 277–281.

Taylor, P. (1988). Technocratic optimism, H.T. Odum, and the partial transformation of ecological metaphor after World War II. *Journal of the History of Biology, 21*, 213–244.

Thom, R. (1974). La linguistique, discipline morphologique exemplaire. *Critique, 30*, 235–245.

Thom, R. (1969). Topological models in biology. *Topology, 8*, 313–335.

Thom, R. (1975). Structural stability and morphogenesis. London: W.A. Benjamin A brief history of automatic control. *IEEE Control Systems*, 16(3), 17–25.

Thom, R. (1983). Darwin, cent ans après. *Rivista di Biologia, 76*(1), 11–22.

Umulis, D., O'Connor, M. B., & Othmer, H. G. (2008). Robustness of embryonic spatial patterning in *Drosophila melanogaster. Current Topics in Developmental Biology, 81*, 65–111.

Vajk, I., Hetthéssy, J., & Bars, R. (2009). In Nof (Ed.), *Control theory for automation – Advanced techniques* (pp. 173–198). Berlin: Springer.

van Nimwegen, E., Crutchfield, J. P., & Huynen, M. (1999). Neutral evolution of mutational robustness. *PNAS, 96*(17), 9716–9720.

Visser, J. A. G. M., Hermisson, J., Wagner, G. P., Meyers, L. A., Bagheri-Chaichian, H., Blanchard, J. L., Chao, L., Cheverud, J. M., Elena, S. F., Fontana, W., Gibson, G., Hansen, T. F., Krakauer, D., Lewontin, R. C., Ofria, C., Rice, S. H., von Dassow, G., Wagner, A., & Whitlock, M. C. (2003). Perspective: Evolution and detection of genetic robustness. *Evolution, 57*(9), 1959–1972.

Voit, E. O. (2013). Biochemical systems theory: A review. *ISRN Biomathematics, 1*, 53.

von Dassow, G., Meir, E., Munro, E. M., & Odell, G. M. (2000). The segment polarity network is a robust developmental module. *Nature, 406*, 188–192.

Wagner, A. (1999). Causality in complex systems. *Biology and Philosophy, 14*, 83–101.

Wagner, A. (2005). *Robustness and evolvability in living systems*. Princeton: Princeton University Press.

Wagner, A. (2008). Robustness and evolvability: A paradox resolved. *Proceedings of the Royal Society B, 275*, 91–100.

Wagner, A. (2014). *Arrival of the fittest: solving evolution's greatest puzzle*. London: OneWorld.

Wagner, A., & Stadler, P. F. (1999). Viral RNA and evolved mutational robustness. *Journal of Experimental Zoology Part B, 285*, 119–127.

Weaver, W. (1948). Science and complexity. *American Scientist, 36*, 536–544.

Weiss, P. (1963). The cell as a unit. *Journal of Theoretical Biology, 5*, 389–397.

Westerhoff, H. V., & Palsson, B. O. (2004). The evolution of molecular biology into systems biology. *Nature Biotechnology, 22*, 1249–1252.

Whitacre, J. M. (2012). Biological robustness: Paradigms, mechanisms, and systems principles. *Frontiers in Genetics, 3*, 67.

Whittle, P. (1996). *Optimal control*. Chichester: Wiley.

Wilkinson, G. N. (1979). In Launer & Wilkinson (Eds.), *Robust inference – The Fisherian approach* (pp. 259–290). New York: Academic.

Wimsatt, W. C. (2007). False models as means to truer theories. In W. C. Wimsatt (Ed.), *Re-engineering philosophy for limited beings* (chap. 6). Cambridge, MA: Harvard University Press.

Woodward, J. (2006). Some varieties of robustness. *Journal of Economic Methodology, 13*(2), 219–240.

Yi, T. M., Huang, Y., Simon, M. I., & Doyle, J. C. (2000). Robust perfect adaptation in bacterial chemotaxis through integral feedback control. *PNAS, 97*, 4649–4653.

Zames, G. (1996). Input-output feedback stability and robustness, 1959-85. *IEEE Control Systems, 16*(3), 61–66.

Zhou, T., & Carlson, J. M. (2000). Dynamics and changing environments in highly optimized tolerance. *Physical Review E, 62*(3), 3197–3204.

Zhou, K., Doyle, J. C., & Glover, K. (1995). *Robust and optimal control*. Englewood Cliffs: Prentice Hall.

Zhou, T., Carlson, J. M., & Doyle, J. C. (2002). Mutation, specialization, and hypersensitivity in highly optimized tolerance. *PNAS, 99*(4), 2049–2054.

Zhou, J. X., Smal, A., Fouquier d'Hérouël, A., Price, N. A., & Huang, S. (2016). Relative stability of network states in Boolean network models of gene regulation in development. *Biosystems, 142–143*, 15–24.

Silvia Caianiello is Senior Researcher at the "Istituto per la Storia del Pensiero Filosofico e Scientifico Moderno" (Institute for the History of Science and Philosophy in Modern Age ISPF) of the Italian National Research Council (CNR) in Naples, and affiliated to the Stazione Zoologica Anton Dohrn, Naples. She is co-editor of the Book Series "Filosofia e saperi. Crossing boundaries between humanities and life sciences" (CNR Edizioni), Vice-Director of the Italian inter-university Centre on epistemology and history of the life sciences "Res viva" (www.resviva.it), and, since 2017, associated to the Zoological Station "Anton Dohrn" in Naples, where she is member of the "Science and Society" Committee. Her research interests and experiences range from History of European Philosophy to History and Philosophy of Life Sciences. Her methodological focus is on conceptual interactions across different disciplines and epistemic approaches. She conducted extensive research on the correlation between representations of time and epistemologies of history from eighteenth to twentieth Century, and on conceptual exchanges between Human and Life Sciences since the nineteenth Century. Her current research field is the history and philosophy of evolutionary theory and evo-devo (Evolutionary Developmental Biology), with particular focus on the breakthrough of systemic and hierarchical approaches to biological organization and evolutionary change.

Chapter 3
Robustness, Mechanism, and the Counterfactual Attribution of Goals in Biology

Marco Buzzoni

Abstract The first part of this paper discusses two important meanings of robustness (robustness as stability as against variations in parameter values and robustness as consilience of results from different sources of evidence) and shows their essential connection with the notion of intersubjective reproducibility. As I shall maintain, robustness in both senses of the term is intimately connected with the notion of scientific experiment. This is the important element of truth of the mechanistic systems approach, which explains events as products of robust and regular systems and processes. In the second part of this paper I shall show that the concept of robustness of a mechanism, if applied to biological systems, is one-sided and incomplete without a heuristic-methodical reference to final causes, even though the assumption of the teleological point of view is justified in biology only to the extent that we use it as a counterfactual artifice, capable of bringing to light causal relations which have a robustly reproducible content. In this way, the reflexive, typically human concept of purposefulness may be employed to investigate living beings scientifically, that is, in an intersubjectively testable and reproducible way, to discover mechanisms in living systems which are robust in both senses of the word.

Keywords Counterfactual attribution of goals in biology · Experiment · Intersubjective reproducibility · Robustness-as-consilience · Robustness-as-stability · Teleology

3.1 Introduction

Confining ourselves to the role that robustness plays in biology as an experimental science, and setting aside robustness as originally introduced for statistical purposes and for model-based simulations (cf. above all Box 1953; Levins 1966, 1993), two main senses of the term "robustness" are usually distinguished (see for example

M. Buzzoni (✉)
Department of Humanistic Studies, University of Macerata, Macerata, Italy
e-mail: marco.buzzoni@unimc.it

© Springer Nature Switzerland AG 2018
M. Bertolaso et al. (eds.), *Biological Robustness*, History, Philosophy
and Theory of the Life Sciences 23, https://doi.org/10.1007/978-3-030-01198-7_3

Calcott 2010), which I shall call respectively robustness-as-stability and robustness-as-consilience:

1. Robustness as stability or insensitivity of output as against variations in parameter values. In this sense, robustness is important both in engineered systems and in living systems because it provides resilience against internal or external perturbations.
2. Robustness as consilience of results resulting from different and independent sources of evidence. These latter are said to be robust if they are supported in a variety of independent ways. It is this sense of robustness that is emphasized by those who espouse the no miracles argument.

According to the first sense that we shall discuss, robustness is usually defined as "a property that allows a system to maintain its functions against internal and external perturbations." (Kitano 2004, p. 826; cf. also Lander 2004, p. 713; Stelling et al. 2004; Clausing 2004, p. 25). In this sense, robustness is very important in engineered systems, as it makes them more resistant to events that cannot *de facto* (and perhaps in principle) be predicted in the early stages of their development: think of a robot that must work in still unexplored portions of nature, for example. This kind of robustness is usually obtained by building redundancy into a mechanism: if one element fails, another element can play its role (cf. Calcott 2010; Clausing 2004).

In this same sense, as resilience against internal or external perturbations, robustness is also a very important property in living systems. For example, the genetic code can be described as a robust encoding of amino acids into codons, or we may say that proteins, developmental pathways, metabolic networks, and tumours are robust against, respectively, translation errors, environmental or genetic disturbances (e.g. "gene knockout experiments"), changes in enzyme efficiency, and various chemotherapies.[1]

The second meaning of robustness, that is, robustness as consilience or coincidence of a variety of different (independent) pieces of evidence, is both a method which is frequently used in everyday life and a venerable concept in the philosophy of science. Apart from some partial anticipation (on this point, cf. especially Wimsatt 2012; Stegenga 2009; Hudson 2014), it is well-known that William Whewell was the first important author to be fully aware of the importance of this concept, which he called "*Consilience of Inductions*":

> We have here spoken of the prediction of facts *of the same kind* as those from which our rule was collected. But the evidence in favour of our induction is of a much higher and more forcible character when it enables us to explain and determine cases of a *kind different* from those which were contemplated in the formation of our hypothesis. The instances in which this has occurred, indeed, impress us with a conviction that the truth of our hypothesis is certain. No accident could give rise to such an extraordinary coincidence. No false supposition could, after being adjusted to one class of phenomena, exactly represent a different

---

[1] Cf. Wilke 2006, p. 695, and Strand and Oftedal 2009. Among the numerous reports published on robustness in biology, cf. Barkai and Leibler 1997; Alon et al. 1999; von Dassow et al. 2000; Kitano et al. 2004; Kitano 2007; Félix and Barkoulas 2015. For the relation of robustness to resilience, see above all Thorén 2014, which is also a useful source of further references.

class, when the agreement was unforeseen and uncontemplated. That rules springing from remote and unconnected quarters should thus leap to the same point, can only arise from *that* being the point where truth resides. [...] I will take the liberty of describing it by a particular phrase; and will term it the Consilience of Inductions.[2]

Newton's theory of universal gravitation is one of the best examples of this sense of robustness. No relation between Kepler's first, second, and third laws about the motion of the planets around the sun was present until Newton's theory of universal gravitation explained all of them at once. Moreover, Newton's theory of universal gravitation did not explain only the perturbations of the moon and planets by the sun and by each other, from which it was originally inferred, but also the apparently independent fact of the precession of the equinoxes (cf. Whewell [1840]1847, Vol. 2, pp. 65–66).

But the most discussed case in the literature is perhaps the measurement of Avogadro's number, that is, the number of molecules per gram-mole of any gas, which Jean-Baptiste Perrin computed by quite different methods and found to be the same, with a relatively good – but not so good as it is sometimes supposed (cf. Perrin [1916], p. 87) – approximation to today's accepted value, that is, $6.02214179 \times 10^{23}$ particles/mole. The fact that all of these measurements essentially agreed with the experimental data within the accuracy of the observations would be an unexplainable coincidence, it would be "miraculous", if each of the measurements referred to an artefact and matter were not really composed of molecules and atoms. If Perrin measured essentially the same Avogadro's number using such a variety of methods, then molecules must be real (for this example, see Psillos 2011; Cartwright 1983; Salmon 1984).

In Sect. 3.2, I argue that, although the distinction between different notions of robustness is necessary in order to avoid confusion (cf. Woodward 2006), we should not run into the opposite error of neglecting important similarities between these different kinds of robustness. Robustness-as-stability and robustness-as-consilience are both intimately connected with the notion of scientific experiment, that is, with the construction of an 'experimental machine' which extends the original operativity of our organic body and whose functioning exemplifies a nomic connection that exists in nature.

This is the important element of truth of the mechanistic systems approach, which explains events as products of robust and regular systems and processes. However, as will be shown in Sect. 3.3, the concept of robustness of a mechanism, if applied to biological systems, is one-sided and incomplete without a heuristic-methodical reference to human purposefulness. On the one hand, a teleological reference is always implicit in the robustness of biological systems, because such robustness is to be considered from the point of view of organisms that have goals and struggle for their survival. On the other hand, the assumption of the

---

[2] Whewell [1840]1847, Vol. 2, pp. 65–66. Entirely ignored in the robustness debate in the philosophy of science, but in my opinion almost equally important, is Bridgman 1927, pp. 56–60, who employed the notion – though not the term – of robustness-as-consilience to the greatest extent, especially as a criterion for proving the physical reality of theoretical entities.

teleological point of view is justified in biology only to the extent that we *use it as a counterfactual artifice, capable of bringing to light robust causal-mechanistic relations that have an objective and intersubjectively testable content.* In this way, the reflexive, typically human concept of purposefulness may be employed to investigate living beings scientifically, that is in an intersubjectively testable and reproducible way, to discover causal-mechanical or experimental relations in living systems which are robust in both senses of the word.

3.2   Robustness and Intersubjective Reproducibility

Even though robustness-as-stability and robustness-as-consilience should be kept distinct in many contexts, it is of the greatest importance for our purpose to make clear the common background against which this distinction stands out. Our question is whether the two senses of robustness are related to each other. Is there something common to both senses of this term, a background against to which the distinction itself may be better understood?

I think that this question should be answered in the affirmative. Robustness in both cases seems to be a property connected with the fact of having grasped something independent of us, something which, in a general sense, may be called objectively real.

In the case of robustness-as-stability this seems relatively clear. The most important point to notice is that the robustness of a mechanism (or of an organism, if regarded in a mechanistic perspective) is intimately connected with the notion of intersubjective reproducibility, which is perhaps the main pillar of scientific experimentation. As many authors already emphasized – from Frege to Poincaré, from Wittgenstein to Popper -, a particular or single event of perceptual awareness (for instance, *my* perception of a blank sheet of paper lying before me) is not only absolutely certain, but also unavoidably subjective and private, because it is not accessible to any other person. As such, it has no right of citizenship either in science or in empirical knowledge in general. Such a perception belongs to a subject, and not to an object, and for this reason it is not intersubjectively testable in principle. As for example Popper rightly noticed, we do not take even our own observations seriously, if they are not in principle intersubjectively testable (Popper 1959[2002], p. 45).

From this point of view, what makes robustness-as-stability so important in science is the fact that it is intimately connected with intersubjectively testable reproducibility. Somewhat at variance with Popper, however, this means that intersubjectively constant and stable reproducibility is, in the last analysis, the most valuable criterion to the independent existence of particular empirical objects (and their properties) and therefore of the truth of propositions which refer to them. On the contrary, to regard our perception as the property of some empirical object is much like conceiving it as a mere hallucination. If someone sees a lion in a room of the house, s/he would perhaps look again and/or ask someone to test whether s/he is seeing the same thing because there is something decidedly strange, if not impossible, in this perception. If no one could find any trace of the lion later on, s/he would know that what s/he had seen was a hallucination, no matter how frightening it was.

In this context, Kant's transcendental deduction might be read in a pragmatic, operationalist, or experimentalist sense, that is, as a claim about the fact that to be an object of possible experience is tantamount to having in principle reproducible properties (an important, but certainly not sufficient, clue is to be found in Kant, *Kritik der reinen Vernunft*, B, 242–243). In this context Kant's idea might be (freely) formulated by saying that *we do not believe that our own observations refer to something real, if they are not in principle reproducible and therefore intersubjectively testable, in relation to the interactions of human beings with the surrounding world.* In other words, the most general condition under which one can assume the existence of an object is that it exhibits properties that are sufficiently constant and reproducible.

In particular, there would be no empirical science whatever without reproducible properties, since it would not be possible to establish law-like connections between our practical or technical interventions on reality and determinate changes in reality. In this sense, to be an object is tantamount to being law-like or (what comes to the same thing) to having in principle reproducible properties. For this reason, robustness is intimately connected with the concept of experiment, that is, with the construction of an 'experimental machine' which extends the original operativity of our organic body and whose reproducible functioning exemplifies a nomic connection that exists in nature.[3]

What I have been saying bears directly on robustness as reproducibility, stability or insensitivity of output as against variations in parameter values, which we have seen to be relevant for engineered and biological systems. But what about robustness-as-consilience? I shall briefly show that what we have said up to this point on robustness-as-stability also applies to robustness-as-consilience. This may be seen from the following two facts: (1) the two meanings of "robustness", though distinct, are in an important respect interdependent; and (2) both meanings are essential to scientific experiment.

First, pace Hudson (2014), it is not true that robustness-as-consilience is less important and less fundamental than robustness-as-stability. The main reason that Hudson gives for supporting this claim is that to underwrite the reliability of a single experimental procedure is the first step one has to take before that of examining whether or not different experimental procedures converge. Now, it must be conceded to Hudson that robustness-as-consilience is reliable only if it is combined, at least to some extent, with robustness-as-stability. However, it is apparent, on a little reflection, that the converse is also true. Take for example sense perceptions in

---

[3] For further arguments in support of this view, see for example Wimsatt (2007, chapter 10), according to whom "robustness has the right kind of properties as a criterion for the real [...]. Furthermore, it works reliably as a criterion in the face of real world complexities, where we are judging the operational goodness of the criterion" (Wimsatt 2007, p. 197).

everyday life. On the one hand – in accordance with Hudson's view –, if a certain intersubjective reproducibility of what is perceived by the sense organs, taken separately, were not presupposed from the beginning, adding a second reproducibility to increase the reliability of the first one would be pointless. On the other hand, this time at variance with Hudson, if a certain intersubjective reproducibility of robustness-as-consilience – that is, of the consistency with which different pieces of evidence point to the same conclusion -, were not given, improving the degree of reproducibility of each piece of evidence would be equally pointless. We feel confident in the reproducibility of the results of any of our interventions on reality (robustness-as-stability) if the different interventions are *stably* consilient, that is, if robustness-as-consilience is itself intersubjectively reproducible to a sufficient degree – sufficient for our purposes. The fact that household objects such as flour, sugar, milk, eggs and currants can be handled with high reproducibility would not be a 'fact' (strictly speaking, they would be absolutely useless to us!) if they were not regarded as stably consilient, that is, as coherently placed into the whole of our (in this case domestic) life. For this reason, pace Hudson, robustness-as-consilience is as fundamental as robustness-as-stability.

Secondly, both senses of robustness are essential ingredients of the notion of scientific experiment. To see how both senses of robustness are intimately linked with one another in this notion, we may adopt Mach's definition of scientific experiment.

As Mach wrote, a scientific experiment is based on the 'method of variation' (*Methode der Variation*), whereby some variables are systematically modified to establish which relation of dependence, if any, holds between them:

> The basic method of experiment is the method of *variation*. If every element could be varied by itself alone, it would be a relatively easy matter: a systematic procedure would soon reveal the existing dependences. However, elements usually hang together by groups, some can be varied only along with others: each element is usually influenced by several others and in different ways. Thus we have to combine variations, and with an increasing number of elements the number of combinations to be tested by means of experiment grows so rapidly (a simple calculation shows this), that a systematic treatment of the problem becomes increasingly difficult and in the end practically impossible. (Mach 1906[1976], pp. 202–203, Engl. Transl. p. 149, translation slightly modified)

As far as robustness-as-stability is concerned, its importance for scientific experiment is evident from the fact that a *reproducible* procedure that reveals "the existing dependences" is part of what makes a good experiment, that is, an experiment able to deliver sound conclusions. From Galilei's experiments on the vacuum to experiments in quantum physics – where we are able to establish relations, not between events, but between the frequency distributions of observed events (on robustness in quantum physics, cf. De Raedt et al. 2014) – the experimenter, on the basis of a hypothetical plan of action, intervenes in a certain way on some aspects (or variables) of the experimental apparatus and notes the corresponding variations of certain others. In all cases, namely, a (perhaps only statistically significant) reproducibility is an elementary condition of scientific significance.

However, in so doing, can the experimenter proceed without any reference at all to robustness-as-consilience? It seems to me that it does not, for the following reasons.

Mach's method of variation holds only on the understood condition that no disturbing force intervenes, and in order to assume this with a degree of certainty that is sufficient for his/her purposes, the experimenter might find himself obliged to go beyond the limited conception of robustness-as-stability in the first of the two senses which have been here considered. As Mach noticed, usually the scientist can vary some elements only along with others because they interact with several others and in different ways. For this reason, we have to "combine variations" (Mach 1906[1976], 203, Engl. Transl. 149), that is to say, we have to combine and to compare different experimental interventions on different experimental apparatuses and laboratories in order to distinguish, in a system of correlated variables, causal from non-causal relationships, and therefore causally dependent variables from causally independent ones.

Varying conditions *in the same* experimental set-up seems clearly to serve the purpose of increasing robustness-as-stability, but varying the experimental apparatus and/or the laboratory is clearly relevant to the aim of increasing the degree of robustness-as-consilience, in the hope that facts already experimentally established (and for this reason already relatively independent) are found to be reproducibly consistent with one another. On reflection, this follows from the fact that natural laws are in principle to be concretely exemplified in the functioning of 'technical machines' or technical apparatuses built and mastered by the scientists, in which that law is present and operates in a controllable form (for instance, a pendulum for the laws of pendulum motion).[4] But if natural laws can only exist as exemplified in concrete technical apparatuses or measuring instruments in which that laws operate in a reproducible form, a change of experimental apparatus and/or laboratory involves a change in the law, and, more precisely, if this leads to success in reproducing the same outcome, it involves an extension of the old law to a new field of phenomena.

One could object to this conclusion on the grounds that two different experimental apparatuses and/or laboratories may instantiate the 'same' law. However, strictly speaking, this sameness would be only seeming: looking more closely it becomes clear that the exemplification of a scientific law in two really different experimental apparatuses (and/or laboratories) implies an extension of its meaning. As Wittgenstein pointed out, we cannot fix a term's intension for all time. Given a series of operations carried out according to a rule, one can never determine univocally what the next application of the rule will be because 'a new decision' is needed at every stage. Actually, each new example of a term or concept modifies its intension, forcing us to reconsider all previous examples in the light of the new one. For the problem under discussion, all this entails that the meaning of a scientific theory cannot be established once and for all by means of univocal definitions, but must be constantly marked out within the development of knowledge especially by means of

---

[4] On this point, cf. Buzzoni 2008, ch. 1.

the theory's experimental exemplification in new particular cases (for more details on this point, see Buzzoni 2008, 31–39).

It may also be argued that there is a huge difference between, on the one hand, the generalization and extension of a law beyond the limiting conditions of one laboratory, and, on the other hand, Newton's physics, which unified, to use Whewell's words, "remote and unconnected quarters" (Whewell [1840]1847, Vol. 2, 65), such as Kepler's Laws, the fact that the planets would slightly disturb one another's motions, etc. This much must be admitted, but this difference, however great it may be, is not a conceptual one: the extensions of important experiments to different apparatuses and/or laboratories and the unification of disparate phenomena differ only in degree and not in kind.

Now we are in a position to conclude that Hudson's objection that robustness-as-consilience reasoning cannot be credited with epistemic warrant if one lacks a minimally reliable observational procedure (which is an expression of robustness-as-stability) is not a fatal one; for what is important in science is not robustness-as-consilience in itself, but only the synergy between robustness-as-consilience and robustness-as-stability, which, on reflection, may be regarded as a robustness of higher order and whose elementary components are the two complementary aspects of technical-experimental reproducibility, taken in its more general sense. Robustness-as-stability, if understood as an aspect of experimental reproducibility, is not an alternative to robustness-as-consilience, but it is one of the most important requirements that must be met by a piece of evidence before entering into relations of consilience, which remains an essential element in the search for more reliable knowledge. What Hudson believes to be "a key part" of his critique of robustness (Hudson 2014, p. 6), is only a reason for maintaining that it is not sufficient to distinguish between two senses of robustness; it is also necessary to connect them with one another.

Thus, robustness as stability of results is only one aspect of a relation, but an aspect that, in the concrete, must be, at least in principle, inseparably bound up with the other aspect, that is, robustness as consilience of different pieces of evidence. For this reason, scientists look for interdependence and mutual growth and development, that is, for synergy, of the two elements upon which the aforementioned senses of robustness are based.

To sum up, robustness-as-stability and robustness-as-consilience are two side of the same experimental and intersubjective reproducibility, which, in this sense, may be plausibly presented as the core of robustness in science. In other words, in principle intersubjective-experimental reproducibility is the most general condition of robustness, which applies to both senses of robustness with which we have been dealing (for a more detailed discussion of this point, see Buzzoni 2017, on which the previous part of this paper is partly based).

So far I have maintained that robustness in both senses of the term is closely connected with the notion of scientific experiment. This applies to all empirical sciences, and must therefore hold also of biology. However, as we shall see in the next section, robustness in the sense just defined is only methodologically necessary, but not sufficient, to characterize robustness in the biological field. The concept of robustness of a mechanism, when applied to biological systems, is insufficient if it is not combined with a heuristic-methodical reference to final causes, even though the assumption of the teleological point of view is justified in biology only to the extent that we *use it as a counterfactual artifice, capable of bringing to light causal relations which have a robustly reproducible content.*

3.3 Robustness and the Counterfactual Attribution of Goals in Biology

The mechanistic systems approach deserves credit for having made evident that robustness as reproducibility, stability or insensitivity of a system as against variations in some parameter values plays an important methodological role not only in engineered systems, but also in the investigation into living systems. According to many exponents of the mechanistic philosophy, scientific explanation is given by describing events as products of robust and regular systems and processes. As Glennan writes, according to this approach events are explained as products of

> robust systems or processes consisting of interacting parts. [...] A mechanistic explanation of the human body's capacity to deliver oxygen to the brain will describe the various parts of the respiratory and circulatory systems that bring oxygen into the lung, transfer it to the blood stream and carry it to the brain. (Glennan 2010, pp. 256–257)

In general, it may be said that, apart from some minor differences, mechanistic approaches agree in conceiving a mechanism as something analogous to the 'experimental machine' that, as I have already mentioned in Sect. 3.2, is the substantial core of scientific experiment. According to the well-known definition of "mechanism" by Machamer, Darden, and Craver, "[m]echanisms are entities and activities organized such that they are productive of regular changes from start or set-up to finish or termination conditions".[5]

In this sense, the concept of "robustness" that we have been sketching – which is essentially connected with experimental testing and whose complementary aspects are stability (or robustness-as-stability) and variety (or robustness-as-consilience) – seems an essential ingredient of the new mechanistic approach.[6]

However, the notion of robustness of the mechanist approach, even understood in this way, is only a methodologically necessary condition, not a sufficient one. The tendency to avoid a definite statement about teleology[7] or the explicitly expressed

---

[5] Machamer, Darden, and Craver 2000, p. 3. On the epistemological and ontological differences between Glennan's and Machamer, Darden, and Craver's concept of mechanism, cf. above all Tabery 2004.

[6] In this context it would be interesting to examine the relationships between robustness, mechanism, and an interventionist account of causation (cf. above all Woodward 2002; Mitchell 2008; Craver 2007, ch. 2; Pâslaru 2009). Unfortunately, there is no space to discuss this issue here.

[7] Glennan 2010, for example, who exploits the mechanism concept to understand explanation in history and in the human sciences, gives no attention to the analysis of such concepts as teleology,

fear for the appeal to top-down causes or interlevel causes in biological explanations (cf. Craver and Bechtel 2007, p. 547) show that the most important exponents of the mechanist philosophy of biology consider teleology and final causes only as useless and/or redundant information, which should be *de facto* ignored or even removed. But, as we shall now see, the concept of robustness, and more generally that of mechanism, is too general and insufficient for interpreting biological knowledge without an implicit and counterfactual reference to final or intentional causes.

It is an undeniable empirical fact that the ability of a rubber to return to its original thickness after it has been compressed shares much in common with the robustness of biological systems. And the hierarchical structures of many physical systems are, like those of biology, only "nearly decomposable" (Simon 1996[1962]), both in the sense that the relations between their elements are only relatively additive and linear and in the sense that they are only relatively modular. In other words, it is not always possible to modify an element of a mechanism without disturbing the causal relationships among the other elements in a way that is not fully predictable and/or controllable: as Mach already pointed out, "each element is usually influenced by several others and in different ways".

Notwithstanding these important similarities, however, it has also turned out as a matter of empirical-historical fact that the teleological assumption was a very fruitful point of view in distinguishing living from lifeless beings, whereas it was not very useful when applied to inanimate objects. As I have argued elsewhere, without the aid of the purposes that *we* project into living beings, there was historically no possibility for biology to have an autonomous subject matter (Buzzoni 2015). In fact, the answer to an experimental question formulated in physico-chemical terms will only contain physico-chemical concepts; it is for this reason that physics and chemistry never talk about living organisms *as such*. Physics and chemistry have no concept that could allow us to distinguish a living being from a lifeless being, or between an organ and another organ. Without a mental teleological reference to a living organism as a whole, which settles preferences and goals, the organic could not be distinguished from the inorganic world – that is to say, the field of biology would not be given at all. What is an organ from the standpoint of physics and chemistry? To put it simply, each organ is thought of as corresponding to a purpose

---

purpose, aim or end (he does not even mention them). A partial exception is Craver 2013, in which a great importance is assigned to the teleological terminology and to the teleological stance. On the one hand, in contrast with the general tendency in the new mechanistic philosophy to remain silent on the issue of teleology, the term "teleology" occurs many times in this paper, while it occurred only once, and then in an insignificant and accidental context, in Craver's book of 2007 (cf. Craver 2007, p. 3). Craver 2013 recognizes that a "teleological feature" is involved in the fact that a mechanism "is explicitly defined in terms of what it does. The mechanism works from beginning to end, where the end is not what the mechanism invariably does but what we think it is supposed to do." (Craver 2013, p. 140) On the other hand, Craver 2013 retains a far from negligible residue of objectivistic views about mechanisms. He writes that an "objective explanation", which should express the objective or "ontic" (Salmon) character of scientific explanations, refers to "an objective portion of the causal structure of the world, to the set of factors that bring about or sustain a phenomenon". For a discussion of this point, see Buzzoni 2016.

of an organism: the sight organ is an organ for seeing; the hearing organ is an organ for hearing, and so on.

It is important to note that this teleological language, insofar as it is spoken for the purpose of empirical or scientific research, is neutral as regards both subjective-mental and objective-ontological (which in turn may be either antinaturalistic or naturalistic) readings of the teleological attribution of goals to reality. Indeed, even an overtly ontological-philosophical view would be compatible with my claim that without the aid of the purposes that we project into living beings there was histori-cally no possibility for biology to have an autonomous perspective from which to explore reality. This claim is in itself not affected by the fact that concepts such as aim, goal or intention are thought of either as corresponding or as failing to corre-spond to something real or existing in nature.

The same proviso must be made in saying essentially the same thing from a dif-ferent standpoint, that is, in saying that the field of biology is only given under the presupposition of the notion of organism, seen as a purposive structure which strug-gles for its survival or organizational maintenance. As Mayr aptly pointed out, it is because the courtship of a male animal can be described from this point of view that it falls within the scope of biology:

> [T]he same event may have entirely different meanings in several different conceptual domains. The courtship of a male animal, for instance, can be described in the language and conceptual framework of the physical sciences (locomotion, energy turnover, metabolic processes, and so on), but it can also be described in the framework of behavioural and reproductive biology. And the latter description and explanation cannot be reduced to theo-ries of the physical sciences. Such biological phenomena as species, competition, mimicry, territory, migration, and hibernation are among the thousands of examples of organismic phenomena for which a purely physical description is at best incomplete if not irrelevant (Mayr 1988, pp. 62–63).

The same is obviously true, *mutatis mutandis*, for biological robustness. For example, if we say that the HIV-1 virus is robust against numerous therapeutic inter-ventions due to a high mutation domain, we tacitly assume general mechanisms *for (viral) survivability* (Kitano 2007, p. 2, who refers to Eigen 1993). In anhydrobiosis, tardigrades, under extreme dehydration, suspend metabolism almost completely and become active again upon rehydration (cf. Crowe and Crowe 2000; Singer and Lindquist 1998), where an implicit but nonetheless fundamental reference is made to 'life' and to a *conatus* to continue to exist.

Another example is provided by the mechanism of phototropism, which is one form of plant movements, with reorientation of growth either towards a light source (positive phototropism) or away from light (negative phototropism). The standard explanation is as follows: a lateral accumulation of the phytohormone auxin (from the Greek αὔξειν, that is, to increase or grow) causes the plant to have elongated cells on the shaded side of the stem (see for example Christie and Murphy 2013). Although there are still some aspects of phototropism that are not completely under-stood as to how photoreceptor activation regulates lateral auxin transport to estab-lish phototropic growth, this explanation has been confirmed and improved by many successive researchers and may now be considered as robustly acquired by science.

Many different mechanisms and submechanisms have been examined in detail: for example, several protein families of auxin transporters have been identified, and the role of exposure to blue and red light have been examined (cf. Fankhauser and Christie 2015). But phototropism (which Darwin called heliotropism) is a biological phenomenon because it is of particular adaptive significance for germinating seedlings that must grow toward the light to increase their survival rate (cf. Darwin 1888, p. 491).

The empirical-experimental fact that living systems stand out against their environments as robust entities is not distinctive of living beings; nor is it the fact that energy and matter must be robustly channeled so as to maintain their structure. What is needed to regard something as a living system is to assume that it is struggling to stay alive or to resist death, an assumption to be supported by the successful prediction of its behavior. In this case, robustness of predictions concerning living systems (in both senses of the term "robustness") is intimately connected with the way in which living systems struggle for their survival, for example by continually assimilating food from their environment and using it to maintain and repair themselves. Without some at least implicit reference to a goal-directed behavior, there are no properties at all which can significantly be ascribed to a living being, robustness included.

Summing up our analysis so far, we can say that, although the mechanistic sense of robustness is an essential ingredient of the corresponding biological one, yet any attempted reduction of the continual flexible restoration of living systems to a mechanism leaves unexplained important empirical and historical facts. On the one hand, it is a historical fact that biological investigations can profitably be pursued by the decomposition of living systems into mechanical structures, and of these into substructures. This (experimentally robust) decomposability is necessary to render scientific explanation possible – in the historically most important sense of the word "scientific". On the other hand, however, although life sciences are in a sense entitled to say, as Bacon did in *De augmentis scientiarum*, that inquiry into final causes is sterile and produces nothing, this decomposability into robust mechanisms cannot get rid *in any sense* of the idea of goal-directed activity (or even, as I shall argue later, of purposeful action). *Mutatis mutandis*, we may paraphrase Jacobi (1815, p. 304) and say that without the assumption of attribution of goals or even intentions to nature we are not able to enter into biology, although we are unable to remain within it, if we accept this assumption without explaining how biology (and biological robustness) gains its operational-experimental value.

Therefore, we have to ask the question whether it is possible to investigate teleology scientifically, taking this last word in the sense of Galilean science, without accepting anthropomorphic final causes in a sense that is irreconcilable with modern science. The short answer to this question is that in biology, on the contrary to physics, attribution of goals or even intentions is a general-methodical assumption that opens the field of biology as well as a powerful heuristic-methodical device to investigate living beings in an intersubjectively testable and mechanistic reproducible way. In the rest of the paper I shall briefly argue in defense of this answer.

The first point upon which it is important to be clear is that, even though to project our inner desires and purposes into outer reality can lead to the greatest mistakes, it is not a mistake in itself. The attribution of goals or intentions to nature is justified in biology to the extent that we do not hypostatize this concept as an ontological quality. And in order to ascribe to goal-directedness only a heuristic value, biological research *must use it as a counterfactual artifice, capable of bringing to light robust causal relations which have an objective and intersubjectively testable content.* More precisely, to ascribe to goal-directedness only a heuristic value by using it as a counterfactual artifice is a procedure that is scientifically harmless since the aim or purpose in question can be bracketed without any loss of objective content (or of its robustness). It is true that we can investigate life only after having projected human teleology into nature, but to do this in strictly heuristic terms exercises no constraints over the empirical content that must be ascertained by the usual experimental means. In this sense, the mechanist approach is entirely right in holding that all knowledge must in the end be reduced to experience in mechanical-experimental terms.

The second point is not less important. It is that, when I use the expression 'attribution of goals,' I do not need to assume any distinction between goal and intention, or between goal-ascription and intention-ascription. There has been a long-standing debate over such a distinction as well as the possibility of providing a translation of intention-ascriptions into goal-ascriptions in which no mention is made of properly human intention, intentionality, or teleology.[8]

A great advantage of the position here defended is that, philosophically speaking, this problem does not need to be solved in order to make our point. Philosophically speaking, we may perhaps remain agnostic: any mere translation of teleological behavior in terms of mechanical feedback ever risks making the analysis of teleological behavior circular, since it invokes terms such as goal which cannot be specified without using teleological concepts. But this is not the interesting point in our context. The interesting point is that biology as biology – that is, as a particular science – ought not to be forced to answer this ontological question, whether positively or negatively, in order to put forward predictions which are experimentally and therefore intersubjectively testable. On the contrary, the biologist qua biologist has to take the same attitude as Laplace and say: « Sire, je n'ai pas besoin de cette hypothèse ».

On the one hand, strictly speaking, purposes are concepts or meanings guiding us in our practical activities. In this sense, the concept of an aim or a purpose characterizes only human actions (or, if you prefer, behaviors). However, it would be radically false to say that this concept plays no part in the explanation of natural phenomena. The objects of biology would not be available if the very general notion of an intentional teleology were not presupposed from the beginning. We could not

---

[8] Among some of the relatively old contributions to this debate, which however are not to be neglected, see for example Rosenblueth et al. 1943, Becker 1959 and 1969, Scheffler 1959, Wright 1968. Of course, the particular issue addressed in these papers, though closely connected, is not to be confounded with the more general issue of "proper functions", which is not a matter that can appropriately be addressed here.

say that cells are nourished by the blood or that a cat attends someone or avoided her/his presence without a reference to standard *human* activities and behaviors which are here applied analogically to plants or animals. This way of speaking would have no meaning in complete isolation from human purposes. Moreover, and more fundamentally, the very nature of scientific experiment demonstrates that teleology and efficient-mechanical causality are not only compatible but that final causes are actually *the condition of the epistemic possibility* of mechanical ones, since without our knowledge of final causes there would be no experiment and therefore no imputation of mechanical causes. In fact, apart from teleology, apart from the intentional and conscious planning of an experimental set-up and apart from the human actions which freely start or 'set in motion' the experimental machine, it would be impossible to establish by experiment causal relations in nature. (This last point requires more qualification than I can reasonably offer in this article. For some more details on this point, see Buzzoni 2015, sect. 2).

On the other hand, however, this is only a necessary, but not sufficient condition, which tests must meet in order to be truly experimental. We need also specific empirical information to give content to our metaphors, which cannot be built up by mere philosophical arguments. Without additional empirical information we would not be able to form hypotheses having factual content, that is, which provide some rule for the anticipation of experience.

It should be clear, therefore, that, in our purely *methodological* context – in which no philosophical assumptions concerning teleology should be made – attribution of goals, attribution of intentions, and attribution of purposes are essentially interchangeable terms. As the attribution of intentions or purposes is associated with an attitude of agnosticism respecting the true nature of (living) beings different from ourselves, intentions and purposes can be ascribed to (living) beings, so to speak, only as a particular version of third person talk, which, in the last analysis, must reduce to causal ascriptions expressed by mechanistic statements.

So understood, a projection of our purposes into outer reality is not only an entirely legitimate move, but is even necessary as a first *methodical* step towards a better and always more robust knowledge of the living world. Therefore, biology may *and* should substitute where possible the unconscious use of the teleological principle *for a fully conscious and methodical-counterfactual use of this principle*. Even though biology is unable either to prove or to disprove the truth of this teleological presupposition, the reflexive, typically human concept of purposefulness or intentionality may be, and strictly speaking must be, profitably employed to discover robust mechanisms in all living beings (where "robust" is to be understood in both senses of the word with which we have been concerned in this paper).

Two simple examples may serve as illustration of these points.[9] Cells communicate with their external environment through particular channels, that is, through proteins that are located in the membrane that encloses the cell and have the capacity of detecting the voltage of the membrane and regulating ion conduction or enzyme activity. In particular, voltage-gated potassium channels alter their confor-

---

[9] Both examples are taken from Buzzoni 2015.

mation in response to changes in the potential of the membrane, thereby allowing or blocking the conduction of ions. Such processes rely on so-called voltage sensors, which are embedded inside the cell membrane and contain an excess of positively charged amino acids which react to an electric field (see for example Lee et al. 2005; Schmidt et al. 2006; Alessandrini et al. 2008).

If we regard the situation from a predator's standpoint, we shall be able to better understand how the physico-chemical mechanisms of these channels work. In order to do this, it is clear that some counterfactual formulation is required. We might say that if the predator were a human being, it could pursue the aim of crossing the critical threshold of the ion-exchange equilibrium in the prey's cells (according to a line of reasoning which is difficult not to consider a thought experiment, a hint which I cannot pursue here). Which chains of cause and effect should the predator trigger in order to achieve this goal? The predator could, for instance, inoculate the victim with venom that interferes with the mechanisms of voltage-gated ion channels in order to impair the prey's nervous system. Now, this hypothesis, according to which ion channels are blocked by toxins, is clearly suggested by a goal-directed or teleological perspective. However, in order to test this hypothesis, we must use the experimental method. According to the manipulability theory, which is here presupposed, causal claims are experimentally elucidated in terms of counterfactuals about what would happen as a result of consciously planned interventions. In order to claim that specific causal chains apply to a particular case, we must have an experimental basis for stating what would happen to the value of a variable if another variable is changed. For instance, it would be advisable (and that is exactly what in fact happened) to test different classes of toxins, whether synthetic or natural, in different circumstances, in order to see whether, how and to what extent they interfere with the external vestibule of the ion conduction pores and work like a plug to block the flow of ions.[10]

---

[10] Wang et al. 2004. It would be instructive to re-read Claude Bernard in the light of what I have been saying, but this is not the place to do this and I shall venture only a hint. Bernard tested poisons (above all curare, but also potassium cyanide and strychnine) in different circumstances, in order to get rid of misleading analogies and to see whether, how and to what extent they differ as regards their way of interfering with the different vital functions of living organisms. He used the poisonous substances as experimental devices, very precise scalpels that interrupt, and thus reveal, important causal correlations (see for example Bernard 1857 and Grmek 1966). As a scientist, he could legitimately ignore the reflexive, typically human concept of purposefulness or intentionality that he was, de facto, using to discover robust mechanisms in living beings. But his philosophical analysis is inadequate. On the one hand, it might be conceded, at least to a certain extent, that he understood that the "vital force" (*force vitale*), which distinguishes living beings from non-living beings, must be regarded, as one critic has said of teleology in Bernard, "as a necessary but subjective principle in biology" (Nils Roll-Hansen 1976, p. 73); and consistently with our view, Bernard writes that the "vital force" is "une cause prochaine ou exécutive du phénomène vital, qui toujours est de nature physico-chimique, et tombe dans le domaine de l'expérimentateur." (Bernard 1878, p. 53) But at other times Bernard uses the phrase "vital force" very ambiguously, if not contradictorily. He writes for example that the vital force must be considered as "une cause première, créatrice, législative et directrice de la vie, et inaccessible à nos connaissances" (Bernard 1878, p 53); where it is not easy to understand how it can be characterised as both "inaccessible to our knowledge" and as "a primary cause – creative, legislative, and directive of life."

As this example shows, the discovery of certain connections between causes and effects in biology depends upon the heuristic fiction of counterfactually considering nature as if it acted intentionally. I would even venture to claim that it is an important part of a biologist's work to try to 'empathize' with the living beings that are the subject matter of her or his research, that is to say, to project him or herself into their roles, counterfactually ascribing to them purposes that, strictly speaking, usually we want to ascribe to human beings alone. Indeed, one can empathize not only with another human being's thoughts or feelings but also with the instinct of an animal searching for food, with the striving of an organism defending itself against a bacterium in order to stay alive, or with a cell regulating its potassium balance, so that one of the most important conditions for its life is maintained. It is only by putting oneself in the position of a predator, of an organism, of a cell, that it is possible to ask what we would do in similar situations to prey on an animal, defend ourselves from a bacterium or prevent the entry of a toxin which would interfere with the chemical equilibrium that keeps us alive.

A second example is this. According to Konrad Lorenz, the horse's hoof – just as any other organ – is adapted to the ground of the steppe and acquired its functional form through ages of encounter or clash (*Auseinandersetzung*) "of reality with reality" (Lorenz 1941/1942, pp. 98–99). On the one hand, this illustrates again that the causal relationship between the horse's hoof and the ground of the steppe can be brought to light in its objective and scientific content only by putting it in relation to the animal's aim or purpose of fleeing from its predators. The particular contents of biological research would not come to light as causal-mechanical relationships within the *mare magnum* of the abstract conceivable conditions of a biological phenomenon if they were not seen against the background of an implicit teleology. In this way, the principle of teleology, which – strictly speaking – is probably only valid within the field of human experience, has a proper use outside this field.

On the other hand, the use of the teleological principle is scientifically harmless since the aim or purpose in question can be bracketed without any loss of robustly reproducible content. That a particular functional form of a horse's hoof is a better (faster and more efficient) way of running can be equally true of an animal that runs away from its predator in order to save its life by flight as of a hypothetical hippomorphous robot that simulates this behavior in an experimental set-up. Or, expressing the same idea in a different way, that a particular functional form of a horse's hoof is better for running can be equally true for one who maintains that animals can be explained entirely through physics and chemistry, and that robots are as intelligent as animals or human beings, as for one who would consider animals as irreducible to physics or chemistry, or even to empirical concepts.

Biological mechanisms are intersubjectively reproducible and law-like patterns extractable from behaviors or processes to which we have ascribed teleological character, though only counterfactually. In other words, mechanisms are what they are in virtue of their relations to a goal-directed or teleological process, and a goal-directed or teleological process is what it is in virtue of mechanisms through which its goal can be attained.

A mechanism (or a machine) must be thought of as a means to a further end in order to gain an intelligible meaning, and conversely the realization of an end is intelligible only in relation to some means, a mechanism or a machine, capable of realizing it. In this sense, the robustness of living beings (and of the investigation into them) must be thought of in the light of the human concept of purposefulness. However, this purposive pursuit of a presupposed end is to be regarded as a kind of Wittgenstein's ladder, which we can throw away after employing, or as a falsework (the so-called centering) of a masonry arch, that is, as a temporary support which takes the entire weight during the building, but that can be removed when the arch is complete. In a similar way, *the heuristic artifice of goal-directedness in biology is the most convenient means to exploit human teleology for scientific purposes*: on the one hand, we run no risk of hypostatizing the teleological moment thanks to the awareness of their fictional status as possible alternative scenarios to any observed or believed reality; on the other, they are only provisionally held and can be dismissed when the connection of means and end is interpreted as a connection of cause and effect about life, which has to have its worth tested experimentally.

3.4 Conclusion

The first part of this paper discussed two important meanings of robustness (robustness as stability as against variations in parameter values and robustness as consilience of results from different sources of evidence) and showed their essential connection with the notion of intersubjective reproducibility. Robustness in both senses of the term is intimately connected with the notion of scientific experiment. This is the important element of truth of the mechanistic systems approach, which explains events as products of robust and regular systems and processes.

However, as shown in the second part of the paper, the concept of robustness of a mechanism, if applied to biological systems, does not exhaust the whole content of what we mean by this term in biology, and it is one-sided and insufficient without a methodical (not ontological) reference to final causes. Contrary to the notion of mechanism as defended by most of the mechanistic philosophers, a 'mechanical' investigation of life mechanisms (or machines), while of fundamental importance in the scientific practice of biology, cannot get rid of teleology and final causes. Pace Lander, no robust calculations may provide the "foundations of a new calculus of purpose" (cf. Lander 2004, p. 713).

We may say that we are not able to enter into biology without teleology, although we are unable to remain within it, if we accept teleology without explaining how biology (and biological robustness) gains, nevertheless, its operational-experimental value. The way out suggested in this paper is as follows. On the one hand, a teleological reference is always implicit in the robustness of biological systems, because such robustness is to be considered from the point of view of organisms that have goals and struggle for their survival. On the other hand, the assumption of the teleological point of view is justified in biology only to the extent that we use it as a counterfactual artifice, capable of bringing to light robust causal-mechanistic relations that have an objective and intersubjectively testable content. In this way, the reflexive, typically human concept of purposefulness may be employed to investigate living beings scientifically, that is in an intersubjectively testable and reproducible way, to discover causal-mechanical or experimental relations in living systems which are robust in both senses of the word with which we have been dealing.

Acknowledgements I presented an earlier version of this paper at the "Interdisciplinary Workshop on Robustness – Engineering Science" (Rome, February 5th – 6th, 2015). Thanks to all those who contributed to the discussion of the paper during and after the conference. Italian Ministry for Scientific Research (MIUR) provided funds for this research (PRIN 2012).

References

Alessandrini, A., Gavazzo, P., Picco, C., & Facci, P. (2008). Voltage–induced morphological modifications in oocyte membranes containing exogenous K channels studied by electrochemical scanning force microscopy. *Microscopy Research and Technique, 71*, 274–278.

Alon, U., Surette, M. G., Barkai, N., & Leibler, S. (1999). Robustness in bacterial chemotaxis. *Nature, 397*, 168–171.

Barkai, N., & Leibler, S. (1997). Robustness in simple biochemical networks. *Nature, 387*, 913–917.

Beckner, M. (1959). *The biological way of thought.* New York: Columbia University Press.

Beckner, M. (1969). Function and teleology. *Journal of History of Biology, 2*, 151–164.

Bernard, C. (1857). *Leçons sur les effets des substances toxiques et médicamenteuses.* Paris: Baillière.

Bernard, C. (1878). *La science expérimentale.* Paris: Baillière.

Box, G. (1953). Non-normality and tests on variances. *Biometrika, 40*, 318–335.

Bridgman, P. W. (1927). *The logic of modern physics.* New York: Macmillan.

Buzzoni, M. (2008). *Thought experiment in the natural sciences.* Würzburg: Königshausen+Neumann.

Buzzoni, M. (2015). Causality, teleology, and thought experiments in biology. *Journal for General Philosophy of Science, 46*, 279–299.

Buzzoni, M. (2016). Mechanisms, experiments, and theory-ladenness: A realist–perspectivalist view. *Epistemologia*. Special Issue of Axiomathes, 26, 411–427.

Buzzoni, M. (2017). Robustness, intersubjective reproducibility, and scientific realism. In E. Agazzi (Ed.), *Scientific realism* (pp. 133–150). Berlin/Heidelberg/New York: Springer.

Calcott, B. (2010). Wimsatt and the robustness family: Review of Wimsatt's re-engineering philosophy for limited beings. *Biology and Philosophy, 26*, 281–293.

Cartwright, N. (1983). *How the laws of physics lie.* Oxford: Clarendon.

Christie, J. M., & Murphy, A. S. (2013). Shoot phototropism in higher plants: New light through old concepts. *American Journal of Botany, 100*(1), 35–46.

Clausing, D. P. (2004). Operating window: An engineering measure for robustness. *Technometrics, 46*, 25–29.

Craver, C. F. (2007). *Explaining the brain. Mechanisms and the mosaic unity of neuroscience.* New York: Oxford University Press.

Craver, C. F. (2013). Functions and mechanisms: A perspectivalist view in. In P. Huneman (Ed.), *Functions: Selection and mechanisms* (pp. 133–158). Berlin: Springer.

Craver, C. F., & Bechtel, W. (2007). Top-down causation without top-down causes. *Biology and Philosophy, 22*, 547–563.

Crowe, J. H., & Crowe, L. M. (2000). Preservation of mammalian cells-learning nature's tricks. *Nature Biotechnology, 18*, 145–146.

Darwin, C. (1888). *The power of movement in plants*. London: John Murray.

De Raedt, H., Katsnelson, M. I., & Michielsen, K. (2014). Quantum theory as the most robust description of reproducible experiments. *Annals of Physics, 347*, 45–73.

Eigen, M. (1993). Viral quasispecies. *Scientific American, 269*, 42–49.

Fankhauser, C., & Christie, J. M. (2015). Plant phototropic growth. *Current Biology, 25*(9), R384–R389.

Félix, M. A., & Barkoulas, M. (2015). Pervasive robustness in biological systems. *Nature Reviews Genetics, 16*(8), 483–496.

Glennan, S. (2010). Ephemeral mechanisms and historical explanation. *Erkenntnis, 72*, 251–266.

Grmek, M. D. (1966). Notes inédites de Claude Bernard sur les propriétés physiologiques des poisons de flèches (curare, upas, strychnine et autres). *Biologie Médicale*, 55, hors série.

Hudson, R. (2014). *Seeing things: The philosophy of reliable observation*. Oxford/New York: Oxford University Press.

Jacobi, F. H. (1815). *Werke, II*. Leipzig: Fleischer.

Kitano, H. (2004). Biological robustness. *Nature Reviews Genetics, 5*, 826–837.

Kitano, H. (2007). Towards a theory of biological robustness. *Molecular Systems Biology, 3*, 1–7.

Kitano, H., Oda, K., Kimura, T., Matsuoka, Y., Csete, M., Doyle, J., & Muramatsu, M. (2004). Metabolic syndrome and robustness tradeoffs. *Diabetes, 53*(Suppl 3), S6–S15.

Lander, A. D. (2004). A calculus of purpose. *PLoS Biology, 2*(6), 0712–0714.

Lee, S. Y., Lee, A., Chen, J. Y., & MacKinnon, R. (2005). Structure of the KvAP voltage-dependent K channel and its dependence on the lipid membrane. *Proceedings of the National Academy of Sciences USA, 102*, 15441–15446.

Levins, R. (1966). The strategy of model-building in population biology. *American Scientist, 54*, 421–431.

Levins, R. (1993). A response to Orzack and Sober: Formal analysis and the fluidity of science. *The Quarterly Review of Biology, 68*, 547–555.

Lorenz, K. (1941/1942). Kants Lehre vom Apriorischen im Lichte gegenwärtiger Biologie. *Blätter für deutsche Philosophie, 15*, 94–125.

Mach, E. (1906[1976]). Erkenntnis und Irrtum. Leipzig, Barth, 2th edition. English (trans: McCormack, T.J.) Knowledge and error. Dordrecht/Boston: Reidel.

Machamer, P., Darden, L., & Craver, C. F. (2000). Thinking about mechanisms. *Philosophy of Science, 67*, 1–25.

Mayr, E. (1988). *Toward a new philosophy of biology: Observations of an evolutionist*. Cambridge: Harvard University Press.

Mitchell, S. D. (2008). Exporting causal knowledge in evolutionary and developmental biology. *Philosophy of Science, 75*, 697–706.

Pâslaru, V. (2009). Ecological explanation between manipulation and mechanism description. *Philosophy of Science, 76*, 821–837.

Perrin, J. (1916). *Atoms*. (D. L. Hammick, Trans.). New York: Van Nostrand.

Popper, K. R. (1959[2002]). *The logic of scientific discovery*. London: Hutchinson (quotations are from the 2002 edition, London: Routledge).

Psillos, S. (2011). Moving molecules above the scientific horizon: On Perrin's case for realism. *Journal for General Philosophy of Science, 42*, 339–363.

Roll-Hansen, N. (1976). Critical teleology: Immanuel Kant and Claude Bernard on the limitations of experimental biology. *Journal of the History of Biology, 9*, 59–91.

Rosenblueth, A., Wiener, N., & Bigelow, J. (1943). Behavior, purpose and teleology. *Philosophy of Science, 10*, 18–24.

Salmon, W. C. (1984). *Scientific explanation and the causal structure of the world*. Princeton: Princeton University Press.

Scheffer, I. (1959). Thoughts on teleology. *The British Journal for the Philosophy of Science, 9*, 265–284.

Schmidt, D., Qiu-Xing, J., & MacKinnon, R. (2006). Phospholipids and the origin of cationic gating charges in voltage sensors. *Nature, 444*(7), 775–779.

Simon, H. E. (1996[1962]). *The sciences of the artificial* (3rd ed.). London/Cambridge, MA: MIT Press.

Singer, M. A., & Lindquist, S. (1998). Thermotolerance in Saccharomyces cerevisiae: The Yin and Yang of trehalose. *Trends in Biotechnology, 16*, 460–468.

Stegenga, J. (2009). Robustness, discordance, and relevance. *Philosophy of Science, 76*, 650–661.

Stelling, J., Sauer, U., Szallasi, Z., Doyle, F. J., 3rd, & Doyle, J. (2004). Robustness of cellular functions. *Cell, 118*, 675–685.

Strand, A., & Oftedal, G. (2009). Functional stability and systems level causation. *Philosophy of Science, 76*, 809–820.

Tabery, J. G. (2004). Synthesizing activities and interactions in the concept of a mechanism. *Philosophy of Science, 71*, 1–15.

Thorén, H. (2014). Resilience as a unifying concept. *International Studies in the Philosophy of Science, 28*, 303–324.

von Dassow, G., Meir, E., Munro, E. M., & Odell, G. M. (2000). The segment polarity network is a robust developmental module. *Nature, 406*, 188–192.

Wang, J. M., Roh, S. H., Sunghwan, K., Lee, C. W., Jae, I. K., & Swartz, K. J. (2004). Molecular surface of tarantula toxins interacting with voltage sensors in K channels. *Journal of General Physiology, 123*, 455–467.

Whewell, W. (1840[1847]). *The philosophy of the inductive sciences. Founded upon their History* (1st ed., London: 1840; 2nd ed., London: 1847). Quotations are from the second edition.

Wilke, C. O. (2006). Robustness and evolvability in living systems. *Bioscience, 56*, 695–696.

Wimsatt, W. C. (2007). *Re-engineering philosophy for limited beings: Piecewise approximations to reality*. Cambridge: Harvard University Press.

Wimsatt, W. C. (2012). "Robustness: Material, and inferential, in the natural and human sciences". In Soler et al., *Characterizing the robustness of science: After the practice turn in philosophy of science*. Dordrecht: Springer, pp. 89–104.

Woodward, J. (2002). What is a mechanism? A counterfactual account. *Philosophy of Science, 69*, S366–S377.

Woodward, J. (2006). Some varieties of robustness. *Journal of Economic Methodology, 13*, 219–240.

Wright, L. (1968). The case against teleological reductionism. *The British Journal for the Philosophy of Science, 19*, 211–223.

Marco Buzzoni is full professor of Philosophy of science at the University of Macerata. Member of the "Académie Internationale de Philosophie des Sciences" (Brussels) and of the "Institut International de Philosophie" (Paris), he is co-editor of "Epistemologia," an annual special issue of the journal Axiomathes. He was a visiting professor at the Universities of Würzburg, Marburg and Duisburg-Essen. He worked on Popper, Kuhn, the relationship between science and technology, the methodology of human sciences, the philosophy of biology and thought experiment. Among his books are *Thought Experiment in the Natural Sciences* (2008); *Science and Technique* (1995, in Italian); *Operationalism and Hermeneutics.* (1989, in Italian); *Semantics, Ontology and Hermeneutics of Scientific Knowledge. Essay on Thomas Kuhn* (1986, in Italian); *Knowledge and Reality in K.R. Popper* (1982, in Italian).

Chapter 4
Multiple Realization and Robustness

Worth (Trey) Boone

Abstract Multiple realization has traditionally been characterized as a thesis about the relation between kinds posited by the taxonomic systems of different sciences. In this paper, I argue that there are good reasons to move beyond this framing. I begin by showing how the traditional framing is tied to positivist models of explanation and reduction and proceed to develop an alternate framing that operates instead within causal explanatory frameworks. I draw connections between this account and the notion of functional robustness in biology and neuroscience. I then examine two cases from systems neuroscience that substantiate my account and show how traditional debates fail to track important features of these cases.

Keywords Multiple realization · Functions · Natural kinds · Explanation · Reduction

4.1 Introduction

Traditionally, multiple realization (MR) has been understood as a thesis about the relation between kinds posited by taxonomic systems in different sciences (e.g. psychology and neuroscience). For instance, a psychological kind, like short-term memory, would be considered multiply realized to the extent that it corresponds to a number of distinct neuroscientific kinds—e.g. differents kinds of synaptic strengthening in hippocampus. This characterization of MR has been heavily influenced by positivist models of explanation, reduction, and the unity of science (Hempel 1942, Nagel 1961, Oppenheim and Putnam 1958), against which early arguments concerning MR (Putnam 1967, 1975; Fodor 1974) were targeted. In this paper, I explicitly reframe MR in terms of causal explanatory frameworks that better capture explanatory practice in the special sciences. Within such frameworks, MR becomes more a thesis about causal structure than about mapping relations between

W. Boone (✉)
Department of History and Philosophy of Science, University of Pittsburgh,
Pittsburgh, PA, USA

 75
M. Bertolaso et al. (eds.), *Biological Robustness*, History, Philosophy
and Theory of the Life Sciences 23, https://doi.org/10.1007/978-3-030-01198-7_4

kinds. This shift in framing exposes connections between MR and the notion of functional robustness in biology and neuroscience.

I proceed as follows. In Sect. 4.2, I show how the traditional framing of MR is tied to outmoded positivist conceptions of explanation and reduction. I then offer an analysis of MR that operates within frameworks of causal explanation that better capture explanatory practice in psychology and neuroscience. In Sect. 4.3, I draw connections between this conception of MR and the phenomenon of functional robustness in biology and neuroscience. I examine in detail two cases of robust functions in neural systems. In Sect. 4.4, I further develop my account by considering and responding to the objection that the account I offer still essentially construes MR to be a relation between kinds. I conclude with brief remarks on the ways this reframing of MR alters the landscape of debate surrounding nonreductive accounts of the mind-brain relation, and the special sciences more generally.

4.2 Multiple Realization and Causal Explanation

The preoccupation with kinds in philosophical discussions of multiple realization has been, in large part, a holdover from now defunct positivist conceptions of explanation and reduction. The deductive-nomological (D-N) model of explanation, of which the Nagelian model of reduction is an extension, maintains that to explain a phenomenon is to subsume it under some law-like regularity (Hempel 1942; Hempel and Oppenheim 1948; Nagel 1961). This conception of explanation thus assigns the explanatory value of theories to their laws, and by fiat to the (natural) kind terms that figure into those laws. Fodor's (1974) seminal argument from multiple realization to the autonomy of the special sciences targeted the Nagelian model of reduction.[1] As a result, Fodor couched MR as a relation between higher- and lower-level kinds, precluding the formation of nomic bridge principles between higher- and lower-level sciences. This general framing has shaped much of the subsequent debate surrounding MR.

The D-N model, however, has proven to be an inadequate account of explanation in the special sciences, particularly psychology and neuroscience (as well as biology, more generally). A primary reason for this is that laws in the traditional sense (qua universal generalizations) do not play a prominent role in either psychology or neuroscience, and relatedly, explanations in neither psychology nor neuroscience proceed by subsuming phenomena under regularities (see, e.g., Cummins 1983a, b, Ch1; Craver 2007). To the contrary, regularities in both psychology and neuroscience provide the targets of explanations, the *explananda*, rather than the *explanantia* (Cummins 1983a, b, 2000).

For instance, the "cocktail party effect" denotes a regularity according to which people are able to single out the sound of their names in a noisy environment

---

[1] Despite the fact that Fodor's title suggests that his target is Putnam and Oppenheim's account of the unity of science. See Shapiro and Polger (2012) for detailed discussion.

(Bronkhorst 2015). Simply citing this effect does little to explain a particular instance of this phenomenon—to do so would be more akin to explaining the sedative properties of opium by appealing to its "dormitive virtue", as famously quipped by Moliere in 1665. Rather, the cocktail party effect characterizes an *explanandum*, and psychology seeks explanations for why this regularity holds. Similarly, gradually depolarizing a neuron to a membrane potential around -40 mV is regularly followed by a rapid depolarization of the cell—the rising phase of the action potential. But again simply citing this regularity does nothing to explain a particular instance of neural depolarization. Rather, the regularity is the target of explanation into the mechanisms of the action potential. A primary function of taxonomic systems is to capture these sorts of regularities within different scientific domains. That is, with respect to explanatory practice in the special sciences, taxonomic systems and the kinds they posit serve more to characterize *explananda* than to provide *explanantia*.[2]

The models of explanation in philosophy of science that have supplanted the positivist framework take causation, rather than subsumption under laws, as the central feature of scientific explanation (e.g. Bechtel 2008; Craver 2007; Salmon 1984; Woodward 2003). While varied in their particulars, what is common to these models is the idea that to explain a phenomenon is to situate it within a causal nexus. Such models fare better at capturing explanatory practice in both psychology and neuroscience. For instance, psychologists look to explain the cocktail party effect by analyzing it in terms of functional subprocesses—e.g. selective auditory attention and speech channel separation. Similarly, neuroscientists have explained the rising phase of the action potential by investigating the workings of voltage-gated Na^+ channels. In explaining how regularities arise, both psychology and neuroscience look to the causal processes that give rise to these sorts of regularities.

For the most part, causal models of explanation stress decomposition of a system in order to explain *how* it operates to give rise to some phenomenon. The mechanistic framework (Bechtel and Richardson 1993; Bechtel 2008; Machamer et al. 2000; Craver 2007) currently provides a dominant framework of explanation via decomposition. According to this framework, roughly, to explain a phenomenon is to decompose it into some set of entities and activities that, appropriately organized, explain how the phenomenon was produced (Fig. 4.1b). Such decompositional explanations, however, only provide half of the story, especially if one is interested in interpolating MR into this framework. That other half consists in upward-directed analyses that explain *what* a system or phenomenon does within some containing system (Fig. 4.1a). Such analyses are closely related to what Craver (2001, 2012) calls "contextual explanations" and the explanatory strategy is similar to Bechtel's (2008) notion "reconstituting a phenomenon."

---

[2] Of course, taxonomic systems also play crucial roles in explanatory practice, but their explanatory value does not consist in capturing nomic regularities or "carving nature at its joints". Rather, the explanatory value of taxonomic systems consists in providing the terms for capturing causal relations between higher- and lower-level analyses. As such, it is those causal relations that do the explanatory heavy lifting in the special sciences, not the kind terms themselves. I revisit this point in more detail in Sect. 4.4, but for now this fast and somewhat loose discussion will do.

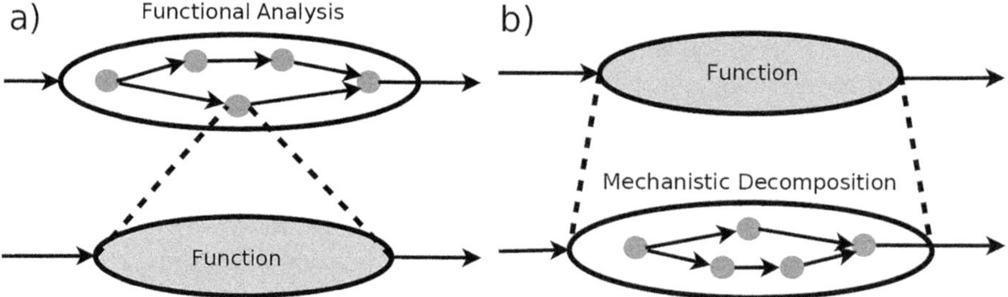

Fig. 4.1 Diagram of the contrast between (**a**) upward-directed functional analyses and (**b**) decompositional mechanistic analyses

Cummins's (1975) account of functional analysis remains one of the canonical ways of capturing this explanatory strategy in both precise and general terms. Cummins's account maintains that functions are ascribed by situating a capacity of a system within an analysis of a capacity of some containing system. In other words functions are attributed relative to the role they play in analysis of other capacities.

> x functions as a φ in s (or: the function of x in s is to φ) relative to an analytical account A of s's capacity to ψ just in case x is capable of φ-ing in s and A appropriately and adequately accounts for s's capacity to ψ by, in part, appealing to the capacity of x to φ in s. (Cummins 1975: 762)

Both decompositional analyses and upward-directed analyses are crucial to understanding functions within a framework of causal explanation. Upward-directed analyses justify functional attributions, explaining *what* a system does within some containing system, while decompositional analyses explain *how* that function is performed by various subsystems. Of course, there is a sort of symmetry between both forms of analysis. A functional analysis can constitute a mechanistic analysis of the function of the containing system, and a mechanistic analysis of a particular function can constitute a functional analysis relative to which the functions of the components of the mechanism are attributed (Piccinini and Craver 2011). Nonetheless, it is useful to keep in mind the distinction between these two forms of analysis in order to interpolate MR into causal-explanatory frameworks.

Juxtaposing functional and mechanistic analyses, MR can be defined as sameness or stability of function (qua causal role in some functional analysis) in spite of difference in the mechanisms performing that function. Mechanisms are individuated in terms of causal relevance. Two mechanisms are distinct just in case they consist in distinct sets of entities and activities that make relevantly different contributions toward explaining the target phenomenon (in this case, some function or capacity).[3] Causal relevance can be understood in terms of manipulation and control

---

[3] Larry Shapiro (2000, 2004) has also developed an account of MR according to which realizations are distinguished on the basis of causal relevance. As such, Shapiro is a progenitor of the move to frame MR in causal explanatory frameworks. However, Shapiro relies on an intuitive notion of causal relevance, and fails to offer a precise criterion for functionality similarity. Instead, in his earlier work on the subject (though cf. Polger and Shapiro 2016), Shapiro accepts Kim's (1992)

(Woodward 2003; Craver 2007, Chap. 3). Thus a feature of a mechanism is causally relevant if manipulating it while holding the other features of the mechanism fixed alters the phenomenon the mechanism is invoked to explain.

Functional sameness or stability can also be cashed out in terms of causal relevance. Here it is the absence of causally relevant differences (relative to the functional analysis of some containing system) that denotes functional stability. That is, a function is stable across multiple instances just in case whatever differences obtain across those instances are not causally relevant to the role of that function within its containing system. Thus MR in this framework amounts to the thesis that there are multiple relevantly different causal pathways that converge on a relevantly stable function.[4] For clarity, this thesis of Causal Explanatory MR can be stated as the joint satisfaction of the following two conditions.

4.2.1 *Causal Explanatory Multiple Realization (CEMR)*

1. Two mechanisms are different realizations of a function just in case there are differences between them that would make a difference to performance of the function they explain under controlled intervention (holding all other aspects of the mechanism fixed).
2. Two instances of a function are relevantly similar just in case whatever differences obtain between them do not make a difference to their roles in explaining the capacity of a containing system under controlled intervention (holding all other aspects of the functional analysis of that system fixed).

This may seem to invite a puzzle. If two mechanisms are really distinct, then there are differences between them that are causally relevant to performance of the function in question. So it would seem that the function cannot be stable across those differences. In other words, it may seem that conditions (1) and (2) are actually mutually incompatible. In essence, this is the same issue that has motivated a predominant thread of MR skepticism due to Larry Shapiro (2000, 2004). Shapiro has argued that proponents of MR face a dilemma when it comes to distinguishing realizations of a functional kind. The dilemma runs as follows.

> Horn one: if two instances of a functional kind do not differ in a way that is relevant to performance of the function in question, then those instances do not properly correspond to distinct realizations of that kind.

---

principle of causal individuation of kinds, which in turns leads him to skepticism regarding MR (more on this below). My account thus can be seen as building on Shapiro's work, offering more precise analysis of both causal relevance and functional similarity by tying both to notions of manipulation and control in causal explanation.

[4] By "causal pathway" here I mean a sequence of steps leading from some causal factor to its effect (in this case, some appropriately specified function). This is in rough alignment with the sequential notion of mechanism laid out in Machamer et al. (2000).

For instance, two waiter's corkscrews that differ only in color are not properly distinct realizations of the kind, corkscrew. It should be clear that the account I've been developing effectively accepts this horn (though I resist framing the issue in terms of kinds—more on this in Sect. 4.4).

> Horn two: if two instances of a functional kind do differ in a way that is relevant to performance of the function in question, then it would seem that there are genuine causal differences between those instances and thus that they correspond to distinct kinds.

The consequent in horn two follows from acceptance of the principle of causal individuation of kinds (Kim 1992). For a mundane example, consider spark-ignition and compression-ignition as relevantly different ways of powering an engine. It would be misleading to say that spark-ignition engines and compression-ignition engines are different realizations of the same kind because the causally relevant difference between them seems to track a difference in kind rather than a difference in realization of the same kind.

The account of MR developed above offers a way around this dilemma. The advance offered is a precise criterion for functional stability that does not get bogged down in issues of kind individuation—i.e. condition (2) of CEMR. Specifically, stating the issue in terms of kinds invites application of a general criteria of "kindhood" (like Kim's principle of causal individuation) that may not actually be relevant to scientific instances of MR. If we instead interpolate Shapiro's second horn into CEMR, it is plain to see that the issue just points to the puzzle outlined above: if two mechanisms are really distinct, then there are differences between them that are causally relevant to performance of the function in question; and so it would seem that the function cannot be stable across those differences. The issue here is just to see how relevantly different causal pathways can converge on a (relevantly) stable function. And the apparent puzzle can be resolved by noting that there may be differences in other causally relevant features of a mechanism that compensate for some particular causally relevant difference to produce a stable function. In such a case, a particular causally relevant difference is sufficient for distinguishing two realizations of a function, while the compensatory differences among other causal factors in turn enable stability (or relevant similarity) of function. This is in fact commonplace in biological systems, as will become clear in the next section.

4.3 Multiple Realization as Distributed Functional Robustness

To this point, I have argued that multiple realization can be understood in causal explanatory frameworks as similarity in what a system does (relative to a functional analysis of some containing system), in spite of differences in how it does it (specified by some set of mechanistic decompositions). This conception of MR is tied purely to the structure of causal explanations rather than to features (e.g. nomicity, causal individuation, projectibility) of kinds that figure into different taxonomic systems (cf. Fodor 1997, Kim 1999, again more on this in Sect. 4.4). The notion of (functional) robustness maps fairly precisely onto this causal explanatory characterization. The aim of this section is to flesh out this connection and to provide empirical examples of robustness that thereby substantiate this account of MR.

In the first place, there are several related notions of robustness that have received substantive attention in both the biological sciences and philosophy of science and should be distinguished. The first, which I will term "methodological" robustness was introduced by Levins (1966) and has been championed in philosophy of science with the work of Wimsatt (1980, 1981, 2007) and more recently Weisberg (2006) and Schupback (2016). Robust in this sense means "accessible (detectable, measurable, derivable, produceable, or the like) in a variety of independent ways" (Wimsatt 2007, p.196). Robustness has also been invoked, with close ties to notions of "stability" and "invariance", that have been cited as criteria on explanatory generalizations that move away from the standard (positivistic) conception of "laws of nature"—qua universal generalizations (e.g. Mitchell 1997; Woodward 2003). The notion of robustness in which I am interested is related to both of these concepts, but is nonetheless distinct in relevant ways.[5] This notion, which I call "functional" robustness, is the robustness of some function or effect produced by a system over variation in or perturbations to the components and properties of that system (Mitchell 2008). This latter notion has been central to recent research in biology and has played a crucial role in systems neuroscience.

Kitano (2004) defines functional robustness as "a property that allows a system to maintain its functions despite external and internal perturbations" (Kitano 2004: 826).[6] The concept is of central relevance to genetic networks in which a large amount of redundancy is built to ensure that systems do not break down in the face of, e.g., minor errors in genetic transcription. Functional robustness is also of central relevance to engineering science in systems in which stable effects have to be maintained in response to a range of environmental perturbations. For instance, the autopilot system in modern airplanes is designed to maintain a flight path against a range of changing atmospheric conditions through compensatory adjustments to various flight mechanisms; similar for cruise control in maintaining a constant speed in automobiles. It is no coincidence that the systems in which functional robustness figures most crucially are also those in which the notion of function has typically been employed—i.e. biological systems and engineered artifactual

---

[5] The main difference being that the notion of functional robustness is a feature of phenomena or systems *in the world*, rather than a feature of models or generalizations *about the world*. Functional robustness is thus, in a relevant sense, a metaphysical property, whereas robustness of models/generalizations is an epistemic property of those means of representing the world. The relation between the concepts is that robust models/generalizations will be necessary to capture certain features of functionally robust systems. This relation seems to be asymmetric in the sense that the referents of robust models/generalizations need not be functionally robust systems.

[6] Kitano uses the term "biological robustness" because he is interested specifically in how the notion of functional robustness applies to genetic networks. For consistency and to keep clear these multiple senses of the term "robustness," I continue to use my more general term, "functional robustness," in reference to Kitano's work and throughout the remainder of the paper.

systems. In such systems there are selective pressures for effects rather than causes, and so there is need for stability in *what* a system does that supersedes stability, and in fact errs toward variation, in *how* it does it.

Some initial distinctions are in order before turning to specific examples of robustness in neural systems. Robustness, in the sense discussed by Kitano and other biologists, should be distinguished from the more general concept of functional stability. For any function there is some normal range of variation in its mechanisms over which it may be stable. For instance spark-ignition engines can combust a range of air-fuel mixture ratios (roughly, between 8:1 and 18:1) that are regulated by a carburetor. The function of the engine is thus stable over this range of ratios. But this form of stability is weaker than or at least distinct from that implied by the concept of robustness of interest to biologists. Functional robustness is a subclass of functional stability that involves some form of reorganization of a system in order to maintain function in the face of perturbations.[7] The concept of reorganization here implies different causal contributions from other components of the mechanism in question.[8]

Here a further distinction can be drawn between (at least) two ways in which reorganization can arise: redundancy and distributed robustness (Wagner 2005). Redundancy occurs when a system maintains function via some redundant mechanism that fills in for a perturbed component. For instance, imagine a spark-ignition engine with a backup carburetor that fills in should the primary carburetor become damaged. In such cases, the redundant part plays the same causal role in the system. As such, functional robustness via redundancy does not qualify as a genuine instance of MR based on the account offered in Sect. 4.2 (due to the lack of causally relevant differences in the mechanisms that explain such functions). By contrast, distributed robustness occurs when many different parts play a range of different causal roles that compensate for effects of perturbations. Though there is no easy analog for engines, it might be something like a spark-ignition engine having the capacity to reorganize itself into a compression-ignition engine and sort out a way of converting gasoline into diesel in response to a carburetor failure. It sounds ridiculous in the context of engines, but something like this seems to be remarkably common in certain biological systems.

In systems neuroscience, the study of robustness very much is a science of multiple realization. Neuroscientists concerned with robustness strive to understand many of the features of the mind-brain relation that motivated early work on multiple

---

[7] Other subclasses of functional stability would include, for instance, stability of a normal range of variation (see above) or stability achieved via redundancy (see below).

[8] A similar notion can be found in the concept of degeneracy (Edelman and Gally 2001). In their definition of degeneracy, however, Edelman and Gally stress *structural* differences in the elements that perform similar functions. Structural differences need not be causally relevant differences, so the relation between my concept of functional robustness and this notion of degeneracy will hang on how one cashes out the notion of structural differences. Further, degeneracy is often tied to the idea that different elements can perform the same function in one context, but perform different functions in some other context. My notion of functional robustness is neutral regarding this sort of multifunctionality.

realization—e.g., the stability of macrolevel regularities to microlevel variation (Putnam 1975; Fodor 1968, 1974), the fact that the same psychological kinds seem to be realized and realizable in different organisms and artificial systems (Putnam 1975), the fact that psychological functions can be stable over changes that occur in the course of development, and the fact that psychological function can be stable over substantial neural damage (Block and Fodor 1972). In spite of the patent relevance of functional robustness to MR, it has received scant attention from philosophers of mind. This is likely due to the fact that the obviousness of the connection has been obscured by the positivistic hangover that has shaped much of the debate on MR. However, with the causal explanatory framing of MR I offered in Sect. 4.2, it is not much work to connect these two concepts.

To see how MR and distributed robustness relate, it will be helpful to first examine some instances of robustness in neural systems. In long-lived organisms—including humans and lobsters (the purpose of this odd association will become apparent)—individual neurons can persist and function properly for decades. By contrast the proteins and receptors that modulate the electrophysiological properties of those neurons are decaying and being replaced on timescales of minutes to hours and days to weeks. As a result, the features that determine a neuron's electrophysiological properties are in a continuous state of flux. And yet those electrophysiological properties are remarkably stable over time. This poses a mystery regarding how this stability is achieved. As Marder and Goaillard (2006) state the problem,

> [E]ach neuron is constantly rebuilding itself from its constituent proteins, using all of the molecular and biochemical machinery of the cell. This allows for plastic changes in development and learning, but also poses the problem of how stable neuronal function is maintained as individual neurons are continuously replacing the proteins that give them their characteristic electrophysiological signatures. (Marder and Goaillard 2006: 563)

The electrophysiological signatures here refer to both the response properties of neurons as well as their intrinsic excitability. These features are determined by proteins and receptors that enable and modulate the flow of ions across the cell membrane. Experimental work has revealed that individual neurons exhibit many-fold variability in their expression of particular ion channels (see Marder and Goaillard 2006, for a review). In spite of this variability in channel density (i.e. ion channels per surface area), those same neurons exhibit remarkably similar electrophysiological profiles. This presents a puzzle. The influx and outflow of ions is what explains the characteristic fluctuations in membrane potential that constitute the electrophysiological properties of a given neuron. So how is it that the channel densities that determine the rates of the influx and outflow of ions can vary while the electrophysiological properties remain stable?

Note that the puzzle encountered here is the same puzzle posed at the end of Sect. 4.2. The problem there was to understand how features that are causally relevant to the performance of a function can vary while that function remains stable. The answer I alluded to was that there can be compensating differences in other causally relevant features that explain this stability. And indeed computational models demonstrate that a variety of combinations of ion channel densities can give rise to similar electrophysiological profiles in model neurons. These results show that very different combinations of channel densities can produce the same intrinsic bursting profiles (Golowasch et al. 2002; Prinz et al. 2003). Taken together with the observed variability in channel density, it can be inferred that neuron electrophysiology is tightly regulated by compensatory mechanisms to maintain target levels of activity. And indeed, the existence of such compensation has been confirmed in genetic knockout experiments (Guo et al. 2005; Nerbonne et al. 2008).

What this all suggests is that the functions of individual neurons provide an instance of the sort of MR outlined in Sect. 4.2. That is, neurons often exhibit stable functions in spite of variation in the mechanisms that allow and explain performance of those functions. Again, mechanisms are individuated on the basis of features that are causally relevant to performance of a function, and functions are specified relative to a functional analysis of some containing system. The densities of ion channels are the primary component features that determine a neuron's electrophysiological profile. And so different combinations of ion channel density distinguish different mechanisms that explain the electrophysiology of a given neuron. It is generally taken for granted in neuroscience that the functions of neurons are determined by their electrophysiological profiles. However, to bring those functions into alignment with the causal explanatory framework from Sect. 4.2, they must be specified relative to some functional analysis of a containing system.

Rarely, and usually only in very simple organisms, do the activities of individual neurons figure directly into explanations of an organism's behavior. To understand how the activities of individual neurons contribute to the behaviors of whole organisms, it is often necessary to first determine the roles those neurons play in intermediate-level causal structures. Specifically, the functions of individual neurons are most often specified relative to their roles in ensembles of neurons—circuits and networks. It is then the functions of these circuits and networks that figure into explanations of simple behaviors. (We do not currently have well-articulated explanations for more complex behaviors in large part because there are likely to be more tiers of intermediate-level causal structure of which we currently have impoverished understanding.) So in order to gain insight into how circuits operate and what functions they perform, neuroscientists look to simpler systems.

The stomatogastric ganglion (STG) of decapod crustaceans is a small network of about 30 neurons in the stomatogastric nervous system that generates and maintains various motions involved in digestion. There are two main functional networks in the STG: the pyloric network and the gastric network. Both networks produce patterned motor outputs that control particular aspects of crustacean digestion. The primary function of the pyloric network is to generate a three-phase motor pattern that traffics food particles through the pylorus in a wave of peristaltic motion. This triphasic rhythm has received extensive attention from systems neuroscientists looking to understand the ways in which activities of individual neurons combine to produce characteristic circuit function. Analysis of the triphasic rhythm has shown that the inference from the functional roles of individual neurons to the functions of neural ensembles is far from trivial. Here again, MR is rife in the structure of these interlevel causal explanations.

Prinz et al. (2004) demonstrated that the pyloric rhythm can be generated by vastly different values of the parameters that define the pyloric circuit. Using a simplified (three neuron) model of the pyloric network, they created a database of all possible combinations of synaptic strength and intrinsic electrophysiological properties of the cells in the circuit. Out of more than 20 million possible combinations of circuit parameters, more than four million sets of those parameters generated rhythms that exhibited the characteristic three-phase signature of the pyloric rhythm—call this the broad criterion. And of those, 11% (just under half a million sets of parameters) satisfied narrowly defined biological criteria derived from in vitro recordings of pyloric rhythms from a large sample of lobster preparations—the narrow criterion. Importantly, the parameter sets that satisfied both the narrow and broad characterizations of the pyloric rhythm included all possible parameter values for both the intrinsic properties of the individual neurons as well as almost all possible values (with one variable having a restricted range) for the synaptic weights between the neurons in the circuit. So no particular component or connection within the circuit dominates circuit function.

Given Prinz et al.'s data, the pyloric rhythm provides a clear instance of multiple realizability of the sort outlined in Sect. 4.2. The intrinsic properties of the neurons comprising the pyloric network and the synaptic weights within the network are precisely the features that are causally relevant to production of the triphasic rhythm. So the different mechanisms that explain the triphasic rhythm correspond to the different sets of parameter values (i.e. particular network configurations) that support the function. It is these specific network configurations that explain *how* the pyloric rhythm is generated in any particular case. But there is no universal answer to this "how" question. That is, there is no single mechanism that is responsible for production of the pyloric rhythm. Just as in the case of electrophysiology of single neurons, tuning of other causally relevant features of the network (other synaptic weights and intrinsic properties of component neurons) allows multiple sets of parameter values to converge on a stable target output (e.g., see Fig. 4.2).

That target output—i.e. the function of the pyloric rhythm—can be specified relative to its role in the functional analysis of crustacean digestion. Thus the rhythm functions to open (and then close) the pylorus and to produce a wave of peristaltic motion to traffic food particles through the pylorus. Prinz et al.'s broad and narrow criteria correspond to two different ways in which the role of the pyloric network in this digestive capacity can be analyzed. The broad criterion specifies relevant similarity simply in terms of production of a three-phase rhythm. There are empirical reasons for thinking this is a reasonable criterion: specifically the motoneuron that mediates between the pyloric network and the pylorus seems to act as a sort of temporal filter, so the relevant information from the network is just the order and timing of the firing of the neurons in the three-phase sequence. The narrow criterion constrains the function to the range of biological variability of circuit output observed

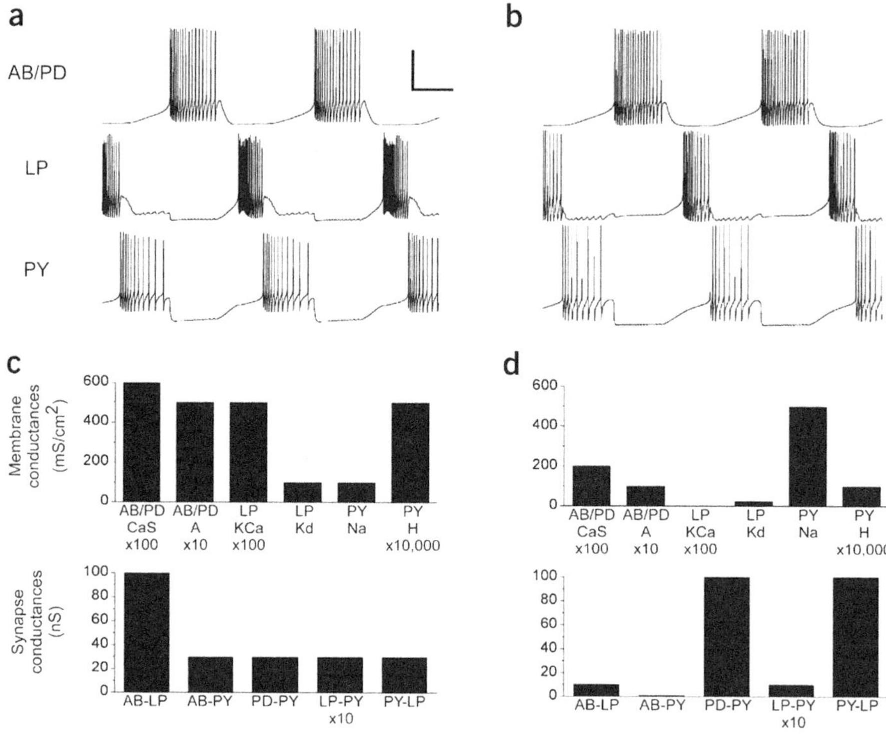

Fig. 4.2 Vastly different sets of parameter values (**c** and **d**) give rise to nearly identical circuit function (**a** and **b**). (From Prinz et al. (2004))

in vitro. Again, there are theoretical motivations—in this case the lack of certainty that order and timing are the only causally relevant features of pyloric network output—to take this as the criterion of relevant functional similarity. These functional analyses specify what the circuit is doing within the organism and as such determine the range of relevant similarity (or acceptable variability) in the output of the circuit.

This can be made more precise by specifically examining the two functional outputs (a, b) and network configurations (c, d) in Fig. 4.2. The two functional outputs are not exactly similar, but they are well within the range of observed variability in in vitro recordings of pyloric network output. On the other hand, the two network configurations—i.e. the two mechanisms realizing those functions—*are* relevantly different. Any controlled intervention changing one of the parameters in configuration (c)—e.g. the KCa conductance (500mS/cm^2) of the LP neuron—to its corresponding value in the second configuration (d)—in this case, completely blocking KCa conductance (0mS/cm^2) of the LP neuron—would cause catastrophic failure of the network rhythm. It is the tuning in other network parameters—i.e. the other causally relevant features of the mechanism—that enables the two networks to produce relevantly similar functions in spite of these differences. Thus the pyloric network provides a clear instance of MR in the sense outlined in the first section: relevant similarity in function in spite of relevant difference in the mechanisms that perform that function.

4.4 Kinds Reconsidered

My central goal to this point has been to provide an analysis of MR, substantiated with empirical examples, that moves away from positivist conceptions of explanation and reduction and operates instead within causal explanatory frameworks. In such frameworks, I have argued MR should be construed as a thesis about the structure of causal explanations rather than a thesis about relations between kinds that figure into different taxonomic systems. One might object that MR in my framework is still, fundamentally, a thesis about the relation between higher- and lower-level kinds. That is, my framing merely offers a different analysis of the kinds involved in putative instances of MR, not a complete abandonment of the concept or utility of kinds in this context. In short my response is that while it is certainly possible to interpolate some notion of kinds into this framework, the relevant shift in the ways kinds are characterized negates much of the philosophical debate that has focused on kinds in the context of MR. The aim of this section is both to develop this objection and to spell out my response in more detail.

Recall that the causal explanatory framework outlined in Sect. 4.2 consists in two parts: (1) realizations of a function consist in the mechanisms that explain how that function is performed; (2) functions are specified as causal roles within a functional analysis of some containing system. In spite of my insistence to the contrary, it would seem there's a natural way to interpolate kinds into this framework. Specifically, the mechanisms that realize functions may be thought to correspond to lower-level kinds, while functions, qua causal roles within some containing system, may be thought to correspond to functional kinds in much the way traditional accounts of MR have assumed. On this modified framing, MR would still amount to the traditional claim that there is a many-one relation between lower-level kinds (mechanisms) and higher-level kinds (functions). Also, notably a causal explanatory framing of type-identity theory could be then couched in this framework as the claim that there is a one-one relation between mechanisms and functions.[9]

To reiterate the challenge, interpolating kinds into CEMR involves (a) identifying mechanisms as lower-level kinds and (b) identifying the functions specified in functional analysis as higher-level (functional) kinds. I'll respond to each of these claims in turn. With respect to (a), identifying mechanisms as kinds (qua members of some scientific taxonomic system) is more problematic than it may appear at first glance. Consider the mechanism of a generic snap mousetrap. That mechanism consists in something like the following. The mousetrap is set by lifting the hammer off

---

[9]I take this to be roughly the view defended, albeit in different terms, by Shapiro and Polger (2012), Polger and Shapiro (2016).

the platform, pulling it back against the force of the spring, placing the holding bar over the hammer/spring, and then engaging (and baiting) the catch that holds the hammer in place. When the catch is released, the potential energy of the spring is converted into kinetic energy causing the hammer to slam down on the other side of the platform. Note that this mechanism is a complex causal process; it is not a *kind* in anything like the traditional philosophical sense, and it is certainly not a simple element of a taxonomic system in terms of which mousetraps might be analyzed.

A taxonomy of the components of a snap mousetrap might consist in a list of elements like: platform, hammer, spring, holding bar, catch. These are the elements in terms of which the function of the mousetrap may be analyzed. But the mechanism itself is a complex of these taxonomic elements, and it is their arrangement and causal coordination that explains how snap mousetraps perform their functions. From the other direction, note that snap mousetraps could be construed as a particular kind in a taxonomy of mousetraps—among others like glue, poison, or electric mousetraps. Generic mapping relations (one-one, one-many, many-one) between this higher-level taxonomic system and the lower-level taxonomic system of mousetrap components do not track anything interesting about explanations of the operations of these different kinds of mousetraps.

This is nothing peculiar to toy examples. The same applies to well-worn scientific examples like the mechanism of the action potential. Action potentials are, plausibly, activity-kinds in cellular-level neuroscientific taxonomy. The subcellular-level taxonomy in terms of which the action potential is explained consists in kinds like: voltage-gated Na^+ channels, voltage-gated K^+ channels, plasma membranes, and Na^+ and K^+ ions. The mechanism itself, of course, belongs to neither of these taxonomic systems. Rather, as in the mousetrap example, the mechanism is a complex causal process that here involves activation of voltage-gated Na^+ when a neuron's membrane potential depolarizes to some threshold, usually between -55 mV and -40 mV, causing a rapid influx of Na^+ ions, and so on.

This may seem like a nitpicking point, but the general framing of MR as an issue of the alignment of taxonomic systems continues to be the default view for many philosophers (see, e.g., Polger and Shapiro 2016).[10] Now, while this all suggests

---

[10] Gillett (2002, 2003, among others) has developed at length a "dimensioned" view of the realization relation. While considerations of space in this chapter prohibit more detailed treatment of Gillett's views, it is worth highlighting that Gillett's dimensioned view also moves beyond the standard framing of MR as a relation between kinds in different taxonomic systems. Gillett instead casts MR as a relation between different individuals constituting part-whole relations—i.e. as a compositional relation. This account gets into controversial (e.g. see Polger and Shapiro 2008) metaphysical territory that I cannot hope to tackle here. But, perhaps more to the point, nothing in my criticism of the standard view of realization hinges on such metaphysical issues. On one hand, MR on my account can be read as a purely epistemic thesis regarding the structure of causal explanations targeting neural, biological, and other complex causal phenomena. So the thesis that MR should not be considered a relation between kinds can be defended independent of deeper metaphysical commitments regarding what MR is a relation between. On another more speculative hand, I will offer that the most natural metaphysical reading of the account of MR I'm developing in this section would be that MR is a relation between different causal *processes*, rather than a compositional relation between distinct individuals. Of course, it would take considerable work to

there are good reasons to resist thinking of mechanisms as kinds in the sense of simple terms in a taxonomic system, one may still object that mechanisms must be kinds because they have scope. That is, mechanisms are not simply token causal processes, but rather are causal process-types that apply across multiple instances. This is borne out, for instance, in both the examples considered above. The mechanism of the snap mousetrap does not just explain how *this particular mousetrap* operates, but rather explains how *snap mousetraps in general operate*. And the same is true mutatis mutandis of the mechanism of the action potential. Presented with such examples it may be tempting to think that mechanisms are actually a sort of functional kind coextensive with the functions they realize. Indeed, the mechanism of the snap mousetrap described above is a sort of functional description; and everything that satisfies that description is a snap mousetrap, and every snap mousetrap satisfies that description. Again, the same seems to be true mutatis mutandis of the mechanism of the action potential.

But here we have to be careful, and thinking in terms of kinds (and generalizing from examples of this form) muddies the waters.[11] Mechanisms and the functions they realize need not be coextensive. To insist that they are would be to rule out causal explanatory MR tout court. The causal explanatory framing of MR outlined in Sect. 4.2 distinguishes the individuation conditions of functions from the individuation conditions of mechanisms: recall that condition (1) lays out the individuation conditions of mechanisms, condition (2) the individuation conditions of functions. The coherence of CEMR thus shows the identification of functions and mechanisms to be conceptually problematic, and the cases of robustness from Sect. 4.3 show that identification to be empirically problematic. So it would seem that be a mistake to construe mechanisms as functional kinds, just as above it proved problematic to construe mechanisms as structural kinds in any straightforward sense. Thus it seems that there is no straightforward way to interpret mechanisms as kinds in any classical sense of the term. And moreover, foisting the concept of kinds onto mechanisms seems to invite confusion regarding the relation between mechanisms and the functions they perform.

We can now take a closer look at (b), the identification of functions specified in functional analysis with higher-level (functional) kinds. On its own, this proposal is not as fraught as (a), but does bear its own pitfalls. The point of maintaining that functions are always attributed relative to a functional analysis of some containing system is to build a significant amount of context-sensitivity into functional attributions. Specifically, functional analyses play the crucial role of determining the relevant grain of generality at which functions are specified.

---

cash out this notion of causal process more precisely and to show how it can be used to skirt the metaphysical controversies raised by the dimensioned view. And that task is beyond the purview of this chapter.

[11] This is one diagnosis of a problem with the type-identity theory that Polger and Shapiro 2008, 2016) defend. They generalize from toy examples like corkscrews and scientific examples that involve quite general characterizations of mental/neural functions to reach the conclusion that functions are, in the vast majority of cases, identical with the mechanisms that realize those functions.

Take hearts as an example. At a most general level the function of a heart can be specified relative to its role in a circulatory system—viz. pumping nutrient fluids. At such a general level, there is no motivation to distinguish between the functions of insect hearts and vertebrate hearts. That is, any organ embedded in a circulatory system that pumps nutrient fluids functions as a heart in this general sense. However, if we perform more fine-grained functional analysis of circulatory systems, and consider the sorts of nutrients those fluids supply (e.g. oxygen) and the ways those nutrients are supplied to body parts (i.e. through open or closed circulatory systems), insect hearts and vertebrate hearts no longer perform the same function. At this grain of functional analysis, however, fish hearts and human hearts do perform the same function. However, if we analyze the functions of vertebrate hearts in terms of their role in blood oxygenation, fish hearts and human hearts no longer perform the same function. In fish circulatory systems, the heart simply functions to circulate blood (via a single pass per circuit), with the blood picking up oxygen from the gills en route to the rest of the organs and body parts. In human circulatory systems, the heart serves a dual function (via two passes per circuit) of circulating deoxygenated blood to the lungs and oxygenated blood to the rest of the body.

The point of these examples is that the relevant function the heart performs changes depending on the way its role within its containing system (the circulatory system) is analyzed. In the context of circulatory systems generally, hearts function to pump nutrient fluids simpliciter; in the context of open circulatory systems, hearts function to pump nutrient fluids (for insects, hemolymph) diffusely throughout the body; in the context of closed circulatory systems, hearts function to pump nutrients and oxygen-transporting red blood cells through a system of blood vessels; and so on. The advantage of tethering functions to functional analyses is that doing so keeps this context in place and encourages clarity regarding the degree of generality at which those functions are specified. Thinking in terms of functional kinds, on the other hand, invites decontextualization of functions ("the heart functions to pump blood"), and encourages lack of clarity with respect to degree of generality.

Further there is a close connection between functional analytic context and the criteria that determine and distinguish between realizations of a given function that risks getting lost when functions are construed as kinds. For instance, are insect hearts genuine realizations of the functional kind, heart, even though they don't "pump *blood*"? What differences between realizations of hearts are causally relevant to their ability to "pump blood"? Are two-chambered hearts and four-chambered hearts two different kinds or different realizations of one kind? These questions are too vague to be determinately answered in the absence of the context provided by some more precise functional analysis of the circulatory systems in which hearts are embedded. Again, the ability to clearly determine and distinguish between realizations gets lost in decontextualized functional attributions—i.e. subsumption under functional "kindhood".

Of course, one could argue that I'm not really giving up the notion of functional kinds, but rather am advocating a radical contextualization of functional kinds. After all, functions, even when tightly coupled to functional analyses, do have scope beyond token instances. Thus, although I may be denying that hearts are a univocal functional kind, what I'm actually advocating is that hearts can be divided into many different functional kinds that correspond to different degrees of generality depending on circulatory system context. There does seem to be something to this. We do distinguish between insect hearts and vertebrate hearts, between fish hearts and mammalian hearts, and these distinctions do seem to track differences in scope, and thus may be construed as tracking differences in kind.

My reply to such a counter is similar to that which arose in the discussion of mechanisms as kinds. On one hand, I can only concede that this sort of stripped-down notion of kinds (qua any predicate with scope) can be applied to my account of functions. On the other hand, I can certainly urge caution in the ways philosophical habits of thought regarding kinds are applied within such an account; and I can further point out that a highly contextualized notion of functional kinds fails to make solid contact with a significant thread of philosophical discussion regarding MR.

On this latter point, I can offer some more specific remarks. Due to the positivist backdrop of most philosophical debates about MR, the focus on functional kinds has centered on their ability to figure into special science laws (rather than causal explanations). For instance, Fodor (1997) argues that functional kinds are vindicated by their role in special science laws (whereas heterogeneous disjunctions are not appropriately nomic, and so lower-level "laws" that attempt to capture higher-level, multiply realized regularities are not in fact laws). By contrast Kim's (1992) MR skepticism is grounded in the claim that scientific kinds must be individuated on the basis of causal powers, which has ties to Shapiro's (2000, 2004) MR dilemma discussed in Sect. 4.2. And further, Kim argues the hallmark of natural laws is that they are projectible generalizations—i.e. a confirming instance of a lawlike generalization of the form "All Fs and Gs" provides reason to believe that Fs will be Gs in all contexts. Kim argues that generalizations involving multiply realized kinds are not projectible in this way, and so MR ought to be rejected.

The issue for both Fodor and Kim in the context of this debate hinges on what criteria one adopts for nomicity or naturalness of kinds. But such criteria fail to gain traction with a highly contextualized notion of functional kinds. Contra Fodor, the generalizations such kinds figure into do not have ambitions for lawlike status; they are confined to their functional analytic contexts. Similarly, their inductive projectibility is confined to functional analytic context; there are no ambitions to project universally. But based on the account I've been developing, none of this should garner pessimism regarding the prospects of MR. One can either give up the prospects for regarding contextualized functions as kinds, or one can accept the characterization as kinds and give up direct contact with these traditional ways of framing MR debates. Once we shift the debate into the context of causal explanations, issues regarding nomicity and lawfulness are exposed as red herrings that the philosophical conversation ought to move beyond.

4.5 Conclusion

The aims of this paper have been largely positive. In the first place, I provided an analysis of MR that moves away from positivist conceptions of explanation and reduction and operates instead within causal explanatory frameworks. In such frameworks, I argued that MR can be construed as a thesis about the structure of causal explanations rather than a thesis about relations between kinds that figure into different taxonomic systems (granted the caveats of Sect. 4.4). My second main aim has been to provide empirical examples that substantiate this notion of MR by drawing connections between MR and functional robustness in systems neuroscience. The traditional philosophical considerations that have surrounded MR (e.g. nomicity, projectibility, causal individuation) fail to adequately track important features of these empirical cases. This should perhaps be unsurprising given that those debates are based on an outmoded framework of explanation and reduction in the special sciences. One might note that tailoring an analysis of MR to these empirical cases is likely to alter the philosophical upshot surrounding MR. Indeed, the connection between MR and robustness that I've articulated invites critical reevaluation of the consequences of MR for nonreductive views of both the mind-brain relation, and the special sciences more generally. My aim has thus been not only to provide a more nuanced account of MR, but also to open a path for future research that may explore the implications of this notion of MR in more detail.

References

Bechtel, W. (2008). *Mental mechanisms: Philosophical perspectives on cognitive neuroscience.* London: Routledge.

Bechtel, W., & Richardson, R. C. (1993). *Discovering complexity: Decomposition and localization as strategies in scientific research.* Princeton: Princeton University Press.

Block, N., & Fodor, J. (1972). What psychological states are not. *Philosophical Review, 81,* 159–181.

Bronkhorst, A. (2015). The cocktail-party problem revisited: Early processing and selection of multi-talker speech. *Attention, Perception, & Psychophysics, 77*(5), 1465–1487.

Craver, C. F. (2001). Role functions, mechanisms, and hierarchy. *Philosophy of Science, 68*(1), 53–74.

Craver, C. F. (2007). *Explaining the Brain.* Oxford: Oxford University Press.

Craver, C. F. (2012). Functions and mechanisms: A perspectivalist account. In P. Huneman (Ed.), *Functions.* New York: Springer.

Cummins, R. C. (1975). Functional analysis. *Journal of Philosophy, 72*(November), 741–764.

Cummins, R. (1983a). *The nature of psychological explanation.* Cambridge, MA: MIT Press.

Cummins, R. C. (1983b). Analysis and subsumption in the behaviorism of Hull. *Philosophy of Science, 50,* 96–111.

Cummins, R. C. (2000). "How does it work?" vs. "What are the laws?": Two conceptions of psychological explanation. In F. Keil & R. Wilson (Eds.), *Explanation and cognition* (pp. 117–145). Cambridge, MA: MIT Press.

Edelman, G., & Gally, J. (2001). Degeneracy and complexity in biological systems. *PNAS, 98*(24), 13763–13768.

Fodor, J. (1968). *Psychological explanation: An introduction to the philosophy of psychology.* New York: Random House.

Fodor, J. (1974). Special sciences (Or: The disunity of science as a working hypothesis). *Synthese, 28,* 97–115.

Fodor, J. (1997). Special sciences: Still autonomous after all these years. *Noûs, 31,* 149–163. https://doi.org/10.1111/0029-4624.31.s11.7.

Gillett, C. (2002). The dimensions of realization: A critique of the standard view. *Analysis, 62,* 316–323.

Gillett, C. (2003). The metaphysics of realization, multiple realization and the special sciences. *Journal of Philosophy, 100,* 591–603.

Golowasch, J., Goldman, M. S., Abbott, L. F., & Marder, E. (2002). Failure of averaging in the construction of a conductance-based neuron model. *Journal of Neurophysiology, 87*(2), 1129–1131.

Guo, Y., Jangi, S., & Welte, M. A. (2005). Organelle-specific control of intracellular transport: distinctly targeted isoforms of the regulator Klar. *Molecular Biology of the Cell, 16*(3), 1406–1416.

Hempel, C. G. (1942). The function of general laws in history. *Journal of Philosophy, 39*(2), 35–48.

Hempel, C. G., & Oppenheim, P. (1948). Studies in the logic of explanation. *Philosophy of Science, 15*(2), 135–175.

Kim, J. (1992). Multiple realization and the metaphysics of reduction. *Philosophy and Phenomenological Research, 52,* 1–26.

Kim, J. (1999). Making sense of emergence. *Philosophical Studies, 95*(1–2), 3–36.

Kitano, H. (2004). Biological robustness. *Nature Reviews Genetics, 5*(11), 826–837.

Levins, R. (1966). The strategy of model building in population biology. *American Scientist, 54*(4), 421–431.

Machamer, P. K., Darden, L., & Craver, C. F. (2000). Thinking about mechanisms. *Philosophy of Science, 67*(1), 1–25.

Marder, E., & Goaillard, J.-M. (2006). Variability, compensation and homeostasis in neuron and network function. *Nature Reviews Neuroscience, 7*(7), 563–574. https://doi.org/10.1038/nrn1949.

Mitchell, S. (1997). Pragmatic laws. *Philosophy of Science, 64*(4), S468–S479.

Mitchell, S. (2008). Exporting causal knowledge in evolutionary and developmental biology. *Philosophy of Science, 75*(5), 697–706.

Nagel, E. (1961). *The structure of science: Problems in the logic of scientific explanation.* New York: Harcourt, Brace, & World.

Nerbonne, J.M., Gerber, B.R., Norris, A., & Burkhalter, A. (2008). Electrical remodelling maintains firing properties in cortical pyramidal neurons lacking KCND2-encoded A-type K+ currents. *The Journal of Physiology, 586*(6), 1565–1579. https://doi.org/10.1113/jphysiol.2007.146597.

Oppenheim, P., & Putnam, H. (1958). Unity of science as a working hypothesis. In *Minnesota studies in the philosophy of science 2* (pp. 3–36).

Piccinini, G., & Craver, C. (2011). Integrating psychology and neuroscience: Functional analyses as mechanism sketches. *Synthese, 183*(3), 283–311.

Polger, T., & Shapiro, L. (2008). Understanding the dimensions of realization. *Journal of Philosophy, CV*(4), 213–222.

Polger, T., & Shapiro, L. (2016). *The multiple realization book.* Oxford: Oxford University Press.

Prinz, A. A., Thirumalai, V., & Marder, E. (2003). The functional consequences of changes in the strength and duration of synaptic inputs to oscillatory neurons. *Journal of Neuroscience, 23,* 943–954.

Prinz, A.A., Bucher, D., & Marder, E. (2004). Similar network activity from disparate circuit parameters. *Nature Neuroscience, 7*(12), 1345–1352. https://doi.org/10.1038/nn1352.

Putnam, H. (1967). Psychological predicates. In W. H. Capitan & D. D. Merrill (Eds.), *Art, Mind, and Religion* (pp. 37–48). Pittsburgh University Press.

Putnam, H. (1975). "Philosophy and our mental life", Chapter 14 of Putnam's *Mind, language and reality: philosophical papers* (Vol. 2). Cambridge: Cambridge University Press.

Salmon, W. (1984). *Scientific explanation and the causal structure of the world*. Princeton: Princeton University Press.

Schupback, J. (2016). Robustness analysis as explanatory reasoning. *British Journal for Philosophy of Science*. https://doi.org/10.1093/bjps/axw008.

Shapiro, L. A. (2000). Multiple realizations. *The Journal of Philosophy, 97*(12), 635–654.

Shapiro, L. A. (2004). *The mind incarnate*. Cambridge, MA: MIT Press.

Shapiro, L., & Polger, T. (2012). Identity, variability, and multiple realization in the special sciences. Chapter 13 of Gozzano, S. & Hill, C.S. (Eds.), *New perspectives on type identity*. Cambridge: Cambridge University Press.

Wagner, A. (2005). Distributed robustness versus redundancy as causes of mutational robustness. *Bioessays, 27*(2), 176–188.

Weisberg, M. (2006). Robustness analysis. *Philosophy of Science, 73*(5), 730–742.

Wimsatt, W. (1980). Reductionistic research strategies and their bases in the units of selection controversy. In T. Nickles (Ed.), *Scientific discovery, Case studies* (Vol. II, pp. 213–259). Dordrecht: Reidel.

Wimsatt, W. (1981). Robustness, reliability, and overdetermination. In M. Brewer & B. Collins (Eds.), *Scientific inquiry and the social sciences* (pp. 124–163). San Francisco: Jossey-Bass.

Wimsatt, W. (2007). *Re-engineering philosophy for limited beings: Piecewise approximations to reality*. Cambridge, MA: Harvard University Press.

Woodward, J. (2003). *Making things happen: A theory of causal explanation*. Oxford: Oxford University Press.

Worth (Trey) Boone is a PhD candidate in the Department of History and Philosophy of Science at the University of Pittsburgh. His research focuses on issues in the metaphysics and epistemology of the mind-brain sciences. He is interested in how recent work in systems and cognitive neuroscience bears on traditional philosophical questions–e.g. multiple realization, reduction, emergence–regarding the relation between the mind and brain. He also has interests in the empirical study of consciousness, and in issues related to attention and representationalism. Before coming to the University of Pittsburgh, he received a BA in philosophy from Lewis & Clark College in Portland, OR, and an MA, also in philosophy, from Simon Fraser University in Vancouver, BC.

Chapter 5
Robustness: The Explanatory Picture

Philippe Huneman

Abstract Robustness is a pervasive property of living systems, instantiated at all levels of the biological hierarchies (including ecology). As several other usual concepts in evolutionary biology, such as plasticity or dominance, it has been questioned from the viewpoint of its consequences upon evolution as well as from the side of its causes, on an ultimate or proximate viewpoint. It is therefore equally the explanandum for some enquiries in evolution in ecology, and the explanans for some interesting evolutionary phenomena such as evolvability. This epistemological fact instantiates general property of biological evolution that I call "explanatory reversibility". In this chapter, I attempt to systematize the explanatory projects regarding robustness by distinguishing a set of epistemological questions. Are they the various expressions of one general project with specific key concepts and methods, or very disparate epistemic projects, unified by the mere homonymy of the term "robustness"? More precisely, are there specific kinds of explanations suited to explain robustness? Finally, how does robustness as an explanandum connect with other explananda in which evolutionists have been massively interested recently such as complexity, modularity or evolvability? After having initially explored various meanings of the concept of robustness and surveyed its instances in biology, I will propose a distinction between mechanical and structural explanations of robustness in evolutionary and functional biology. Then, among the latter, I will highlight the class of "topological explanations," and the subclass of explanations based on networks, as a major explanatory tool to address robustness. Focusing on evolutionary issues, I will eventually address the "explanatory reversibility" of robustness and consider its relation to key evolutionary concepts that are also explanatorily revertible such as modularity, evolvability and complexity.

P. Huneman (✉)
Institut d'Histoire et de Philosophie des Sciences et des Techniques,
CNRS/Université Paris I Sorbonne, Paris, France

© Springer Nature Switzerland AG 2018
M. Bertolaso et al. (eds.), *Biological Robustness*, History, Philosophy
and Theory of the Life Sciences 23, https://doi.org/10.1007/978-3-030-01198-7_5

5.1 Introduction

The robustness of a system, namely – roughly speaking – its ability to maintain itself with respect to some range of perturbations (at a given timescale), is a salient feature of biological and ecological systems. Physiologists such as Claude Bernard (1859) emphasized in the late nineteenth century the capacity of organisms to maintain constant key variables that describe their functioning (Canguilhem 1977), which Cannon later called "homeostasis" (Cannon 1932). And developmental biologists know that many developmental processes are still likely to produce functional adults, even though the normal environments, or some genes, are disturbed: Waddington famously labeled this property "canalization" (Waddington 1940). In ecology, a striking feature of communities is that the set of species one may find in an area seems to be quite stable along several decades, even though we know that "struggle for life" continues and all species evolve, speciate, go extinct, etc. Community ecologists have been focusing on the reasons for this stability since the inception of their discipline in the 1930s (Ives and Carpenter 2007; Cooper 2004). Hence, robustness of systems is a primary explanatory focus for biology and ecology. From the viewpoint of an evolutionary biologist, indeed, the reasons why this robustness evolves constitute an intriguing question, and many theoretical views and models have been framed to address this question (e.g. Wagner 2005a, b, de Visser et al. 2003).

In this chapter, I'll attempt to systematize those explanatory projects through a set of epistemological questions. Are they all the expression of one general project with specific key concepts and methods, or are they very disparate epistemic projects, unified only by the homonymy of the term "robustness"? More precisely, are there specific kinds of explanations suited to explain robustness? And then, which questions do they precisely address? Finally, how does robustness as an *explanandum* connect with other explananda in which evolutionists have been massively interested recently such as complexity, modularity or evolvability? The major aim of this paper is to sketch a conceptual cartography of the science of biological robustness.

First I'll characterize robustness and survey its widespread distribution in biology. In Sect. 5.3 I'll propose a distinction between mechanical and structural explanations of robustness. Among the latter, I'll highlight the class of "topological explanations," and the subclass of explanations based on networks, as a major explanatory tool to address robustness.

In Sect. 5.4, I'll partition the biological questions addressed to robustness into 'evolutionary' and 'functional' questions, then focus on evolutionary issues and highlight relations between these projects. I'll indicate that robustness can be either an *explanans* or an *explanandum*, and tie this feature to a general property of evolution, which I call "evolutionary reversibility" – namely, that robustness can be what explains something related to evolution, or something whose evolution is in needs of explanation. Under this label I'll compare robustness to plasticity and to dominance, and then show how evolvability and robustness are correlated. In the last section, I'll consider how some properties – modularity and complexity – can sometimes be proxies for robustness, and ask how they can be related between themselves.

5.2 Characterizing and Situating Robustness

Robustness intuitively means the fact that a system is able to remain the same under perturbation. This crude characterization allows some varieties of robustness according to what "remains" means, and to what "the same" means. The former especially concerns timescales, the latter concerns the features against which robustness is assessed, and how they are defined. As an example of the first aspect: a forest after some forest fires may be very different, with many fewer trees, yet on a larger timescale it may recover the same level of species abundance; its robustness is thereby assessed at the longest timescale. As to the "sameness" enveloped in the above characterization, it focuses on some specific parameters of interest. In biology, what remains the same is often related to survival and reproduction – many things could change through environmental disturbances, but the organism retains its ability to live and reproduce, and we would call that robustness.

5.2.1 Sameness

There are several ways to capture this idea of sameness. The first one focuses on functions, and states that a system is robust when it's able to remain functional, or achieve some key functions, in the face of perturbations of a certain range. Kitano's (2004) review paper uses this functional definition. Of course the range of perturbations should always be specified; few systems would remain the same after a nuclear holocaust but that's not the kind of perturbations relevant to the definition of robustness; yet, for instance, if someone is interested in the robustness of the Earth System (Lenton 2016) or of the biosphere, that could be a possible perturbation to consider.

The second approach partitions the set of variables likely to describe the system into low-level and high-level variables, or micro- and macro-variables. In this perspective, the robustness of the system would be defined by its ability to keep values of the high level variables relatively constant even if the low level variables vary in relation to perturbations. For instance, in an ecosystem where the composition of the species community is a macro-variable, and the abundance of each species a micro-variable, ecologists call "persistence" the property, which is such that the composition of the community remains unchanged in response to perturbations, even though the abundances may fluctuate a lot. This persistence thereby fits the second characterization of robustness.

Those two definitions are not exclusive and often they overlap, because functional properties can easily be understood as supervenient – since a same function can be realised by distinct mechanisms, but a difference in functions entails a difference in mechanisms – and then be considered as macro-properties, or as high-level properties (the realizing mechanisms being the micro-level properties).

5.2.2 *"Remain"*

Independently of the previous distinction, the way a system is said to "remain" the same gives rise to two families of robustness, which one could call *stability* and *resilience*, both of which are of interest to biologists and ecologists. Stability concerns constancy, functional or structural, in the face of perturbations. Resilience is about the capacity for returning to an equilibrium state after a transient out-of-equilibrium period where the system may behave very differently, or organizes itself very differently from the way it initially did. Each aspect of robustness is the focus of a research tradition in ecology: a longstanding issue in the field is the stability-diversity debate, namely the problem of testing and validating the widespread intuition that diversity begets stability (Ives and Carpenter 2007; Pimm 1984), and on the other hand functional ecologists have devised formal notions of resilience (Holling 1973) to characterize ecosystems and better understand current issues such as adaptation to habitat fragmentation and climate change as effects of anthropic forcing on ecosystems.

5.2.3 *Robustness Across Levels and Scales*

Biology and ecology are organized into levels of organization: genes, cells, multicellular organisms, species, communities, and ecosystems... Eldredge (1985) famously distinguished two kinds of hierarchy, one about ecological entities likely to interact, others about genealogical entities that reproduce. "Ecological hierarchy" comprises therefore cells, organisms and communities, while "genealogical hierarchy" comprises genes, cells, species and clades. Interestingly, robustness is widespread at all levels of those two hierarchies. Against the ancient view that each gene yields a specific protein product, recent molecular biology has shown that individual genes often don't make much difference to the outcome of the cell when they are knocked down, which indicates a certain form of robustness. Genes indeed generally work in networks, such as the Gene Regulatory Networks studied after Davidson's work (1986), which are such that the alteration of some nodes in the network, hence of some genes, often doesn't trouble the end-product. Genomic systems are robust systems. As Romano and Gray (2003) made clear, even deleting a massive part of the gene regulatory network of gene *Endo 16* in sea urchins doesn't significantly change the phenotypic product of it. This is not proven by

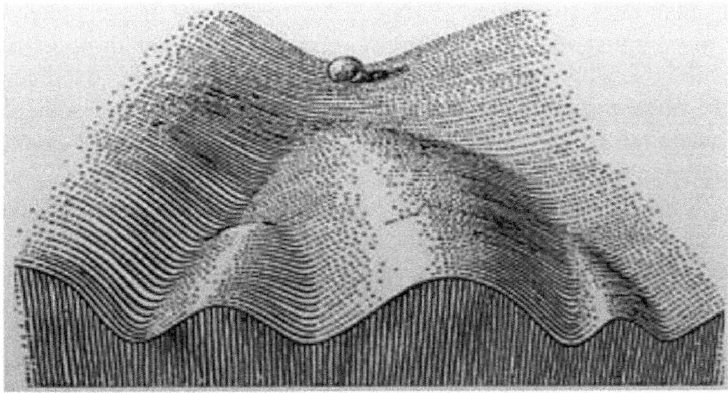

Fig. 5.1 Canalization, according to Waddington (1940). The rolling ball is the developing organism or cell, the pathways are the fates. This represents what has been called an "epigenetic landscape"

experimental biology, but through evolution: species *Strogylocentrotus purpuratus* and *Lytechinus variegatus* are different, the latter diverged from the former 16 millions years ago. But with regard to their skeleton they are phenotypically equivalent, meaning that the two GRNs are functionally equivalent – yet the GRN of the latter has been massively lost in evolution.

How is this possible? Many genes are duplicated in the genomes, which means that deleting an allele does not have much effect because another gene will do the same thing at another locus. Such redundancy therefore yields robustness at the genomic level. Alternatively, Wagner (2005b) calls "distributed robustness" the fact that a genomic network overcomes the deletion of several genes and keeps producing the same outcome, even though there is no attested redundancy.

Canalization (Waddington 1940), also concerns the resistance of a phenotypic outcome across changes of genetic bases (Fig. 5.1). It takes place during development of an organism, or a cell, not at the level of the functioning of an already developed entity. Canalization is observed through controlled experimental alteration of development. The structure of the developmental process is such that many of these changes don't change the endpoint of the developmental pathway.

But at the level of organisms, a way for systems to remain the same and continue to ensure functions related to survival and reproduction in the face of fluctuations or perturbations, is what has been called "plasticity". This word has several meanings in biology, depending upon the field where it is used. For example, population genetics, developmental biology or behavioral ecology all use the term in different ways: in population genetics, plasticity can be seen as a property of genes described by the "reaction norm" (Woltereck, see Sarkar 1999), namely a function mapping the expression of a gene to various environments; in behavioral ecology, it is the organism's property of phenotypic plasticity, namely the capacity of producing adaptive phenotypes in relation to distinct environments; and in developmental biology, the terms refers to a property of cells, which may "choose" distinct developmental fates (totipotency) (Nicoglou, ms.) Even if plasticity denotes a capacity to vary that seems to contradict robustness, one could argue that *phenotypic plasticity*, namely, the capacity for organisms, with a given fixed genotype, to display the phenotype that is most adaptive to their current environmental conditions, is a way for the organism to remain highly functional in the face of intergeneration environmental change, and this fits a definition of robustness.

Finally, as I mentioned it, at the level of ecological community robustness (in the mode of community stability, and in the mode of resilience) is pervasive.

Robustness seems thereby to be a cross-level distinctive property of living, evolutionary and ecological systems. Of course, an obvious reaction to this statement would be that if those systems were not robust, they would have disappeared. In other words their existence proves that they should be robust; but that's not so much an answer as a fact in need of explanation. This being said, one should ask how robustness is ensured, and why it is the case that it is widespread at all levels.

Robustness is not, however, distinctive of biology. Many physical systems are robust, and the mathematical tools to address stability and robustness have indeed been elaborated for dealing with them (especially non-linear dynamics, theory of ergodicity, or control theory). The robustness of biological systems may be distinct from the robustness of physical systems in general, and possibly evolution by natural selection would play a role in this difference, but this is left aside in this chapter (see Lesne 2008 on this difference). My focus is the systematic relationships between various questions regarding robustness in evolutionary biology and ecology. I'll therefore start by considering two strategies to explain robustness; they are very general and not especially tied to biology; those two strategies are the unraveling of robustness mechanisms, and the identification of robustness-enhancing structural properties (which can be of two sorts, topological or non-topological).

5.3   Three Families of Explanations of Robustness

5.3.1   *Mechanisms*

The nature of scientific explanation, especially in biology, is often understood by philosophers of science in the framework of what some call 'neomechanicism' (Machamer et al. 2000; Glennan 1996). This view claims that to explain means to unveil a mechanism that produces the *explanandum*, understanding the term 'mechanism' as a set of entities that have specific activities, and which is properly organized (Craver and Darden 2013). Identifying those entities, their activities and the organization is the core program of a scientific explanation. Mechanist views have proven very accurate to account for many research programs in molecular biology (e.g. Darden 2006) or neuroscience (e.g. Craver 2007). To explain robustness in this sense means to unravel the mechanism underlying the system's capacity to remain the same in the case of perturbations or changes.

Consider experimental physiology in the nineteenth century; a key issue then was to understand the various *regulations* that constitute the organism's life. Regulation was indeed one of the major problems of nineteenth century physiology, after Lavoisier's experiments, and led to Bernard and Cannon's views on homeostasis and the *milieu intérieur*. The nervous system was considered by many physiologist of this century as the main regulator (Sherrington, Bell, etc.).[1] For this reason Bernard consistently experimented on the parasympathic nervous system (e.g. Bernard 1858). Regulation is a kind of process aligning the organism's features on a set of parameter values that are beneficial or optimal for it, and adjusted to the environment and the lifestyle. Regulation processes support the robustness of the organisms, since they allow the main biological parameters to be approximately constant (temperature, level of glucose in organism, etc.), so that vital functions can be performed and life continues. Regulators are typically realized by mechanisms that are the target of the biologists' research (hormonal or nervous).

In the twentieth century, after Metchnikoff and Pasteur established the field of immunology, a major vector of robustness with respect to perturbations instantiated by heterogeneous substances ('antigens') was considered to be the immune system. Clearly, it provides a major contribution to the robustness of metazoan organisms, or even to all organisms, as argued by Pradeu (2012).

So, in this context, to explain robustness of a system is indeed to unveil a mechanism. And most of these mechanisms are instantiating negative feedbacks or sets of those, since such feedbacks are by definition allowing the system to correct the disturbed value of some fixed parameters.

5.3.2 Topology and Mathematical Structures in General

But generally speaking, in science, another kind of explanation does not typically unravel mechanisms, but relies on formal properties of systems. Several instances have been described, e.g. 'mathematical explanations' (Batterman 2010; Baker 2009), 'statistical explanations' (Lange 2013), variance explanation (Walsh 2015), 'topological explanations' (Huneman 2010; Woodward 2013; Jones 2014). A general characterization of those explanations has been proposed by Huneman (2017) under the label 'structural explanations'. In those explanations, mathematical facts do not describe or represent features or mechanisms that account for the phenomena investigated, but explain by themselves since they account for why the *explanandum* should take place. To put it bluntly, mathematical properties are here explanatory to the extent that they constitute a constraint for any mathematical descriptions of the mechanisms at stake, and for this reason they account by themselves for the *explanandum*.

Among these explanations, topological explanations are such that the *explanandum* is explained by showing how topological properties of a space related to the

---

[1] See Canguilhem (1977) for a historical account.

system entail properties that correspond to the *explanandum* in the system. By 'topological' one intends either topology stricter sensu, i.e. the science of invariance through continuous transformations– or graph theory, and network properties in general (both being merged in 'topological graph theory', Gross and Tucker 1987). The latter are of pervasive use in ecology (trophic networks, interactions networks, see Pimm 2002 or recently Kéfi et al. 2016), and in genetics (e.g. Gene Regulatory Networks, Davidson 1986, Revilla-i-Domingo et al. 2003).

Some explanations of robustness belong to this family, and therefore don't pinpoint any mechanisms for robustness. Instead, they characterize some topological property whose possession by a system entails a specific invariance regarding perturbations, and therefore a sort of robustness (Huneman 2015a). A simple example consists in some ecological explanations in trophic or interaction networks. As indicated elsewhere (Huneman 2010), those explanations have played an interesting role in advancing the traditional debate in ecology regarding the relations between stability and diversity. Ecologists from the early twentieth century (Cooper 2004) shared the intuition that the more a community – namely, a set of species in a common space, interacting (e.g. competing and preying) – is diverse, the more it is stable. The intuitive sense of stability, here, is that over a reasonably long (compared to our lives) timescale the same species more or less remain there, with more or less the same abundances (that is, number of individuals). Yet this has been neither rigorously empirically documented, nor theoretically modeled. And when May (1974) tried to model it, he came up with the challenging fact that, in mathematical theory, a community of species which is very diverse tends not to be more stable, but, rather, less stable.

Networks were already widely used to model trophic interactions in communities (e.g. Yodzis 1989). So May has shown that in these networks, the increase in the number of nodes does not by itself begets stability. But May's result was not a purely negative one; on the contrary, it led ecologists to reflect what 'stability' and 'diversity' should mean. Indeed, as seen above, there can be various kinds of robustness, one being persistence, another being the constancy of some variables such as biomass (Tilman 1996), another one is what many ecologists meant by 'stability', namely the reliability of abundances, and then resilience is another aspect.

On the other hand, the simplest meaning of 'diversity' is the number of nodes in a network. However, May just assumed random networks, namely, networks where the degree of nodes are randomly distributed. However, this clearly may not always be the case in ecology. And, as graph theory makes clear, random networks are one specific kind of network, but there are also many other types of networks characterized by the value of specific mathematical parameters: path length between nodes, distribution of degree (i.e. the number of edges related to a given node), clustering coefficient, etc. (Strogatz 2001; Watts 2003). Thus, the stability-diversity debate moved onto a question about what kinds of connectivity pattern between nodes would be such that stability would be yielded, and reinforced by increasing the amount of nodes (Dunne et al. 2002a, b, Montoya and Solé 2002, etc.).

A very simple explanation of the intuition that some diversity begets some stability therefore came from graph theory. Suppose that a network is scale-free,

Fig. 5.2 Scale-free
network

namely, there are many nodes poorly connected, and very few nodes very highly connected, and then more precisely the distribution of degrees follows a power-law. (Fig. 5.2).

Now, given this definition, when a random species goes extinct the probability that a highly connected node disappears is very low. Therefore random species extinction very probably doesn't affect the general pattern of connection in the network, and hence the functioning of the community.

This explanation of stability is clearly a topological explanation, in which the topological property of being a small world entails the *explanandum*, namely the stability regarding random species extinctions. It's not a toy example, because many trophic networks are indeed "truncated scale-free" networks (Dunne 2006), so one can apply such explanation. And in these explanations, the nature of interaction – be it predation, mutualism, or else – does not play any role; neither do the many processes that go on in the networks. As such, it clearly contrasts with the feedback model, which constituted above a paradigm of mechanistic explanation of robustness, for instance in the paradigm of 'regulators' above.

Actually, two categories of networks are highly studied in graph theory: the scale-free ones, and the so-called "small-world" networks (the latter are such that two nodes are always very close to another, but the whole network is highly clustered). Small-world networks are indeed ubiquitous in nature: in the brain, in the gene regulatory networks, in the signaling networks that support the chemical activity of the cell, or in social life (Watts and Strogatz 1998). In the case of ecology (Montoya and Solé 2002), this kind of networks also supports a topological explanation of stability: suppose that a species B goes randomly extinct, and interacting with A. If B is in the same cluster as A, A is still not very affected because the interactions it had though B with other species in the cluster are maintained, since by definition all nodes in the cluster are highly connected; and if B is in another cluster, then because of the property of short path length between two random nodes, there exists another node C in the cluster where B was to which A is close, and therefore A is still connected to this cluster, and interacts with the nodes related to C. The functioning of the network is therefore mostly unaffected.

Hence, the property of being a small-world network also entails some stability in ecological networks. This explanation as well as the explanation resorting to scale-free networks is a topological explanation: it deduces the stability property from the topological property of the network of connections, without considering the dynamics, the causal interactions between species, etc. If the network were a network of interactions in general rather than a trophic network, it would make no difference; the nature of interactions is not relevant, each interaction just defines an edge in the network, whose global properties are what is explanatory for the kind of robustness one is interested in. In other words, a given ecological network is stable, whether it's a network of predation relations, or a network of mutualism. What yields its robustness is just the formal property – e.g. in the scale-free case the relative number of small and large hubs – no matter what all the temporal mechanisms are, which can occur in the community.

Those topological explanations of robustness are pervasive well beyond ecology, precisely because at all levels of biological reality, one finds networks: signaling networks in cells (Sameer et al. 2012), GRN in the genome (Revilla-i-Domingo et al. 2003), neural networks of various sorts (Bassett and Muldoon 2016) in the brain. In each case, some properties of those networks are likely to yield robustness of the system, therefore the availability of such kind of explanations is pervasive.

All networks indeed have a topology, which can be roughly described as a global connectivity pattern describable by some variables (connectance, degree of nodes, clustering degree, etc.; Strogatz 2001). The properties such as scale-freeness or small-worldliness define equivalence classes between networks: if a network is scale-free, then deleting some random node will turn it into another scale-free network, therefore, the property defines invariance across perturbations of the network. Hence, topology is crucial to capture the robustness of the network as invariance through a given range of perturbations. Moreover, the dynamics on the network will be more or less constrained by the topology of the network, because this topology induces some states that, once reached, are not likely to cease, just because of the invariance features involved by the properties under focus (on these constraints see Huneman 2015a).

But more generally, the very idea of topology, as a study of invariance through continuous transformations, is essentially fit to explain robustness as a specific kind of invariance through perturbations (Huneman 2015a, b). Therefore, one can easily understand why topological explanations of robustness, either under the form of graph theoretical explanations, or under the form of topology *stricto sensu* (on continuous manifolds), are all over the place in science. Once a system is such that a space attached to it (e.g. a network, a phase space) possesses a topological property, this property defines a specific kind of invariance regarding some continuous transformations, so it entails for the system a specific disposition of invariance regarding

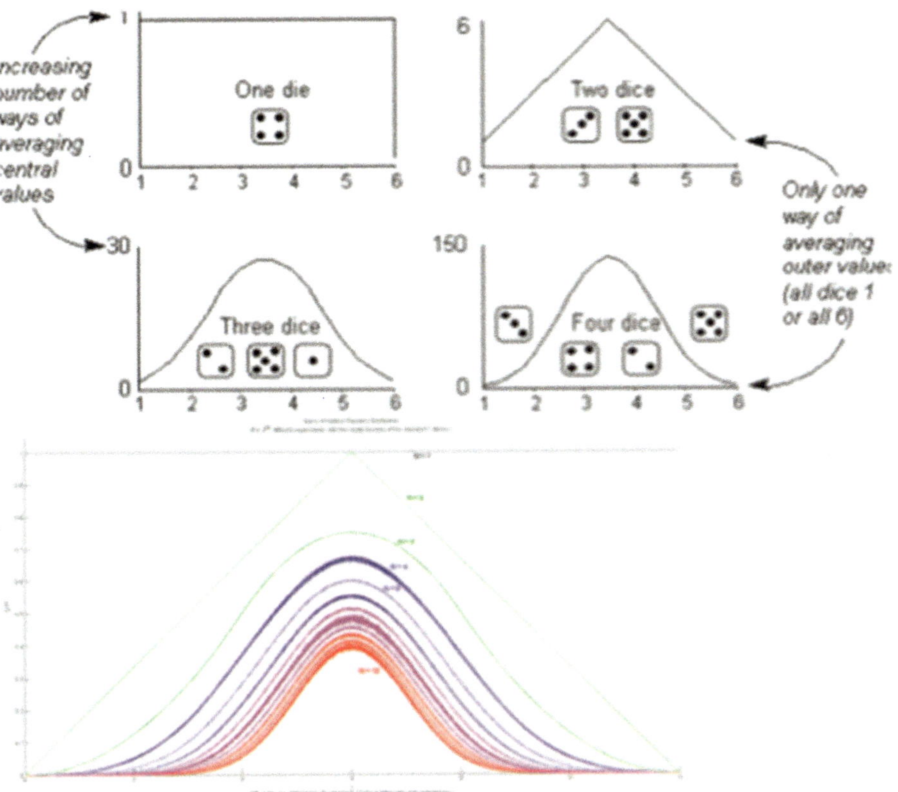

Fig. 5.3 The Central Limit Theorem: each die instantiates a random distribution. Adding die tossing, hence aggregating identical random variables, tends to a normal distribution

to a range of perturbations. Hence it yields some robustness, and explaining this robustness will thereby pertain to a topological explanation.

5.3.3 Normal Distributions

Nevertheless these explanations are not the only ones, among structural explanations (*sensu* Huneman 2017), that provide explanatory schemes alternative to mechanistic explanations of robustness in terms of feedbacks. When we try to understand the robustness of a system regarding stochastic fluctuations, one often turns to an explanation that relies purely on probability theory, namely, explanations using the Central Limit Theorem[2] (CLT).

---

[2] See Huneman 2017, §3, for an account of explanations relying on the Central Limit Theorem as paradigmatic of structural explanations.

This theorem states that when aggregating a set of random variables with finite equal mean and variance, one gets a normal distribution. Hence, if one considers a system's outcome made up by the aggregation of many independent forces or processes, this outcome will be described by a normal law (Fig. 5.3). For instance, given that one can consider human size as yielded by a set of factors, namely many different alleles on hundreds of loci, and then variable environmental inputs, so that it appears as a result of many independent random variables, its distribution must appear as normal. Moreover, the requisite of equal mean and variance can be relaxed, and substituted by the so-called Lindeberg condition, which (roughly said) stipulates that no variable makes a contribution incommensurably higher than the others. So the conditions for getting a normal distribution when there is aggregation of independent variables are very lightly demanding, and the pervasiveness of the normal laws in nature is so explained. As Frank (2009) explains, this is also the explanation of why many patterns in nature are normal, since many patterns in nature result from combining independent mechanisms yielding comparable distributions (size in a population, measures taken by various observers, etc.).

Then a significant consequence of the CLT is the following: suppose a system that produces a phenomenon is such that sources of noise are regularly added to the phenomenon, namely stochastic fluctuations of environment. Then this can be seen as the addition of a random variable to the variable describing the outcome of the system, and, assuming the Lindeberg condition, the outcome will still be a normal distribution. In other words, the normal law produced by the system is *not* affected by the perturbations. This means that the system is robust, and the explanation wholly relies on the CLT: we have a structural explanation of robustness.

As an example, consider what has been studied since the 2000s under the name "stochastic gene expression"; such research uses extensively those kinds of explanations. This research domain emerged when biologists got techniques to track at the level of individual cells the expression of genes (Elowitz et al. 2002). Then it turned out that different cells in an organism, even if they have the same genes, and even if they are functionally equivalent (the same organ, the same tissue; etc.) do not equally express the gene. The amount of gene product seems to vary stochastically across the cells. Before those findings most of the studies assumed that all cells with the same genes were phenotypically identical. But they were often predictively correct because of the fact that aggregating gene expression in all cells of a given organ begets through the CLT a normal distribution, peaked around the mean value. For many purposes therefore, a group of cells of this nature would be equivalent to a group of cells that identically produce the gene product at this average level. The CLT accounts for the fact that even though gene expression is stochastic in an individual cell, the population of genes could be successfully handled like a population of expressively identical genes. It robustly produces a predictable amount of gene product at the level of organ or tissue, in spite of thermodynamic and chemical fluctuations in and around each cell.

5.4 Robustness as Explanandum in Evolutionary Biology, and the Explanatory Reversibility Proper to Evolutionary Biology

5.4.1 Explanatory Reversibility: Robustness, Plasticity, Dominance

Once we have considered this typology of explanations of robustness, and emphasized that some explanations can be structural explanations, I turn to the cartography of the enquiries about robustness in biology.

Mayr (1961) suggested a useful partition between biological questions, which is often used currently when one intends to order or classify biological studies: some questions aim at "proximate causes", namely causes pertaining to the lifetime of the individual organism under focus (disciplines like physiology, cell biology or molecular biology fall in this category); other questions aim at "ultimate causes" namely causes that lie within past populations of the same species as the organism under focus. The former set of questions is called "functional biology" by Mayr, the second "evolutionary biology".[3]

Along these lines, the investigations about robustness can be partitioned into inquiries about the *proximate mechanisms* of robustness and inquiries about the *ultimate causes* of robustness, which therefore belong to evolutionary theory. In this second perspective, robustness is seen as something that occurred in evolutionary history; then the researchers wonder why, and they model some possible histories in order to suggest an answer. If robustness arose through natural selection, it means that it was selected for some reason; but it may also have been a correlated effect of the evolution of a selected trait; or even just a trait that in some species or clade arises so to say 'for free', because the developmental processes in those species can't do otherwise. An example of the latter view is the following: Uri Alon and colleagues (Alon 2007) have shown the existence of recurrent motifs in gene networks, likely to be interpreted as logical transformations on chemical inputs, and highly conserved across many clades. This seems to support the hypothesis that they are here because of natural selection. However, Solé and Valverde (2006) argued that those motifs can be easily produced by self-organization processes and therefore may not be yielded by selection. So here, the worry would be that, even though robustness can easily be seen as advantageous to biological systems, in many cases it did not evolve by natural selection but can be a very regular and spontaneously produced result of basic living structures.

---

[3] This may be sometimes confusing because evolutionary biology includes many questions about functions, and there is even a way of making sense of the very concept of functions in terms of evolutionary facts, which is widely shared by philosophers under the label "etiological view of function" (Wright 1973; see Huneman 2013, for a current overview). However in the present context the label "functional" should not be misleading.

However, the evolutionary problem of robustness once again should be divided since robustness may mean two very different things: one is the robustness of the system in the face of environmental perturbations, seemingly the object of this paper. But the kind of robustness Waddington (1940) talked about under the name "canalization" is not about the environment; rather, it's robustness *in the face of genetic mutations*. His idea is that even though the genes involved in developmental processes are slightly changed, altered by some somatic or germinal mutations, or if their expression changes (according to thermodynamic fluctuations), the process still leads to the same typical adult outcome. Similarly, the robustness of a Gene Regulatory Network demonstrated by evolution, as in the case mentioned in Sect. 5.2 about the GRN of gene *Endo 16* in sea urchins, only concerns the loss of genes, hence instantiates mutational and not environmental robustness.

Thus, given that both mutational robustness and environmental robustness exist, and assuming that robustness has been selected, the pending question is: was it for environmental perturbations, or for buffering against genetic fluctuations? And, whatever the choice we make on the basis of available evidence, how did the other aspect of robustness evolve? In a review paper about the topic De Visser et al. (2003) argued that mutational robustness has been selected for, and environmental robustness arose as a by-product, which is then maintained by natural selection.[4] In this view, the concept of robustness is explanatorily dual: mutational robustness is an *explanandum* of natural selection, and environmental robustness is in turn an *explanandum* of mutational robustness.

But this duality does not exhaust the explanatory status of robustness in evolution. In effect, once the question of the explanation of robustness has been addressed, we remain with the question of *what does robustness itself explain* in evolution? What role does robustness play, and how does it differ from the evolution of systems that are much less robust? This second question does not equate with the former, since even though robustness is there, say, for advantages regarding buffering against mutational fluctuations, yet the question remains: *What does it do in evolution?*, Or, in other words, how is biological evolution with robust systems different from what would be biological evolution with non-robust ones? Intuitively, there would seem to be a substantial difference but the question is how to exactly characterize it.

This last difference introduces us to a very general aspect of evolutionary biology, which I call "explanatory reversibility". It means that a feature of biological systems that has been explained evolutionarily can also be investigated from the viewpoint of *what it contributes to explain at the evolutionary level*. This reversibility is characteristic of Darwinian evolution in general. Very generally, parameters that are used to describe a population likely to evolve, such as the mutation rate, the genetic make-up, the dominance and recessivity, can in turn be addressed from an evolutionary viewpoint: for example, mutation rates, that are fixed in many population genetics models, can be studied as an evolving variable (Lynch 2010; Sniegowski et al. 2000; Denamur and Matic 2006). For instance, there are phylogenetic differences between clades regarding mutation rates, which are stud-

---

[4] For an alternative view of the evolution of mutational robustness see Van Nimwegen et al. (1999).

ied directly, and Taddei et al. (1995) have shown that in bacteria a stress system governs the mutation rates, making the genome more prone to mutate when stressed. Recombination, which is given with sexual reproduction and is a key feature of many evolutionary models to the extent that Mendelian inheritance with diploid assumes recombination, can be itself investigated as something resulting from evolution, since sex is a result of evolution. Since the 1980s a huge literature has been devoted to making sense of the evolution of sex, namely of its origin and its maintenance (Maynard Smith 1978; Williams 1975; Gouyon et al. (2015) for an overview). The evolution of recombination is itself an object of study (Barton 2010).

This reversibility in evolutionary *explananda* has been noticed for a very long time. In a 1932 conference, Fisher distinguished the issue of the "genetic bases of evolution" – i.e., what he, Sewall Wright, Haldane and few others were doing, namely, capturing the conditions of evolution from modeling a Mendelian population – from the question of the evolutionary bases of genetics, which he formulates as follows: "Can genetical phenomena be explained in terms of known evolutionary causes?" (Wright 1932) The latter was very poorly studied. Here Fisher says that the very general properties of Mendelian systems assumed when modeling populations are themselves the result of evolution. Another research program, which he only sketches there, would thereby consist in elaborating an evolutionary history of those key parameters of any model in population genetics.

At his time, Fisher only considered the evolution of the difference between two kinds of alleles in Mendelian settings, namely dominant and recessive. Thus, his question about the evolutionary basis of genetics first concerned the issue: "why do dominance and recessivity exist"? And he had a selectionist answer for that, whose main rationale was the following:

- First, organisms are highly integrated entities; hence most of the mutations would be detrimental, to the effect that by altering one trait it would touch upon many other traits and finally, be highly likely to harm the organisms. In the conference he notes: "That the vast majority of mutations should be deleterious is a perfectly natural consequence from the view that the organism is maintained in a highly adapted condition by natural selection, for *a highly adapted condition can mean nothing else than one which is more easily injured than improved by a change in its organization*". (Fisher 1932) This fact is shown formally in his "geometric model" (Martin and Lenormand 2015; Orr 2000).
- Hence, dominance appears as a kind of robustness. When a genetic system includes the feature of dominance/recessivity, even if in one parent a gene is altered, however it requires two copies to be effective when it's a recessive gene, so the organism is buffered against mutational harm. Thus, dominance and recessivity, so to speak, evolved to buffer against such harm. That's how Fisher depicts it in his text: "the recessiveness of mutations is itself a consequence of a prolonged evolutionary process, by which each species reacts to the unfavorable mutations with which it is persistently peppered." (Fisher 1932) In turn, the dominance vs recessivity system is a key explanatory parameter in population genetics.

Explanatory reversibility concerns many of the major evolutionary parameters. For some, like dominance, most of the biological questions were raised about *its effect* in evolution; for robustness, things are different: robustness is rarely a parameter in models, rather, it is often *something whose evolutionary history is investigated;* but some time biologists also ask about its causal role in evolution is, especially for the last two decades (the major reviews by Kitano, Wagner, and De Visser et al. appeared in the years 2000).

Plasticity is another property that crucially displays explanatory reversibility. It is often taken for granted in genomic systems, for instance under the form of "reaction norm", which has been used to model phenotypic plasticity in quantitative genetics since the 1930s, starting with Woltereck and Schmalhausen (Sarkar 1999). But from the 1970s and 1980s population geneticists began to enquire about why plasticity should evolve, thereby beginning to consider it as an *explanandum* rather than an *explanans*. Thus in the 80s, they disputed whether plasticity would be as such a trait targeted by natural selection (Via and Lande 1985), or whether it would evolve as a by product of selection for traits that are more or less plastic (Scheiner 1993).[5]

5.4.2   Robustness as Explanans/Explanandum, Evolvability and Topological Explanations

Turning to what robustness explains, or what role it plays in evolution, the issue often hinges on the relation between robustness and *evolvability*. First, a robust system, at any level (cell, organism, genome) is characterized by a great amount of self-identity. Therefore, it seems *prima facie* that robustness opposes variability; populations with less variability are less likely to evolve by natural selection, since variation is the fuel of natural selection. Fisher's fundamental theorem of natural selection – even though controversial, both regarding its interpretation and its validity (Okasha 2008; Edwards 1994) – says that the intergeneration variation in phenotypic value due to natural selection equals additive genetic variance, which entails that poor variation (genetic, at least) equals very little possible selection. Thus, it seems problematic that robustness could play any facilitating role in evolution; rather, robustness seems to contrast with plasticity, which appears as a facilitator of evolution, since it provides populations with resources in phenotypic difference that are by definition adaptive. West-Eberhard (2003) extensively documented the positive role plasticity plays in evolution.

But the rigidity of robust systems, their lack of variation potential, is not the last word on the role of robustness in evolution. A recent model, elaborated by Stadler, Schuster, Fontana and colleagues (Stadler et al. 2001; Stadler and Stadler 2004; Schuster et al. 1994; Fontana et al. 1999, Cupal et al. 1999, Schuster 2002) relying on topological explanations, shows how robustness as such may enhance the capacity of systems to evolve and therefore yields what's called "evolvability" (Wagner and Altenberg 1996; Minelli 2017). The model first considers RNA sequences, understood as genotypes, and their relation with phenotype and fitness. The phenotype is the functional protein, namely, the folded protein; but given that proteins do

---

[5] See Nicoglou (2015) for an account of those controversies.

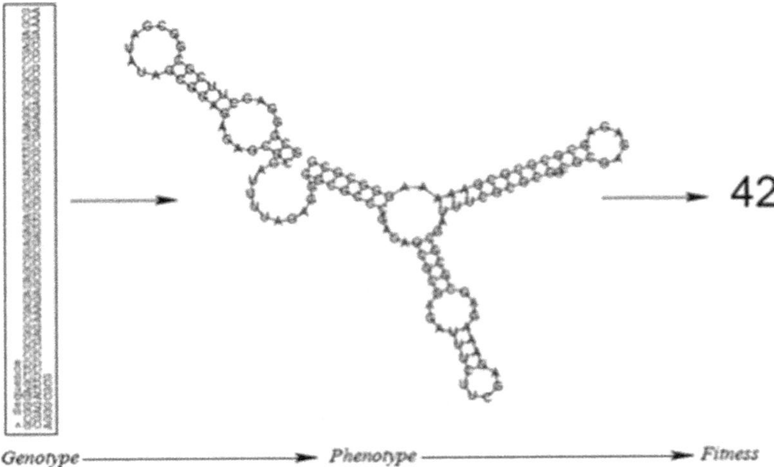

Fig. 5.4 Mapping RNA Genotype, Functional protein (phenotype) and fitness; After Stadler and Stadler (2004)

Fig. 5.5 Network of genotypes, each node being a RNA sequence. An edge between two nodes represents a distance of one mutation. Nodes of the same color have the same fitness. Black subnetworks are neutral networks

interact by receptors on their surface, several proteins having a different sequence may have the same functional profile and therefore the same fitness (Fig. 5.4).

The model therefore considers the network made up by the possible genotypes (each genotype is a node); the distance between two genotypes is defined by the amount of mutations by which they differ (Fig. 5.5). Then the space of genotypes is mapped onto phenotypes and the fitness. Several genotypes are likely to have the same fitness. Stadler, Schuster and colleagues define what they call "neutral networks", namely, subsets of the network of genotypes in which all genotypes have the same fitness, and all genotypes in one of such subset are neighbors, namely they are separated by one mutation step.

Now if one picks a genotype located deep inside a neutral network, each mutation transforms it into a genotype in the same subset, and so a genotype of equal fitness. This means that this genotype is fairly robust against mutations. The position of the genotype within the neutral network explains its robustness, therefore we here have an explanation based on graph theoretical properties, i.e., a topological explanation. Such explanation also concerns cases of distributed robustness in networks, mentioned earlier (Sect. 5.2: if gene networks are considered as genotypes in such model, then being situated deep within a neutral network means that all mutations of genes only move the genotype a few steps away, and therefore it remains in its neutral network, which leaves the fitness unchanged.

Moreover, suppose that the neutral network N comprises genotypes of fitness W, and that one neighbor of a genotype G1 at the external boundary of N has fitness $W' < W$, and another genotype G2 at another external boundary of N has a fitness $W'' > W$. By a series of one-step mutations, the genotype G1 could enter the neutral network and then circulate within it without being selected against, and finally mutate into G2, which is a higher fitness genotype ($W'' > W > W'$).

Subsequently, the larger the neutral network N, the higher the amount of neighboring genotypes of various finess values such that the system could explore the genotype space without being counterselected (since it would remain in the neutral network). The more numerous and the larger those neutral networks are, the more the system is capable of evolving and increasing in fitness, because those opportunities for genotypes to explore subspaces of higher fitness will increase; namely, the more evolvable it is. Hence the structure of the space of possible genotypes, which includes the sizes and amount of neutral networks, accounts for the evolvability of the population.

In this model, two interesting facts are established: there can be a topological explanation for mutational robustness in terms of genotype networks; and the reasons for robustness are at the same time reasons for the evolvability of the system (i.e. the population of genotypes), therefore evolvability and robustness are not in principle antagonistic.

5.4.3 Summing Up

To sum up this examination of the moving explanatory status of robustness in evolutionary biology, the following explanatory picture of robustness ensues from the models considered here:

Mutational Robustness is an explanandum *of natural selection*

- Mutational robustness has been selected for;

Environmental robustness is an explanandum *of mutational robustness*

  - Environmental robustness is a by-product, which is maintained by natural selection;

Mutational robustness is the explanans *of evolvability*

  - Evolvability occurs via neutral spaces in mutation-robust gene networks.

5.5 Robustness and Other Reversible Explananda of Evolutionary Biology

5.5.1 *Robustness, Dominance and Plasticity: A Comparison*

A major feature of evolutionary concepts is the explanatory reversibility sketched above, and it is highly relevant for robustness, which is addressed through distinct explanatory projects. As indicated above, dominance and plasticity share the same fate. The last question I consider here is the systematic relation between the revertible *explananda*.

There are many hypotheses on the evolution of dominance. Fisher's was interesting for us here, because to some extent his key idea is that the dominance/recessivity feature *provides robustness* against mutational fluctuations, since for harmful mutations it takes generally two alleles and not one to be phenotypically deleterious.

Robustness may not contradict plasticity either, notwithstanding the first impression of an opposition between constancy and variation that they instantiate. As indicated above, robust systems do not change in the face of perturbation (either environmental or mutational), while plastic systems are modified along with environmental changes. This said: mutational robustness does *not* contradict in principle plasticity, since plasticity is defined with keeping genotypes fixed. As to environmental robustness, one could suggest the following hypothesis about its compatibility with plasticity: plasticity and robustness may be two ways of achieving the same result at different timescales. When the environment changes, plasticity ensures constant fitness through phenotypic environment-sensitive variability without changing genotypes, so that the relation between fitness and genotypes can remain constant at the expense of the relation between phenotype and genotype. This entails that at the level of the population, some constancy may be realized through natural selection – namely, the population of genotypes keeps its fitness constant in the face of environmental fluctuations. Plasticity is therefore a short-time strategy to realize robustness of genetic or genomic systems at the larger timescale. Sultan and Stearns (2005) insist in the same spirit that plasticity and robustness should not be opposed. In this perspective, plasticity appears as a developmental-scale mechanism to achieve robustness at the evolutionary scale.

5.5.2 Complexifying the Explanatory Cartography: Robustness, Modularity and Complexity

Another feature of genetic systems, which has been addressed both as an *explanans* and as an *explanandum*, is modularity[6]: at many levels biological systems are modular: the cellular architecture, the existence of developmental and cognitive modules (Winther 2011) and even the architecture of the genome indicate modularity. Therefore the evolution of modularity is a relevant question, and for two decades authors have looked at various advantages of modularity. By definition, modular systems can function when one module is altered – so clearly there is a kind of robustness entailed by modularity, which provides in turn a selective advantage. Moreover, modularity allows systems to vary one module and explore some possibilities while they remain functional: to this extent, modularity also entails evolvability, as highlighted by Wagner and Altenberg (1996).

A caveat here: I just formulated the general form of an explanation of modularity by natural selection; this is not a priori valid for all instances of modularity, and actually many authors have challenged this hypothesis in precise cases, because there are structures that allow systems to exhibit modularity somehow 'for free', through self-organization. For instance, Solé and Valverde (2008) make such a case regarding the modular architecture of metabolic networks in the cell.

Notwithstanding this caveat, and assuming the adaptationist or selectionist perspective, modularity is hypothesized as *evolved by natural selection* for its robustness-enhancing properties, but it also contributes to *explain the course of evolution* by entailing evolvability. This provides an idea of the complex relation between those key concepts within the explanatory structure of biology.

Thus, because of the robustness it confers to systems, modularity can be taken as a proxy for robustness; but modularity is not the only way to ensure robustness. The neutral networks explanation I described above is an alternative explanation of mutational robustness. And dominance, according to Fisher, is also an explanation of robustness that is alternative to modularity. Hence robustness is a more general concept, which relates to several proxies – such as modularity – that have themselves a complex situation within the general explanatory picture of evolutionary biology, since they too are likely to be *explanans* and *explananda*.

Robustness, modularity and evolvability therefore have subtle explanatory ties, depending upon whether one is taken rather as *explanandum* or as *explanans*, and whether robustness is considered directly or through a proxy such as modularity.

The last notion that is theoretically crucial and often invoked in those contexts because it is conceptually related to those three notions of robustness, modularity and evolvability, is "complexity". This concept is notoriously difficult to define;

---

[6] Many definitions of « module » exist, and for a given system, many partition into 'modules' are possible, as Winther (2011) argues. Here, I take modules following the famous definition by Simon (1969), namely, within a system of interacting elements, a subset of elements that do interact within itself more than with other elements. Networks provide a way to identify modules, by running clustering analyses and pinpointing the major clusters.

algorithmic definitions have been given (e.g. 'Kolmogoroff complexity') but they are rarely in use in biology, and one sees hardly how to implement them easily in this field.[7] Complexity is often captured either in structural terms, for instance as the amount of types of parts in a system, or in functional terms, for example in terms of the amount of the functions performed, or the structural complexity of the mechanisms achieving them. In their book about evolution of complexity, Brandon and McShea (2011) reduce complexity to structural complexity in a deflationary manner, suggesting that any appeal to function will deprive the concept of its operationality.

In any case, multicellular organisms are complex under many definitions of complexity, and some trends in complexity increase in evolution can be detected as McShea (2005) has shown (irrespective of the causes of these trends). Fisher (1930) already clearly saw the relation between complexity and robustness: complex systems have many part types and many relations between those part types – which supports the functionality of the system and ultimately its survival and reproduction. Therefore, as complex, organisms are very likely to be altered by either mutational or environmental changes; the more complex they are, the more fragile they are (Orr 2000). Hence natural selection will favor devices that temper this fragility, namely, devices or features that promote robustness – which for Fisher were first and foremost genetic properties such as dominance and recessivity. Hence complexity evolves, and by evolving it creates a selective pressure in favor of robustness, hence various types of robustness at many levels have evolved among many clades, especially in metazoans.

To conclude, let's consider the general relation between modularity, robustness and complexity. By complexity, one intends that systems have parts that are not wholly modular. In formal terms, a signature of complexity is in general non-linear or non-additive interactions, which by definition oppose modularity (the contributions of modules A and B in a modular subsystem are by principle additive, otherwise affecting A would affect the contribution of B).

So complexity in biological organisms sets a limit to modularity, which is on the other hand favored by selection as a proxy for robustness. But complexity also requires some robustness, from an evolutionary viewpoint. There seems to be a conundrum here: complexity favors robustness and impedes modularity, which itself begets robustness (Fig. 5.6). So, how can we understand the coexistence of modularity and robustness in biological complex systems?

The hypothesis I suggest here starts by noticing the pervasiveness of one specific kind of network at all levels of biological hierarchy (which is full of networks, as indicated at the beginning: Alon 2007, Watts 2003, etc.) – namely, 'small world networks' (Watts and Strogatz 2008). Recall that by definition those networks are highly clustered and they have a short path between any two nodes (Fig. 5.7).

Networks like this are more robust, as I indicated when dealing with stability in ecology – and they are also highly modular (the modules being the clusters) but, because of the fact that all nodes are close to one another, they are complex, in the

---

[7]A tentative way to do that is Huneman (2015b).

Fig. 5.6 Explanatory and facilitating relations between modularity, robustness and complexity

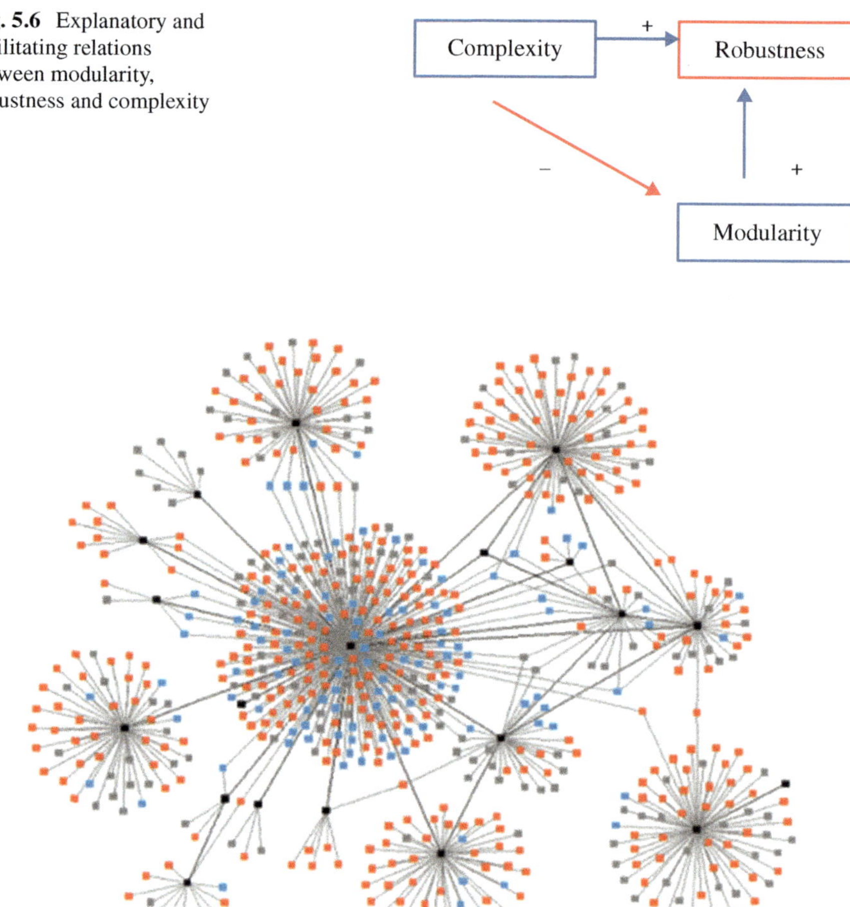

Fig. 5.7 Example of small world network

sense of achieving a high degree of holistic integration. The pervasiveness of these networks at all levels of the biological hierarchy suggests that evolution by natural selection favored them. A reason for this advantage would arguably be the fact that they provide robustness to complex systems without having to drop in complexity. One can therefore argue that they have been overwhelmingly selected because they implement a *trade-off between modularity and complexity.* Or to say it differently, this is a kind of network, which, for a given amount of complexity, minimizes the fragility due to the decline in modularity.

Hence, what appears here is a nuanced picture of the relations of complexity, modularity, robustness and evolvability as *explananda.*

5.6 Conclusion

Finally the concept of robustness in evolutionary biology is at the heart of a set of complex relations within the explanatory picture of biology. It can be circumscribed by reiterating the following distinctions, through which I tried to grasp its explanatory roles: they include distinctions characterizing the nature of robustness (1–3), and then two distinctions characterizing the explanatory strategy through which it is addressed (a,b).

- Distinction 1. Types of robustness defined by **nature of the invariants**

  State variables; functional properties

- Distinction 2. Types of robustness defined by **perturbation under focus**

  Mutational/environmental

- Distinction 3. Types of robustness defined by **timescales**

  Short, individual timescale; lineage timescale.

- Distinction a. Types of robustness defined by **explanatory variety**

  Mechanistic explanation vs. structural explanation

- Distinction b. Types of robustness defined by **explanatory role**

  Explanans or *explanandum.*

The major claim of this chapter is that no concept of biological robustness makes precise sense except if situated within this explanatory cartography, by specifying, for each of these pairs of notions, which element is operational within the scientific investigation of biological robustness one undertakes.

To end this examination, a last issue (not addressed in this chapter) consists in the relations between the meaning of robustness in biology and another notion of robustness put forth by evolutionary ecologist Richard Levins (1966), which became central in the epistemology of modeling, namely the notion of 'robust models', or robustness of theorems across models. This latter concept does not characterize (biological) systems themselves – like the concepts of robustness addressed here – but the *models* of these systems. However, even though there is no synonymy between them, they still should entertain complex relations. This is left to another investigation.

Acknowledgements The author thanks Anya Plutynski, Nick Jones, Carl Craver and Matteo Mossio for helpful comments and discussions on the arguments presented in this paper. Many thanks to the editors of the volume, whose careful reading improved the manuscript. I also thank Andrew McFarland for his thorough language-check. I am finally grateful to an anonymous reviewer for constructive criticism. This work was possible thanks to the grant ANR--13-BSH3-0007 Explabio and the LIA CNRS Paris "Montreal ECIEB."

References

Alon, U. (2007). Network motifs: Theory and experimental approaches. *Nature Reviews Genetics, 8*, 450–461.

Baker, A. (2009). Mathematical explanation in science. *British Journal for the Philosophy of Science, 60*, 611–633.

Barton, N. (2010). Mutation and the evolution of recombination. *Philosophical Transactions of the Royal Society B, 365*, 1281–1294.

Bassett, D., & Muldoon, S. (2016). Network and multilayer network approaches to understanding human brain dynamics. *Philosophy of Science, 83*(5), 710–720.

Batterman, R. (2010). On the explanatory role of mathematics in empirical science. *British Journal for the Philosophy of Science, 61*, 1–25.

Bernard, C. (1858). *Leçons sur la physiologie et la pathologie du système nerveux*. Paris: Baillière.

Brandon, R., & McShea, D. (2011). *Biology's first law*. Chicago: University of Chicago Press.

Canguilhem, G. (1977). La formation du concept de régulation biologique aux XVIII et XIXe siècles. In *Idéologie et rationalité dans l'histoire des sciences de la vie*. Paris: Vrin.

Cannon, W. B. (1932). *The wisdom of the body*. London: Norton.

Cooper, G. J. (2004). *The science of the struggle for existence: On the foundations of ecology*. New York: Cambridge University Press.

Craver, C. (2007). *Explaining the brain*. New York: Oxford University Press.

Craver, C., & Darden, L. (2013). *In search for mechanisms: Discovery across the life sciences*. Chicago: University of Chicago Press.

Cupal, J., Stadler, P., & Schuster, P. (1999). Topology in phenotype space. In J. Giegerich (Ed.), *Computer science in biology* (pp. 9–15). Dordrecht: Springer.

Darden, L. (2006). *Reasoning in biological discoveries: Essays on mechanisms, interfield relations, and anomaly resolution*. Cambridge: Cambridge University Press.

Davidson, E. H. (1986). *Gene activity in early development*. Orlando: Academic.

de Visser, J. A. G. M., Hermisson, J., Wagner, G. P., Ancel Meyers, L., Bagheri-Chaichian, H., Blanchard, J. L., & Chao, L. (2003). Evolution and detection of genetic robustness. *Evolution, 57*, 1959–1972.

Denamur, E., & Matic, I. (2006). Evolution of mutation rates in bacteria. *Molecular Microbiology, 60*, 820–827.

Dunne, J. (2006). The network structure of food webs. In M. Pascual & J. Dunne (Eds.), *Ecological networks: Linking structure to dynamics in food webs*. Oxford: Oxford University Press.

Dunne, J. A., Williams, R. J., & Martinez, N. D. (2002a). Network structure and biodiversity loss in food webs: Robustness increases with connectance. *Ecology Letters, 5*, 558–567.

Dunne, J. E., Williams, R. J., & Martinez, N. D. (2002b). Food web structure and network theory: The role of connectance and size. *PNAS, 99*, 12917–12922.

Edwards, A. W. F. (1994). The fundamental theorem of natural selection. *Biological Reviews of the Cambridge Philosophical Society, 69*(4), 443–474.

Eldredge, N. (1985). *Unfinished synthesis: Biological hierarchies and modern evolutionary thought*. New York: Oxford University Press.

Elowitz, M. B., Levine, A. J., Siggia, E. D., et al. (2002). Stochastic gene expression in a single cell. *Science, 297*, 1183–1186.

Fisher, R. (1930). *The genetical theory of natural selection*. London: Methuen.

Fisher, R. (1932). The evolutionary modification of genetic phenomena. *Proceedings of the 6th International Congress of Genetics, 1*, 165–172.

Fontana, W., Stadler, P. F., Bornberg-Bauer, E., Griesmacher, T., Hofacker, I. L., Tacker, M., et al. (1999). RNA folding and combinatory landscapes. *Physics Review E, 47*, 2083–2099.

Frank, S. A. (2009). The common patterns of nature. *Journal of Evolutionary Biology, 22*, 1563–1585.

Glennan, S. (1996). Mechanisms and the nature of causation. *Erkenntnis, 44*, 49–71.

Gouyon, P. H., Vienne, D., & Giraud, T. (2015). Sex and evolution. In T. Heams, P. Huneman, G. Lecointre, & M. Silberstein (Eds.), *Handbook of evolutionary thinking in the sciences* (pp. 499–502). Dordecht: Springer.

Gross, J. L., & Tucker, T. W. (1987). *Topological graph theory*. Reading: Wiley Interscience.

Holling, G. (1973). Resilience and stability of ecological systems. *Annual Review of Ecology and Systematics, 4*, 1–23.

Huneman, P. (2010). Topological explanations and robustness in biological sciences. *Synthese, 177*(2), 213–245.

Huneman, P. (Ed.). (2013). *Functions: selection and mechanisms*. Dordrecht: Springer.

Huneman, P. (2015a). Diversifying the picture of explanations in biological sciences: Ways of combining topology with mechanisms. *Synthese*. https://doi.org/10.1007/s11229-015-0808-z.

Huneman, P. (2015b). Redesigning the argument from design. *Paradigmi, 33*(2), 105–132.

Huneman, P. (2017). Outlines of a theory of structural explanations. *Philosophical Studies*. https://doi.org/10.1007/s11098-017-0887-4.

Ives, R., & Carpenter, J. (2007). Stability and diversity of ecosystems. *Science, 317*(5834), 58–62.

Jones, N. (2014). Bowtie structures, pathway diagrams, and topological explanation. *Erkenntnis, 79*, 1135.

Kéfi, S., Miele, V., Wieters, E. A., Navarrete, S. A., & Berlow, E. L. (2016). How structured is the Entangled Bank? The Surprisingly simple organization of multiplex ecological networks leads to increased persistence and resilience. *PLoS Biology, 14*(8), e1002527.

Kitano, H. (2004). Biological robustness. *Nature Review Genetics, 5*, 826–837.

Lange, M. (2013). Really statistical explanations and genetic drift. *Philosophy of Science, 80*(2), 169–188.

Lenton, T. (2016). *Earth system science: A very short introduction*. Oxford: Oxford University Press.

Lesne, A. (2008). Robustness: Confronting lessons from physics and biology. *Biological Reviews, 83*, 509–532.

Levins, R. (1966). The strategy of model building in population biology. *American Scientist, 54*, 421–431.

Lynch, M. (2010). Evolution of the mutation rate. *Trends in Genetics, 26*(8), 345–352.

Machamer, P., Darden, L., & Craver, C. (2000). Thinking about mechanisms. *Philosophy of Science, 67*(1), 1–25.

Martin, G., & Lenormand, T. (2015). The fitness effect of mutations across environments: Fisher's geometrical model with multiple optima. *Evolution, 69*, 1433–1447. https://doi.org/10.1111/evo.12671.

May, R. M. (1974). *Stability and complexity in model ecosystems*. Princeton: Princeton University Press.

Maynard Smith, J. (1978). *The evolution of sex*. Cambridge: Cambridge University Press.

Mayr, E. (1961). Cause and effect in biology. *Science, 134*, 1501–1506.

McShea, D. (2005). The evolution of complexity without natural selection: A possible large-scale trend of the fourth kind. *Paleobiology, 31*(2), 146–156.

Minelli, A. (2017). Evolvability and its evolvability. In P. Huneman & D. Walsh (Eds.), *Challenging the modern synthesis*. New York: Oxford University Press.

Montoya, J. M., & Solé, R. V. (2002). Small world patterns in food webs. *Journal of Theoretical Biology, 214*, 405–412.

Nicoglou, A. (2015). The evolution of phenotypic plasticity: Genealogy of a debate in genetics. *Studies in History and Philosophy of Biological and Biomedical Sciences C, 50*, 67.

Nicoglou, A. (ms). *Plasticity in biology*.

Okasha, S. (2008). Fisher's "fundamental theorem" of natural selection: A philosophical analysis. *British Journal for the Philosophy of Science, 59*(3), 319–351.

Orr, H. A. (2000). Adaptation and the cost of complexity (PDF). *Evolution, 54*, 13–20.

Pimm, S. L. (1984). The complexity and stability of ecosystems. *Nature, 307*, 321–326.

Pimm, S. (2002). *Food webs* (2nd ed.). Chicago: University of Chicago Press.

Pradeu, T. (2012). *The limits of the self: Immunology and biological identity*. Oxford: Oxford University Press.

Revilla-i-Domingo, I., Domingo, R., & Davidson, E. (2003). Developmental gene network analysis. *International Journal of Developmental Biology, 47*, 695–703.

Romano, L. A., & Gray, G. A. (2003). Conservation of endo 16 expression in sea urchins despite evolutionary divergence in both cis and trans-acting components of transcriptional regulation. *Development, 130*(17), 4187–4199.

Sameer, S., Bajikar, S. S., & Janes, K. A. (2012). Multiscale models of cell signaling. *Annals of Biomedical Engineering, 40*, 2319–2327.

Sarkar, S. (1999). From the Reaktionsnorm to the adaptive norm: The reaction norm, 1909–1960. *Biology and Philosophy, 14*, 235–252.

Scheiner, S. M. (1993). Genetics and evolution of phenotypic plasticity. *Annual Re- view of Ecology and Systematics, 24*, 35e68.

Schuster, P. (2002). A testable genotype-phenotype map: Modeling evolution of RNA molecules. In M. Lässig & A. Valleriani (Eds.), *Lecture notes in physics, 585* (pp. 55–81). Dordrecht: Springer.

Schuster, P., Fontana, W., Stadler, P. F., & Hofacker, I. (1994). From sequences to shapes and back: A case study in RNA secondary structures. *Proceedings of the Royal Society of London Series B, 255*, 279–284.

Simon, H. (1969). *The sciences of the artificial*. Cambridge: MIT Press.

Sniegowski, P. D., Gerrish, P. J., Johnson, T., & Shaver, A. (2000). The evolution of mutation rates: Separating causes from consequences. *Bioessays, 22*, 1057–1066.

Solé, R., & Valverde, S. (2006). Are network motifs the spandrels of cellular complexity? *TREE, 21*(8), 419–422.

Solé, R., & Valverde, S. (2008). Spontaneous emergence of modularity in cellular networks. *Journal of the Royal Society Interface, 5*, 129–133.

Stadler, B., & Stadler, P. (2004). The topology of evolutionary biology. In C. Ciobanu (Ed.), *Modeling in molecular biology. Natural computing series* (pp. 267–286). Dordrecht: Springer.

Stadler, P., Stadler, P., Wagner, G., & Fontana, W. (2001). The topology of the possible: formal spaces underlying patterns of evolutionary change. *Journal of Theoretical Biology, 213*(2), 241–274.

Strogatz, S. (2001). Exploring complex networks. *Nature, 410*, 268–276.

Sultan, S. E., & Stearns, S. C. (2005). Environmentally contingent *variation*: Phenotypic plasticity and norms of reaction. In B. Hallgrimsson & B. Hall (Eds.), *Variation*. Boston: Elsevier Academic Press.

Taddei, F., Matic, I., & Radman, M. (1995). Cyclic AMP-dependent SOS induction and mutagenesis in resting bacterial populations. *Proceedings of the National Academy of Sciences of the United States of America, 92*, 11736–11740.

Tilman, D. (1996). Biodiversity: Population versus ecosystem stability. *Ecology, 77*(2), 350–363.

Van Nimwegen, E., Crutchfield, J., & Huynen, M. (1999). Neutral evolution of mutational robustness. *Proceedings of the National Academy of Sciences of the United States of America, 96*(17), 9716–9720.

Via, S., & Lande, R. (1985). Genotype-environment interaction and the evolution of phenotypic plasticity. *Evolution, 39*, 505–522.

Waddington, C. (1940). *Organisers and genes*. Cambridge: Cambridge University Press.

Wagner, A. (2005a). *Robustness and evolvability in living systems*. Princeton: Princeton University Press.

Wagner, A. (2005b). Distributed robustness versus redundancy as causes of mutational robustness. *Bioessays, 27*, 176–188.

Wagner, G., & Altenberg, L. (1996). Complex adaptations and the evolution of evolvability. *Evolution, 50*(3), 967–976.

Walsh, D. (2015). Variance, invariance and statistical explanation. *Erkenntnis, 80*(3), 469–489.

Watts, D. (2003). *Six degrees of separation*. New York: Norton.

Watts, D. J., & Strogatz, S. H. (1998). Collective dynamics in "small-world" networks. *Nature, 393*, 440–442.

West-Eberhard, M. J. (2003). *Developmental plasticity and evolution*. Oxford: Oxford University Press.

Williams, G. C. (1975). *Sex and evolution*. Princeton: Princeton University Press.

Winther, R. (2011). Part-whole science. *Synthese, 178*, 397–427.

Woodward, J. (2013). II–Mechanistic explanation: Its scope and limits. *Aristotelian Society Supplementary, 87*(1), 39–65.

Wright, L. (1973). Functions. *Philosophical Review, 85*, 70–86.

Wright, S. (1932). « The roles of mutation, inbreeding, crossbreeding, and selection in evolution » *Proceedings of the sixth international congress on genetics* (pp. 355–366).

Yodzis, P. (1989). *Introduction to theoretical ecology*. New York: Harper & Row.

Philippe Huneman   research director at the Institut d'Histoire et de Philosophie des Sciences et des Techniques (CNRS/Université Paris 1 Sorbonne) in Paris, is a philosopher of biology. After having worked and published on the constitution of the concept of organism and Kantian metaphysics, he currently investigates philosophical issues in evolutionary theory and ecology, such as the emergence of individuality, the relations between variation and natural selection in evolutionary theory, the role of the concept of organism, and the varieties of explanations in ecology, as well as the history of the modern Synthesis, and of neutral theories in ecology and evolution. He edited seven books on history and philosophy of biology – among them Functions: selection and mechanisms (Synthese Library, Springer, 2012), From groups to individuals. Evolution and emerging individuality (MIT Press, 2013, with F. Bouchard), Challenging the Modern synthesis: development, inheritance and adaptation (Oxford UP, 2017, with D. Walsh) – and authored two.

Chapter 6
Robustness and Autonomy in Biological Systems: How Regulatory Mechanisms Enable Functional Integration, Complexity and Minimal Cognition Through the Action of Second-Order Control Constraints

Leonardo Bich

Abstract Living systems employ several mechanisms and behaviors to achieve robustness and maintain themselves under changing internal and external conditions. Regulation stands out from them as a specific form of higher-order control, exerted over the basic regime responsible for the production and maintenance of the organism, and provides the system with the capacity to act on its own constitutive dynamics. It consists in the capability to selectively shift between different available regimes of self-production and self-maintenance in response to specific signals and perturbations, due to the action of a dedicated subsystem which is operationally distinct from the regulated ones. The role of regulation, however, is not exhausted by its contribution to maintain a living system's viability. While enhancing robustness, regulatory mechanisms play a fundamental role in the realization of an autonomous biological organization. Specifically, they are at the basis of the remarkable integration of biological systems, insofar as they coordinate and modulate the activity of distinct functional subsystems. Moreover, by implementing complex and hierarchically organized control architectures, they allow for an increase in structural and organizational complexity while minimizing fragility. Finally, they endow living systems, from their most basic unicellular instances, with the capability to control their own internal dynamics to adaptively respond to specific features of their interaction with the environment, thus providing the basis for the emergence of minimal forms of cognition.

Keywords Regulation · Control · Functional integration · Organization · Autonomy · Cognition

L. Bich (✉)
IAS-Research Centre for Life, Mind, and Society, Department of Logic and Philosophy of Science, University of the Basque Country (UPV/EHU), Donostia-San Sebastian, Spain
e-mail: leonardo.bich@ehu.es

© Springer Nature Switzerland AG 2018
M. Bertolaso et al. (eds.), *Biological Robustness*, History, Philosophy and Theory of the Life Sciences 23, https://doi.org/10.1007/978-3-030-01198-7_6

6.1 Introduction

One of the characteristics that differentiate living systems from physicochemical ones is the capability to self-produce and self-maintain by means of continuous exchanges of matter and energy with the environment. While doing so, organisms exhibit a remarkable robustness in how they respond to external perturbations and manage internal variations in such a way as to maintain their viability. They do not just produce, modify and maintain their components in order to persist: another distinctive feature of life — often included in definitions employed in origins of life and synthetic biology (e.g. Ruiz-Mirazo et al. 2004; Damiano and Luisi 2010; Bich and Damiano 2012a) — is the adaptive capability to constantly oppose the thermodynamic tendency towards disintegration, and to counteract potentially destabilizing interactions with the environment.

From this perspective, robustness is a crucial and ubiquitous biological property, that allows living systems to adaptively cope with variation. It is generally regarded as the capability of a system to maintain its functions and performances despite perturbations, or under uncertainty in general (Kitano 2004; Stelling et al. 2004; Chen 2008). It is often associated with degeneracy in the realisation of functions, and characterized in terms of flexibility of behaviours or of steady maintenance of some property (Mitchell 2009).

Physiological mechanisms and behavioral strategies by which organisms achieve robustness and maintain their viability as integrated wholes, are implemented at all levels of biological organization, and they can range from distributed network properties to more complex mechanisms, which rely on different degrees of modularity. An open question, which requires careful theoretical scrutiny, is whether mechanisms responsible for achieving robustness do not just assure the survival of living systems under changing internal and external conditions, but also play a fundamental, inherent, role already in the realization of biological organizations and of relevant biological properties. A closely related issue concerns whether complex forms of robustness – relying on hierarchical or modular mechanisms – are already necessary at the level of minimal life or, instead, they are later additions that ensured the survival of living systems in more variable environments than the ones where life might have originated. This chapter tackles the open question — and aims to provide conceptual tools to address the issue related to minimal life — from an organizational standpoint. In doing so, it regards robustness as a system property, and analyzes it in relation to the whole organization of the organism and the environmental conditions in which it operates.[1]

A candidate theoretical framework to address these issues is that of *biological autonomy*. Historically inspired by the work of Immanuel Kant (Kant 1790) and Claude Bernard (Bernard 1865), among others, on the internally self-determined

---

[1]An alternative way to address biological robustness, closer to engineering approaches, would be to focus on individual behaviours and mechanisms, and on the maintenance of specific functions or performances.

organization of living systems, this perspective has been developed in Systems Theory, Cybernetics and Theoretical Biology starting from the 1960s (Bich and Damiano 2008; Bich and Arnellos 2012; Mossio and Bich 2017). The main contributions to this line of research include: Jean Piaget's work on organizational closure and thermodynamic openness (Piaget 1967); Robert Rosen's theory of M/R-Systems and his formal model of minimal biological organization (Rosen 1972, 1991; Letelier et al. 2006); Humberto Maturana and Francisco Varela's theory of autopoiesis (Varela et al. 1974; Maturana and Varela 1980) and its applications to origins of life and synthetic biology (Luisi 2006); Tibor Ganti's Chemoton Theory (Ganti 1975, 2003a, 2003b); Stuart Kauffman's auto-catalytic sets and his theory of autonomous systems (Kauffman 1986, 2000); and, more recently, the notion of basic autonomy and the organizational approach developed by Alvaro Moreno and collaborators (Ruiz-Mirazo and Moreno 2004; Moreno and Mossio 2015).

The notion of biological autonomy is grounded in the idea that living systems are metabolic self-producing systems able to self-maintain and keep their network organization invariant through the continuous exchange of matter and energy with the environment. According to this perspective, living systems are "endogenously active" (Bechtel 2008), due to their distinctive thermodynamic nature. They realize a specific kind of internal organization — defined by the notion of *organizational closure* (Varela 1979; Rosen 1991; Bich and Damiano 2008; Mossio and Moreno 2010) — where not only the very existence and activity of the constituents depend on the network of processes of transformation that they realize but, in addition, they collectively promote the conditions of their own existence through their interaction with the environment. The thermodynamic nature of biological organization, which combines at its core endogenous activity with essential interaction with changing environments, implies also that the system is required to harbor an internal dynamical variability to enable different viable responses to a variety of environmental perturbations. This is one of the reasons why robustness plays a crucial role in the characterization of living systems from their most basic instances.

By adopting this perspective, this chapter advocates the view that complex mechanisms implemented by living systems to achieve robustness are essential for an understanding of life and of some of its distinctive features. Specifically, by leaning on a theoretical characterization of the concepts of biological control and signal (Sect. 6.2), it focuses on the organization and role of those specialized regulatory mechanisms that contribute to the robustness of living systems by coordinating compensatory responses (Sect. 6.3). Regulatory mechanisms are characterized as forms of higher-order control, exerted over the basic regime responsible for the production and maintenance of the organism, in response to specific signals and perturbations (see also Bich et al. 2016). The thesis defended in Sect. 6.4 is that while enhancing robustness, regulatory mechanisms play also a fundamental role in the realization of an autonomous biological organization. By enabling the integration and coordination of the basic biological functions, they contribute to the construction of biological identity. They also allow living systems to overcome bottlenecks of complexity in the transitions from basic self-maintaining (bio)chemical networks to increasingly more sophisticated organizations. Moreover

(Sect. 6.4.2), regulatory mechanisms lay the basis for the emergence of minimal forms of cognition already at the unicellular level, when a system becomes capable to internally generate operational meanings — expressed through self-regulatory loops — associated with environmental variations.

6.2 Basic Concepts: Stability, Control and Signal in Autonomous Systems

Robustness, as applied to biology, is a general and flexible concept that includes diversified properties and phenomena, related to how organisms or their parts cope with variation. It admits different approaches depending on the level of description and goals: the engineering approaches, centered on the maintenance of individual functions and performances; those approaches interested in local properties such as the persistence and stability of specific molecules or classes of molecules (see for example Pascal and Pross 2016); and system-oriented approaches that address robustness as a property related to (and emerging from) collective self-production. This chapter, which aims to contribute to an understanding of the fundamental adaptive properties common to living organisms, takes the system-oriented path, and addresses robustness in the context of biological autonomous organizations.[2]

In this scenario, robustness is usually characterized as a network property, mostly related to forms of stability (Kitano 2007), and described dynamically in terms of properties of attractors or of the capability to shift between attractors. When mechanistic aspects are taken into account, robustness is often understood in terms of feedback loops, and of their capability to make attractors more stable or to facilitate the establishing of new steady states when the systems' dynamics are displaced from the initial ones.[3]

These properties are very general and widespread, and not specifically biological: they can be exhibited by many types of natural and artificial networks. A case of special interest for its proximity to biology is that of out-of-equilibrium self-maintaining chemical systems: *dissipative structures* emerging and maintaining themselves under specific external boundary conditions, like Bénard cells, tornados,

---

[2] As argued by Cornish-Bowden (2006), among others, most engineering approaches "often seem to imply little more than reductionist biology applied on a large scale" while a "systemic approach to biology ought to put the emphasis on the entire system". A somehow similar categorization is proposed by Dupré and O'Malley (2005), who distinguish between 'pragmatic systems biologists' – who find it useful to refer to some systems properties in terms of interactions and of integration of data – and 'systems-theoretic biologists' – who focus on the investigation of general systems principles. An interesting case is that of Robert Rosen, whose contributions include both engineering (Rosen 1967, 1970) and system-oriented approaches (Rosen 1972, 1985, 1991), and who saw the former as inadequate to capture the distinctive features of living organisms *as systems* (Rosen 1991).

[3] For different ways to achieve biological robustness through stability see for example Rosen (1970), a classic textbook in dynamical systems theory.

whirlpools etc. (Nicolis and Prigogine 1977). They share some properties with living systems, such as stability and out-of-equilibrium self-maintenance but, importantly, they lack the capability of self-production, and the internal organization and functional differentiation that are distinctive of biological systems.

Let us consider the most basic instances of robustness in biological systems. It is possible to think of prebiotic systems and of the earliest (and simplest) forms of life as distributed biochemical networks of processes of production capable of generating and maintaining stable dynamic regimes, encapsulated by a compartment of their own making. An example is the basic scenario centered on metabolic self-production provided by the theory of autopoiesis (Varela et al. 1974; Maturana and Varela 1980), among similar approaches. The emphasis is on the network capability of the system as a whole to achieve a stable regime and employ distributed compensations for perturbations in such a way as to maintain the global organization of the system stable. In such a scenario robustness could be achieved, for example, through the capability to recover after the loss of most of its catalysts (Piedrafita et al. 2010).

A more detailed model of minimal biological organization is represented by Tibor Ganti's Chemoton (Ganti 1979, 2003b), a system characterized by an internal differentiation between three coupled chemical subsystems — a metabolic cycle, a template subsystem and a compartment — working like the cogwheels of a clock.[4] This model describes a non-hierarchical and very fragile basic biological network, exhibiting a certain (although low) degree of functional differentiation, and a minimal robustness that allows it to be viable under a limited range of environmental conditions (Bechtel 2007), and to cope with stochastic variation (Van Segbroeck et al. 2009).

In this scenario, a minimal form of biological robustness is achieved. It can be understood in terms of *dynamic stability*[5] of the self-producing and self-maintaining distributed network organization that puts together the basic living system. This organization realizes highly distributed endogenous patterns of compensations that respond to variation in such a way that the system remains within its viability region. The system simply "absorbs" the effects of perturbations or internal variations as a network, by compensating them through internal reciprocal adjustments between tightly coupled subsystems, while the whole dynamics is maintained in the initial attractor — or shifts to a new one.[6]

---

[4] The coupling between these subsystems is realised by means of supply and demand of metabolites necessary for the production of the components in the different subsystems (*metabolic complementarity*). The subsystems provide the necessary substrates for the internal processes of production taking place in the others and, in turn, consume the metabolites supplied by the others.

[5] Dynamic stability is the capability to counterbalance the displacement of the system from a certain initial state, provoked by a perturbation, and end up in the same final state (Rosen 1970). It is a widespread property in the natural world, instantiated by any system whose dynamic behavior is characterized by the presence of at least one stable attractor. When realized collectively, dynamic stability is a distributed property of a whole network of reactions - it cannot be attributed to any single transformation, or to a partial subset of transformations.

[6] See Bich et al. (2016) for a discussion of dynamic stability in models of minimal living organizations.

Implicit in these accounts is the idea that prebiotic and early living systems were characterized by a distributed organization capable to achieve a minimal form of robustness as stability under very specific (almost invariant) and favorable environmental conditions.[7] Only later, an increase in the scope and efficiency of mechanisms related to robustness would have allowed living systems to survive in more variable environments. *Vice versa*, according to this view, today when the environments of current living systems are particularly favorable and almost invariant, many regulatory mechanisms would not be needed anymore. This is the case of many intracellular endosymbionts, which live under very stable conditions within the cellular internal environment of the host organism. They tend to undergo a reduction of their organizational complexity and genome size, and they lose several constitutive and compensatory mechanisms. An example is *Buchnera aphidicola BCc* — an endosymbiont of the cedar aphid *Cinara cedri* — which lost some of its regulatory mechanisms and also the capability to realize the full pathway for the synthesis of the amino acid tryptophan and the co-enzyme riboflavin necessary for its own functioning and that of the host. The production of these components is achieved only by sharing biosynthetic pathways with other endosymbionts in the same host (Lamelas et al. 2011).

Dynamic stability is an important property to understand and model some aspects of life and its robustness, but it is not an exclusively biological property. It is also crucial to explain the emergence of several relevant phenomena in chemistry and at the edge of biology, such as oscillatory behaviors in reaction networks (Semenov et al. 2016). The paradigmatic case of dynamic stability in these infra-biological domains is represented by dissipative structures, where a high number of microscopic elements spontaneously self-organize and realize a stable global, macroscopic ordered configuration in the presence of a specific flow of energy and matter in far from thermodynamic equilibrium conditions (Nicolis and Prigogine 1977). These dynamic structures show the capacity of reacting conservatively to a certain range of perturbations, due solely to distributed network properties.

Yet, unlike these cases at the edge of biology, dynamic stability is not the only way actual living systems can achieve robustness. As pointed out by Hiroaki Kitano, the study of biological robustness should not be limited to dynamic stability, considered as a distributed property. Robustness should be understood, instead, as a more general phenomenon.

> Whereas homeostasis and stability are somewhat related concepts, robustness is a more general concept according to which a system is robust as long as it maintains functionality, even if it transits through a new steady state or if instability actually helps the system to cope with perturbations. [...] Examples of extreme robustness under harsh stress conditions show that organisms can attain an impressive degree of robustness by switching from one steady state to the other, rather than trying to maintain a given state. [...] Robustness is also not identical to stability. Some species gain robustness by increasing instability in a part of its system (Kitano 2007, 1–2).

---

[7]These favorable conditions might also include the presence of most of the necessary building blocks for early living systems, as claimed for example by the heterotrophic models of the origin of life (see for example Mansy et al. 2008).

A theoretical account of biological robustness needs to take into consideration not only stability but also more complex ways of achieving robustness. In particular, it should take into consideration modularity and hierarchy – two aspects which, according to Kitano, still require a proper formal characterization (Kitano 2004, 2007).[8] Modularity may enable functional differentiation and, therefore, different contributions of subsystems to the maintenance of the more comprehensive system that harbors them. Hierarchy allows to modulate and coordinate such contributions in an efficient and precise way, and to specify in each circumstance how to ensure the viability of the whole system.

As a matter of fact, biological systems can locally produce distributed network responses. However, one of their distinctive features is that they are internally organized in such a way that different subsystems contribute in different ways to the maintenance of the system (Montévil and Mossio 2015; Moreno and Mossio 2015; Mossio et al. 2016). These features are absent in infra-biological thermodynamically open systems (see also Moreno and Ruiz-Mirazo 2009; Mossio et al. 2009; Arnellos and Moreno 2012).

One of the consequences of being internally organized and functionally differentiated is that the interactions between the components of a system can take place in ways that are qualitatively different from one another, and they can play distinct causal roles within the system. This differentiation might be invisible to a network characterization, focused on patterns of interconnectivity and numbers of connections, rather than types of interactions. As a consequence, alternative and more complex way to achieving robustness that are distinctively biological might be passed over.[9] This is a serious conceptual issue if we consider that in all known instances of living systems, biological robustness is not achieved exclusively through reciprocal adjustments between the coupled subsystems of a distributed network, and indeed all living systems employ also other forms of direct control of the basic dynamics that imply a more active role of the system organization in handling variation (Bich and Damiano 2012b).

With the aim to tackle this issue, the following sections will not dwell upon robustness as a property of stable networks, but on mechanisms that are more demanding in terms of organizational requirements, such as hierarchical regulatory

---

[8] It does not imply that different forms of robustness, based respectively on dynamic stability, and modular and hierarchical control mechanisms cannot coexist in the same system or subsystem.

[9] This issue is closely related to the debate on the relationship between mechanistic and network descriptions. Whereas mechanisms describe distinguishable parts which play different specific tasks, network descriptions focus on global properties and patterns of connectivity (Moreno et al. 2011). Finding ways to bring together network and mechanistic descriptions, apparently irreducible to each other, is on the of the challenges faced by complex systems theory. An attempt to develop an heuristics to move between the two descriptive approaches has been proposed by Bechtel (2017a, 2017b). The basic idea is that clusters of interconnections in a network description are possible candidates for the parts of the mechanistic description of the same system. The limit of this approach is that patterns and network configurations derived from numbers of connections do not provide the same kind of information on the system as the identification of different types of contributions (e.g. in metabolism a complex hierarchy involving, metabolites, enzymes acting on metabolites, energy currencies, kinases acting on enzymes, etc.), but a complementary one.

mechanisms. However, this approach needs to face some conceptual difficulties related to the specific object of study. Basic living systems, such as bacteria, are characterized by fluid biochemical machineries with low internal structural differentiation, realized by highly distributed interactions where each component interacts with — or is indirectly influenced by — many others. These features favor network descriptions and seem to make it particularly difficult to provide a characterization of robustness in terms of different functional contributions to the viability of the system. This scenario, therefore, requires the introduction of theoretical tools to operationally distinguish between distinct kinds of interactions in the systems. These tools can then allow us to address issues related to complex biological robustness such as control and regulation, to operationally identify which components or subsystems act as controllers or regulators, and to distinguish them from those other subsystems that are the targets of their activities.

6.2.1 Control and Signal

Let us proceed stepwise by first clarifying some basic concepts such as control and signal in the framework of autonomy.

Autonomous biological systems constantly produce, transform, repair and replace their own components, and maintain themselves through exchanges of matter and energy with the environment. Unlike dissipative structures, which are mostly and largely determined by external boundary conditions, they do not emerge spontaneously under appropriate environmental conditions, but they contribute to determine their own conditions of existence (Mossio and Bich 2017). In order to maintain themselves in far-from-equilibrium conditions, they need to exert some *control* over their underlying thermodynamic processes which would otherwise proceed toward equilibrium.

Control is generally defined as the capability to modify the dynamics of a system toward a certain state (e.g. parameters acting upon variables, enzyme upon concentrations of metabolites, etc.[10]). It is an *asymmetric interaction*. In biological systems control is exerted by some molecules or supra-molecular structures, generated and maintained by the system itself, which act as *constraints*[11] on thermodynamic (matter/energy) flows.[12] A biological system is capable of generating some of the (internal) constraints that control its dynamics in such a way that they collectively

---

[10] See for example Rosen 1970; Hofmeyr and Cornish-Bowden 1991.

[11] Given a particular thermodynamics process P, a molecular configuration C acts as a constraint upon P if: (1) at a time scale characteristic of P, C is locally unaffected by P; (2) at this time scale C exerts a causal role on P, i.e. there is some observable difference between free P, and P under the influence of C (Mossio et al. 2013, 164. A more detailed characterisation can be found in Montévil and Mossio 2015).

[12] See Pattee 1972; Bich et al. 2016; Winning and Bechtel 2018, for a discussion of control in biological systems. See Arnellos et al. 2014 and Veloso 2017, for a discussion of inter-cellular control mechanisms in development.

achieve self-maintenance. These constraints are involved in at least three main kinds of control mechanisms. One is *kinetic control* (e.g., catalysis), which specifies the rates of diverse synthetic pathways: e.g. an enzyme that harnesses (catalyzes) a chemical reaction (without being directly affected by it). A second one is *spatial control*, which defines the spatial scale of the system, the selective passage of molecules, and thereby, the concentrations of its components. Examples are selectively permeable boundaries and diffusion barriers, which avoid the dilution of certain key compounds and keep their concentration above critical threshold values. A third kind is *template control* (e.g. DNA, mRNA[13]), responsible for constraining the sequences of amino acids in proteins.

The distinctive feature of biological systems is that the constraints which exert these basic types of control are organized in such a way that they are mutually dependent for their production and maintenance, and collectively contribute to maintain the conditions at which the whole network can persist (Moreno and Mossio 2015). This continuous operational integration puts together the *constitutive regime* (C) responsible for the basic self-production and self-maintenance of a living system.[14] This basic regime involves at least two different kinds of interactions — *processes* and *constraints* acting on processes — and it is characterized by the strict *stoichiometric coupling* between the subsystems involved. Responses to variation are essentially governed by changes in concentrations (both of the substrates that take part in the processes, and of the molecular structures that carry out control tasks on those metabolites). In this context, robustness is achieved as a network property: variations in concentrations affecting a given process or subsystem can propagate throughout the system, producing the change of one or several other processes and control subsystems which, in turn, compensate for the initial one. As a result, the system can be regarded as stable.

The activity of control constraints can be affected in different ways.[15] In the context of the constitutive regime alone, the most basic ways to affect the activity and rates of control subsystems are through direct molecular interactions, changes in the concentrations of the substrates and products of the processes upon which the constraints act[16] and, finally, through variations in the processes responsible for the synthesis of the control components themselves, resulting in changes in the concentration of the control components in the system.

All these are constitutive interactions governed by stoichiometry. However, there are additional ways in which the activity of control subsystems can be affected. These additional modalities open new spaces of possibilities for the action of control subsystems (new degrees of freedom) and are responsible for the com-

---

[13] See for example Mossio et al. 2016 for a characterisation of the role of mRNA as a constraint.

[14] See Nghe et al. (2015) and Ruiz-Mirazo and Moreno (2004) for a discussion of the some of the key elements for the origin of these self-producing and self-maintaining networks in prebiotic conditions.

[15] The requirement for a structure to be a constraint is to be locally unaffected by the process they are harnessing. But a constraint can be affected by other interactions in the system.

[16] For example, the effects described by the law of mass action.

plexification of biological organizations. It is the case of *signal molecules*. In living systems they play a fundamental role in the coordination and integration between subsystems in response to both internal variation and interactions with the environment. The distinctive features of signals in general, as argued by Haven Wiley, are the capability to trigger a response in the targets without providing the energy for it, and the fact that their causal power is insufficient for determining the response. The response, therefore, depends to a large degree on the properties of the receiver (Wiley 2013).

In the framework of autonomy described above, signals are not part of constitutive processes. They are usually by-products of biosynthetic processes and, more importantly, their role *as signals* does not consist in participating as metabolites in production processes. They *do not act as* constraints either, but they usually *interact with* control constraints by giving rise to activated or inhibited molecular complexes. A paradigmatic case is constituted by cyclic AMP, a ubiquitous signal molecule in cells, which forms molecular complexes with regulatory proteins like kinases — responsible for the activation or inhibition of enzymes — and also with other regulatory proteins such as CAP (catabolic activation protein), which interacts with the promoters of the genes coding for the enzymes involved in the metabolism of glucose.

6.3 Biological Regulation

Another way control constraints can be affected or modulated beyond constitutive interactions is hierarchically, by the direct action of other specialized constraints in the system. These internally produced, second-order control subsystems provide a living system with the capability to act on its own constitutive dynamics. They can be functionally recruited into regulatory mechanisms (R) that modulate constitutive control constraints on the basis of internal and external signals in such a way as to maintain the overall viability of the system.[17] These higher-order architectures open a new space of possible control operations. They enable the realization of more complex mechanisms that contribute to enhancing the robustness of the system, *beyond and on top* of the constitutive ones, which instead are ruled by changes in concentrations and embedded in the basic self-maintaining network.

In this new organizational architecture, the functional role of a regulatory subsystem is to modulate the basic constitutive network, by shifting between distinct metabolic regimes available to the system in relation to changes in environmental conditions. It does so in such a way that the new metabolic/constitutive regimes brought forth by regulatory shifts should be capable of coping with the new environmental conditions and internal variation, thus extending the range of perturbations

---

[17]For a more detailed discussion of biological regulation as second-order control see Bich et al. (2016).

or stimuli to which the system may respond in a rapid and efficient way, as well as enriching the sphere of dynamic functional behaviors available.

One of the difficulties that arise when addressing control and regulation, is to find an operational/naturalized way to distinguish between the regulator and what is regulated. Regulatory control cannot be regarded as a straightforward extension of the collective control that enables the dynamical stability of the constitutive regime. It is not the consequence of either a different way to wire constraints and processes, or of the introduction of additional functional nodes in the basic self-maintaining network. In these cases, the result would still be a constitutive network.

A more complex organizational architecture is required in order to realize regulation. Its distinctiveness and the difficulty in characterizing it, stem from the fact that a regulatory subsystem (R) is part of the living system, and like the other parts is produced and maintained by the processes integrating the constitutive regime (C). Yet, to act as a second order controller of C, R needs to work according to a different logic than the one characteristic of C, which is based on the strict (stoichiometric) coupling between control subsystems.[18] R needs to be able to exhibit some independence — i.e. a *dynamical decoupling* — from what takes place in C, in order to freely and asymmetrically modulate the activity of the control constraints in C on the basis of signals. This capability can be achieved when R exhibits additional degrees of freedom with respect to the controllers in C. Some of these new degrees of freedom make R sensitive to activation and inhibition signals, while not being directly dependent on the state of C. Others endow R with the effector capability to modulate the activity of the controllers in C (Bich et al. 2016).

The decoupling required in order for R to be operationally distinct from C, and capable to independently modulate the latter, is realized when the operations of the regulatory subsystem R are neither specified nor directly determined by the metabolic activity of C: i.e. they are 'stoichiometrically free' from the latter (Griesemer and Szathmáry 2009). More specifically, the activation and operation of a regulatory subsystem R is not directly dependent on its concentration (or variation of concentration) — that is, on its production by C — even though C guarantees its presence in the system (Bich et al. 2016). Instead, the activation of R is triggered by signals, and its operations depend on its internal organization and on the structure of its functional parts (in particular molecular geometries that are complementary to those of the controlled constraints in C). In such a way the activity of R is operationally distinct from C, and R can act as a dedicated regulatory controller of C. The functional regulatory loop realized by a control architecture so built, takes place in three steps: (1) a signal or perturbation activates R, which (2) in turn modulates C and brings forth a new constitutive regime. Finally, (3) the modification of C enables the system to cope with the specific variation which triggered the regulatory response.

---

[18] As argued by William Bechtel, "Although stoichiometric linkages between reactions are effective for insuring linkages between operations, they do not provide a means for varying the reactions independently. Such independent control can only be achieved by a property not directly linked to the critical stoichiometry of the system" (Bechtel 2007, 229).

This general theoretical model applies to autonomous systems at different levels of organization. Let us now consider two examples of how living systems — basic ones such as bacteria and multicellular systems like mammals — rely on mechanisms of regulation to adaptively respond to the composition (or variation in the composition) of their internal and external environments.

Some of the most basic mechanisms of regulation are present at the very core of biological machinery, such as protein synthesis. Organisms need to be able to determine which proteins/enzymes to produce on the basis of their internal state, of the availability of specific amino acids, and of the characteristics of the environment. To do so they need to activate, inhibit and modulate the synthesis of specific proteins according to their needs. The regulation of protein synthesis can take place at many different steps of this process: at the level of DNA through the control of transcription, which involves longer time scales; at the level of RNA through RNA processing control, RNA transport and localization control, translational control, mRNA degradation control, etc.; and at the post-translational level by modulating, at shorter time scales, the folding and activity of proteins — e.g. through forms of allosteric control, phosphorylation, etc.

Let us consider for example how bacteria modulate the metabolism of tryptophan through a regulatory control of the transcription step. Bacteria are able to produce this amino acid when it is not available in the environment. The genes responsible for the five enzymes which contribute to its synthesis are grouped together into one operon. A repressor protein exerts a regulatory control upon the promoter of the operon, by repressing it in presence of tryptophan in the cell. Two molecules of this amino acid act as the signals that allosterically activate the repressor protein. When tryptophan is present in the environment, the repressor protein is activated and blocks the endogenous production of this amino acid by repressing the synthesis of the enzymes responsible for it. In absence of tryptophan, instead, the repressor protein is in an inhibited state, and the cell can start synthesizing the enzymes responsible for the production of the amino acid. In this example the regulator R (the repressor protein) is dynamically decoupled from the constitutive regime C, i.e. from the enzyme-coding operon and the metabolism of tryptophan. The activity of the repressor protein does not depend on variation in its own concentration — which is determined by the rates of C[19] — but, rather, on its (stoichiometrically free) structural affinities with the signal molecules and with the promoter sequence.

The second example concerns the regulation of glucose concentrations in mammals through glycogen synthesis during fast and food uptake.[20] The metabolic path-

---

[19] The concentration of repressor proteins is usually low (proportional to the number of copies of the promoter sequence it regulates) and does not undergo variation in concentrations to bring forth its regulatory effect. The *lac*-operon system, a more complex example of genetic regulation, which coordinates the metabolism of two sugars (lactose and glucose) through the second-order control of the transcription step, follows the same logic (see Bich et al. 2016 for an analysis of regulatory decoupling in this latter case).

[20] Specifically, the example relies on the analysis of glycogen synthesis in the rat gastrocnemius muscle provided in Schafer et al. (2005).

way that in mammal cells connects glucose to G6P (through the GT/HK subsystem) and then to glycogen (through the GSase subsystem) is characterized by homeostatic capabilities, that in the fasting state of low glucose level can compensate slight variations in glucose through negative feedbacks and other distributed responses (Schafer et al. 2005).[21] Yet, the unregulated metabolic pathway alone is not able to cope through a network response with the strong increase in glucose that takes places during food uptake.

Therefore, additional mechanisms such as the release of insulin are implemented to prevent glucose from reaching dangerously high levels, by activating and increasing glycogen synthesis sufficiently and quickly enough to cope with the rise in plasma concentration of glucose. "Larger-scale flux changes operating on a more intermediate time scale will involve an external detector/effector (e.g. pancreas/insulin) that stimulates both up- and downstream subsystems, thereby maintaining excellent internal and external homeostasis despite increased flux" (Schafer et al. 2005, 69).

The release of insulin by the pancreas is triggered by signals in presence of high concentrations of glucose, and it leads to the "coordinate activation of glucose transport, hexokinase and glycogen synthase" (Schafer et al. 2005, 67)[22] which allows metabolizing the exceeding sugar. In virtue of the action of the pancreas/insulin regulatory subsystem, dynamically decoupled from the intracellular metabolism of muscular cells, the system as a whole can reach a new regime that is able to cope with the perturbing variation — the absorption of glucose through food.

This example shows how distributed network properties (dynamic stability) and regulatory mechanisms coexist and interact to enhance the organism's robustness, and what is the crucial role played by regulation in integrating and coordinating some core metabolic functions to maintain the viability of the whole system. The main point is that, whereas intracellular stability network mechanisms that compensate for slight variations in glucose concentrations are always at work during fast, they do not guarantee the viability of the system in response to a strong increase of glucose that occurs during food uptake. During food uptake, regulatory mechanisms intervene by releasing insulin, and they bring forth a coordinate response of the organism's metabolism to transform glucose into glycogen. This case also shows that regulation does not depend solely on genetic control — which requires more time to exert its effects. When quick responses are needed, they can be achieved by the coordinated control of the activation state of several enzymes and subsystems in a pathway.

---

[21] See also J. S. Hofmeyr and Cornish-Bowden (2000) for an analysis of this type of stoichiometrically dependent responses.

[22] One of the effects of the release of insulin is the rapid change of the phosphorylation state of the enzyme glycogen synthase (GSase), which alters its kinetics. It is a clear case of second order control interaction, in which a regulatory subsystem acts on a first order control constraints (GSase) by activating (or inhibiting) it.

6.4 Regulation at the Crossroads Between Identity, Complexity, and Cognition

A possible approach to the study of the nature and evolution of biological robustness is to consider dynamic stability as the default property of self-producing and self-maintaining prebiotic and minimal living systems. According to this view, additional and more complex regulatory mechanisms were later developed beyond the constitutive regime C. These mechanisms act on top of C, and they allow the system to cope with a wider range of variations and to improve the specificity of its responses to perturbations. Regulation, therefore, contributes to extend the viability space of the already dynamically stable biological system.[23]

In this view, the requirements for the emergence of biological regulation would also roughly coincide with fundamental transitions in the process of evolution of robustness, leading from basic stability to full-fledged adaptive regulation. They would include the emergence of: (a) organizational self-production and self-maintenance; (b) functional differentiation; (c) network stability; (d) multistability; and (e) sequence dependent components that enable stoichiometric freedom by means of interactions based on complementary molecular geometry.

In particular, steps (d) and (e) would be crucial for the transition towards regulation. Multistability enables the realization of distinct constitutive regimes which, in turn, may allow the system to remain viable under different conditions. As it has been recently shown, bistability and related properties such as oscillations, can arise already in simple autocatalytic networks of relevant organic reactions such as those involved in origins of life (Semenov et al. 2016). The problem that arises at this step is how to govern multiple stability precisely and efficiently: a distributed system characterized by multiple possible stable regimes can be more and more fragile the higher the number of attractors. Yet, the existence of more than one viable regime is the basis upon which specialized second-order control subsystems can functionally act and bring forth transitions in C, compatibly with internal and external conditions. One way to look at multistability is to regard it as a starting point, capable to generate a new *adjacent possible* (Kauffman 2000) in the evolution of robustness. The additional requirement for the development of regulatory capabilities is then the presence in the system of geometry-dependent molecular complexes that can be recruited by the system to operate as dynamically decoupled switches that control the shifts between attractors in C.

It is therefore reasonable to think that regulation is dependent on the existence of a full-fledged constitutive network already capable to exhibit stability (and multistability) and, consequently, that it is both logically and historically secondary to more basic forms of biological robustness. A possible weakness of this idea it that the

---

[23] Another way to express the idea is that systems can be alive under very special stable conditions, without regulatory mechanisms, by relying only on the constitutive network. Regulation would then become necessary only when the system is immersed in changing environments and develops a higher internal differentiation.

stoichiometric couplings that characterize the constitutive regime are very fragile: anything disturbing the delicate balance of metabolites or creating a new branching in the metabolic pathways would easily break the system apart. Moreover, some further considerations that will be discussed in Sect. 6.4.1. might make us question this idea and admit the possibility that mechanisms responsible for complex form of robustness might have already been necessary for the realization of a constitutive regime and for the emergence of some of the fundamental properties of biological systems.

6.4.1 *Functional Integration and Organizational Complexity*

Functional integration can be characterized as the degree of interdependence between the (functional) subsystems that are necessary to realize and maintain the system that harbors them. As argued in Sect. 6.2.1, the integration into one coherent system of several types of kinetic, spatial and template control constraints is the basis for the realization of a self-maintaining and self-producing biochemical machinery. The way these subsystems are put together constitutes the specific organization of the constitutive regime of the system. This organization is what determines its *identity* as a living system, characterized by functional autonomy, cohesion, and asymmetry with respect to the environment. There are different ways in which the subsystems of a basic autonomous organization can operate in a functionally integrated fashion (Bich 2016). The simplest ones are constituted by (1) metabolic complementarities — in which the products of one subsystem become the substrates for the processes controlled by another, like in the case of the chemoton — or (2) forms of cross control — in which control constraints are the direct products of the process harnessed by other constraints in the network (e.g. cross-catalysis).

Yet, the integration of subsystems into a basic constitutive network cannot be achieved by simply coupling pre-existing molecular or supra-molecular complexes and recruiting them as control constraints. As pointed out by Shirt-Ediss, "the first systems with the ability to robustly maintain themselves far-from-equilibrium [...] could only have existed as such if they were encapsulated and their internal (proto-metabolic) organization and compartment were tightly integrated" (Shirt-Ediss 2016, 85). This tight integration requires a matching between the features of the subsystems involved: for instance, the composition of the membrane and the position of its molecular machineries to meet the demands of metabolism; the synthesis of the right components by metabolism to be used in compartments to achieve the required permeability. As it has been recently argued, "the encapsulation of a self-maintaining chemical system has far-reaching organizational implications since its viability imposes significant changes on both parts (compartments and metabolic networks) in order to enable a functional coupling between them" (Moreno 2016, 10).

A minimal constitutive network might not be enough to realize a living autonomous system, and surely not to achieve more complex forms of biological organization, not only because of its overall fragility. The basic functional subsystems, in fact, need not only to be matched. Also their activity and rates need to be functionally coordinated in order to achieve integration, to avoid conflict, and to realize robustness at the system level under changing internal and external conditions.[24] The functional coordination necessary to realize and maintain this integrated organization might already require control hierarchies such as those realized by regulatory mechanisms.

As a matter of fact, all current living systems employ forms of hierarchical control to modulate the relations between their constitutive subsystems in such a way that they are capable to coordinate their basic functions and achieve integration. In bacteria, for example, membrane channels are not only activated directly by concentration gradients, but also by protein phosphorylation triggered by specific signals from the environment and the metabolism (Karpen 2004; Kulasekara and Miller 2007). In turn, signals from the membrane can trigger the regulation of gene expression, which can affect metabolism in a way that is compatible with the state of the membrane (Stock et al. 1989). The coordination between genome and metabolism is exemplified by genetic regulatory mechanisms such as the tryptophan operon, described above, which modulates metabolism through protein synthesis, compatibly with the composition of the internal and external environments of the living system. The *lac*-operon provides an even more interesting case: the regulatory mechanisms that govern the diauxic shift between glucose and lactose metabolisms, coordinate two different functional regimes within metabolism itself according to the state of the internal and external environments (Jacob and Monod 1961; Bich et al. 2016). Bacterial chemotaxis is another example in which the coordination between two functional subsystems — metabolism and the flagellum responsible for movement — is achieved, in different ways, through the activity of a regulatory mechanism (Eisenbach 2004; van Duijn et al. 2006; Eisenbach 2007; Alexandre 2010; Bich and Moreno 2016). These examples show that regulation plays a role in achieving the functional integration necessary for the continuous realization of the constitutive regime of biological autonomous systems, starting from their simplest instances in the prokaryotic world.

A further remark concerns organizational complexity. As argued above, a basic constitutive regime already exhibits a relevant internal complexity that allows it to harbor the functional differentiation necessary for its production and maintenance. Such degree of complexity cannot be supported reliably by means of network properties alone in presence of variations, insofar as generating a compensatory effect depends on propagating changes through many local interactions. As argued by

---

[24] Different subsystems might present different internal norms of operation, and need to be coordinated to ensure both their compatibility and their joint functional contribution to the maintenance of the system. Another relevant feature of biological cells is that they cannot synthesise all the possible molecules at the same time, due to energetic and spatial limits. Therefore, they have to exert some control over biosynthetic processes in order to produce the necessary components at the right times. Other types of functional resources need coordination as well. Let us think of the interplay between metabolic regimes relying on different carbon sources, such as lactose and glucose. Without the proper regulation (realised for example by the *lac*-operon subsystem) these regimes would compete for basic catalytic resources in the cell.

Wayne Christensen (2007), in this context achieving reliability and specificity of responses becomes an issue. The time required for the responses depends on both the size of the system and the degree of complexity found in it, and, as argued above, the internal complexity required for a constitutive network is already high. Additional problems derive from the difficulties in generating multiple differentiated global states and in reaching the appropriate one for a given perturbation. The lack of specificity in the responses is responsible for the increased fragility of the systems when its internal complexity rises, unless additional switch mechanisms devoted to the selective modulation of the basic dynamics are in place. What regulation does is precisely to make it possible to overcome this bottleneck of complexity, by endowing the system with the capability to induce the appropriate collective pattern of behavior in a more rapid and efficacious way. The responses so produced are specific, and they are not negatively affected by the size of the system.

Moreover, evolutionarily speaking, the adaptation of constitutive networks to new environmental perturbations would require each time a modification of the organization of the core constitutive network of the system, which would more likely drive the system to disruption than to the complex responses. A modification of those subsystems dedicated to handling internal changes, such as switches, instead, provides more reliable solutions (see Kirschner and Gerhart 2005), that do not only provide further viability, but also enable the increase of complexity.

The individual and evolutionary capability to maintain or enhance robustness at the system level — by extending the range of internal and external variations the system is capable to cope with, and by enabling internal differentiation without loss of viability — are the results of a trade-off between stability and complexity. Dynamically stable distributed systems are fragile under qualitatively different conditions, and unspecific in their responses. To implement those differential and specific responses necessary to maintain a complex organization and to respond to a variety of changing conditions, internal variability (increased degrees of freedom) under the coordination of regulatory mechanisms is fundamental, and it is realized at the expenses of the intrinsic stability of the constitutive regime and of its subsystems.[25]

In sum, the increase of robustness at the system level goes hand in hand with increases of complexity and of internal variability, and it can take place only through the action of regulatory mechanisms. Novel and higher levels of regulation need to be invented to ensure the maintenance of an increasingly complex constitutive (or low-level) part of the system. As the complexity of the system increases, the corresponding regulatory subsystem is bound to become increasingly necessary for the continuous maintenance of the basic organization, from minimal living systems to more complex forms of life.

---

[25] The idea of improving adaptivity and robustness by increasing the degrees of freedom available to the systems' dynamics at the expense of distributed stability of the network can be found already in Piaget's school (Meyer 1967). For a recent discussion of the interplay between organisation and variation in biology see instead Montévil et al. (2016).

6.4.2 Minimal Cognition

The capability of living systems to adaptively cope with their interactions with the environment by means of regulatory mechanisms is also closely related to the question of the origin and characterization of minimal cognition.[26] Traditionally, the authors that first developed the framework of autonomy — in particular Piaget (1967) and Maturana and Varela (1980) — defended the view according to which cognition consists basically in the viable interactions that the organism can enter with the environment without losing its identity, and the internal modifications it undergoes in this process. This approach considers cognition in its minimal form as coextensive with life or coinciding with the interactive dimension of life. The general idea underlying this thesis is that what for physical systems would be just external influence, in living system is adaptively integrated and transformed into a "meaningful interpretation" (Heschl 1990, 13). Another way to formulate it, is that since autonomous systems — and, therefore, all living beings — are capable of "enacting a meaningful world", they would also *ipso facto* be cognitive agents, at least in a minimal sense (Maturana and Varela 1980; Varela et al. 1991; Bitbol and Luisi 2004; Bourgine and Stewart 2004). According to this perspective, then, the adaptive behavior of minimal organisms such as bacteria is already a cognitive phenomenon.

A different view is supported by those who argue that even though some of the properties exhibited by minimal living systems are important aspects of cognition, they are not sufficient to capture cognition. One of the weaknesses of the thesis that cognition emerges already with unicellular life, it has been argued, is that by dissolving cognition in broader biological phenomena, then it would become difficult to understand the nature, function, and evolutionary history of cognition as a specific phenomenon (Moreno et al. 1997). According to these latter approaches, it is increased *behavioral capacities* (Christensen and Hooker 2000) or a higher degree of organizational complexity — namely, a nervous system decoupled from metabolism, and with its own *distinctive norms* (Barandiaran and Moreno 2006) — which are the primary discriminating dimensions of cognition.

Bechtel, among others, sustains a position closer to the former one, and provides a theoretical and heuristic justification for addressing cognition already at the level of bacteria: "Evolution is a highly conserved process, and the mechanisms developed in our common ancestors with these species provide the foundation for many of our cognitive activities. Since these organisms lack some of the complications that have evolved in us, research on them can help reveal key features of our cognitive mechanisms" (Bechtel 2014, 158). Other supporters of minimal cognition at the unicellular level argue that the decoupling between metabolism and cognition

---

[26] See Bich and Moreno (2016) for a more detailed discussion of minimal cognition in relation to biological regulation.

takes place already in bacterial chemotaxis, and propose an account of minimal cognition based on sensory-motor activity (van Duijn et al. 2006).[27]

The account of regulation proposed here provides a generalized theoretical model to understand the most basic decoupling responsible for minimal forms of cognition: the decoupling between adaptive regulation and constitutive metabolism. The adaptive capability provided by regulation is not specific of sensory-motor activity, but rather exhibits a more fundamental logic common to all regulatory mechanisms dedicated to the interaction with the environment, such as for example the tryptophan and *lac*-operons (Bich et al. 2016). When applied to cognition, therefore, the notion of decoupling from constitutive metabolism needs to be addressed in this more encompassing context, which transcends sensory-motor activity to include other forms of (*sensory-effector*) adaptivity. The central idea is that one of the essential aspects of cognition that can be analyzed at the basic level of biological organization, is that cognitive agents should be able to distinguish between some specific features of their interaction with the environment and to act accordingly, in such a way as to maintain their viability. And this fundamental requirement for cognition can be met only in presence of regulatory mechanisms (Bich and Moreno 2016).

When a mechanism of regulation is at work, the environment is not only a source of indistinguishable perturbations, but also of specific and recognizable ones. The crucial point is that the regulated system reacts in a distinctive way: it does things according to what it distinguishes (what activates the regulatory subsystem) in its interactions with the environment. In presence of regulatory mechanisms perturbations do not directly drive the response of the system. It is the regulatory subsystem, activated by specific perturbations, which modulates the constitutive one. It does so in such a way that the system as a whole becomes able to cope with the environmental perturbations which triggered the regulatory response: the organism eats a new source of food, or secrets chemicals to neutralize a lethal substance, etc.

In this context, an environmental perturbation becomes a specific and recognizable interaction because of the nature of the relation it holds with the regulatory subsystem. The response of the system is the result of the evaluation operated by the regulatory subsystem (activation-plus-action). The regulatory subsystem establishes "classes of equivalence" (Rosen 1978) in the environment according to how the variation activates it and triggers the regulatory action, so that such categories are actually employed by the system to modify its own internal dynamics in a viable way. Therefore, perturbations achieve an endogenous, operational, significance for the system: the interactions with the environment become more than just a source of indistinct noise, but are converted into a world of endogenously generated (naturalized) significances.

This approach still leaves open the question whether these cognitive relevant properties are sufficient for cognition — and bacteria should be considered as fully-fledged cognitive — or a new source of normativity needs to emerge on top of

---

[27] See Godfrey-Smith (2016) for a critical discussion of this account of minimal cognition focused on sensory-motor activity.

the one related to the metabolic viability of the cell. This latter view is supported for example by the advocates of the emergence of cognition with the nervous system (Barandiaran and Moreno 2006).

A related question is whether a theory of minimal cognition should account for the simplest instantiation of those features we ascribe to human or higher animal cognition — and depict its evolution as continuous from the appearance of the nervous system — or consider minimal cognition as a distinct category in itself, specified by a first decoupling from the constitutive regime and capable to generate behavioral capabilities analogous to those found in multicellular systems. In the latter case the evolution of cognition would be understood as a discontinuous process characterized by the emergence of several decouplings. However, whichever position is adopted, the adaptive regulatory mechanisms that confer robustness to the system, by realizing different types of decoupling, play a relevant, if not crucial, role in the origin of cognition.

6.5 Final Remarks

What is the relationship between robustness and autonomy in living systems? An answer can be found by looking at biologically distinctive ways to achieve robustness beyond and on top of general network properties. The advantages of pursuing an organizational approach to this question, is that it allows us to identify and analyze robustness mechanisms by focusing on different types of contributions to the realization and maintenance of a biological system: for instance, processes, constraints, control, signals and regulation. This approach also provides an understanding of robustness, functional differentiation, modularity and integration in the encompassing context of a living cell or of a multicellular organism, rather than in relation to individual properties or performances of a system, or of parts of it. Moreover, the thermodynamic nature of the notion of constraint, as used in this approach,[28] can be useful in the attempt to lay a bridge between robustness and thermodynamics, one of the open issues in the field of robustness (Kitano 2007).

From this theoretical standpoint, this paper has advocated the view according to which mechanisms related to robustness — and, specifically, regulatory ones — play a fundamental role at the core of biological organization. They contribute not only to enhancing the viability of the system under changing conditions. They also make it possible to coordinate and integrate the basic constitutive functions of a living system, and to overcome bottlenecks of complexity. Moreover, they are the basic requirements for the emergence of minimal forms of cognition or of cognitively relevant properties. Some questions remain open, among which: whether or not such mechanisms played a role in the origin of life, and whether a viable minimal biological system integrating metabolism, compartment and template can be actu-

---

[28] See Pattee (1972, 1973), Kauffman (2000), Umerez and Mossio (2013), Moreno and Mossio (2015), Winning and Bechtel (2018).

ally realized without regulatory mechanisms, although in very special or controlled environmental conditions. A negative response to the latter question would then require a revision of our theoretical models of minimal life.

Acknowledgements This project has received funding from: the European Research Council (ERC) under the European Union's Horizon 2020 research and innovation programme – grant agreement n° 637647 – IDEM; from the Ministerio de Economia, Industria y Competitividad (MINECO), Spain ('Ramon y Cajal' Programme RYC-2016-19798 and research project FFI2014-52173-P); and from the Basque Government (Project: IT 590-13).

References

Alexandre, G. (2010). Coupling metabolism and chemotaxis-dependent behaviours by energy taxis receptors. *Microbiology, 156*(8), 2283–2293.

Arnellos, A., & Moreno, A. (2012). How functional organization originated in prebiotic evolution. *Ludus Vitalis, XX*(37), 1–23.

Arnellos, A., Ruiz-Mirazo, K., & Moreno, A. (2014). Organizational requirements for multicellular autonomy: Insights from a comparative case study. *Biology and Philosophy, 29*(6), 851–884.

Barandiaran, X., & Moreno, A. (2006). On what makes certain dynamical systems cognitive: A minimally cognitive organization program. *Adaptive Behavior, 14*(2), 171–185.

Bechtel, W. (2007). Biological mechanisms: Organized to maintain autonomy. In F. Boogerd, F. Bruggerman, J. H. Hofmeyr, & H. V. Westerhoff (Eds.), *Systems biology: Philosophical foundations* (pp. 269–302). Amsterdam: Elsevier.

Bechtel, W. (2008). *Mental mechanisms: Philosophical perspectives on cognitive neuroscience.* New York: Routledge.

Bechtel, W. (2014). *Cognitive biology: Surprising model organisms for cognitive science.* In Proceedings of the 36th annual conference of the cognitive science society (pp. 158–163).

Bechtel, W. (2017a). Explicating top-down causation using networks and dynamics. *Philosophy of Science, 84*, 253–274.

Bechtel, W. (2017b). Systems biology: Negotiating between holism and reductionism. In S. Green (Ed.), *Philosophy of systems biology: Perspectives from scientists and philosophers* (pp. 25–36). New York: Springer.

Bernard, C. (1865). *Introduction à l'étude de la médecine expérimentale.* Paris: Balliére.

Bich, L. (2016). Systems and organizations: Theoretical tools, conceptual distinctions and epistemological implications. In G. Minati, M. R. Abram, & E. Pessa (Eds.), *Towards a Post-Bertalanffy Systemics* (pp. 203–209). New York: Springer.

Bich, L., & Arnellos, A. (2012). Autopoiesis, autonomy and organizational biology: Critical remarks on "Life After Ashby". *Cybernetics and Human Knowing, 19*(4), 75–103.

Bich, L., & Damiano, L. (2008). Order in the nothing: Autopoiesis and the organizational characterization of the living. *Electronic Journal of Theoretical Physics – Special Issue Physics of Emergence and Organization, 4*(16), 343–373.

Bich, L., & Damiano, L. (2012a). Life, autonomy and cognition: An organizational approach to the definition of the universal properties of life. *Origins of Life and Evolution of Biospheres, 42*(5), 389–397.

Bich, L., & Damiano, L. (2012b). On the emergence of biology from chemistry: A discontinuist perspective from the point of view of stability and regulation. *Origins of Life and Evolution of Biospheres, 42*(5), 475–482.

Bich, L., & Moreno, A. (2016). The role of regulation in the origin and synthetic modelling of minimal cognition. *Biosystems, 148*, 12–21.

Bich, L., Mossio, M., Ruiz-Mirazo, K., & Moreno, A. (2016). Biological regulation: Controlling the system from within. *Biology and Philosophy, 31*(2), 237–265.

Bitbol, M., & Luisi, P. L. (2004). Autopoiesis with or without cognition: Defining life at its edge. *Journal of the Royal Society Interface, 1*(1), 99–107.

Bourgine, P., & Stewart, J. (2004). Autopoiesis and cognition. *Artificial Life, 10*(3), 327–345.

Chen, C. (2008). Review on robustness in systems biology. *Journal of Biomechatronics Engineering, 1*(2), 17–28.

Christensen, W. (2007). The evolutionary origins of volition. In D. Spurrett, H. Kincaid, D. Ross, & L. Stephens (Eds.), *Distributed cognition and the will: Individual volition and social context* (pp. 255–287). Cambridge, MA: The MIT Press.

Christensen, W. D., & Hooker, C. (2000). An interactivist-constructivist approach to intelligence: Self-directed anticipative learning. *Philosophical Psychology, 13*, 5–45.

Cornish-Bowden, A. (2006). Putting the systems back into systems biology. *Perspectives in Biology and Medicine, 49*(4), 475–489.

Damiano, L., & Luisi, P. L. (2010). Towards an autopoietic redefinition of life. *Origins of Life and Evolution of Biospheres, 40*(2), 145–149.

Dupré, J., & O'Malley, M. (2005). Fundamental issues in systems biology. *Bio Essays, 27*, 1270–1276.

Eisenbach, M. (Ed.). (2004). *Chemotaxis*. Singapore: Imperial College Press.

Eisenbach, M. (2007). A hitchhiker's guide through advances and conceptual changes in chemotaxis. *Journal of Cellular Physiology, 213*, 574–580.

Ganti, T. (1975). Organization of chemical reactions into dividing and metabolizing units: The chemotons. *Bio Systems, 7*, 15–21.

Ganti, T. (1979). *A theory of biochemical supersystems*. Baltimore: University Park Press.

Ganti, T. (2003a). *The principles of life*. Oxford: Oxford University Press.

Ganti, T. (2003b). *Chemoton theory*. New York: Kluwer Academic/Plenum Publisher.

Godfrey-Smith, P. (2016). Individuality, subjectivity, and minimal cognition. *Biology and Philosophy, 31*, 775–796. https://doi.org/10.1007/s10539-016-9543-1.

Griesemer, J., & Szathmáry, E. (2009). Ganti's chemoton model and life criteria. In S. Rasmussen, M. Bedau, L. Chen, D. Deamer, D. Krakauer, N. Packard, & P. Stadler (Eds.), *Protocells. Bridging nonliving and living matter* (pp. 481–513). Cambridge, MA: The MIT Press.

Heschl, A. (1990). L = C. A simple equation with astonishing consequences. *Journal of Theoretical Biology, 145*, 13–40.

Hofmeyr, J. H., & Cornish-Bowden, A. (1991). Quantitative assessment of regulation in metabolic systems. *European Journal of Biochemistry/FEBS, 200*(1), 223–236.

Hofmeyr, J. S., & Cornish-Bowden, A. (2000). Regulating the cellular economy of supply and demand. *FEBS Letters, 476*(1–2), 47–51.

Jacob, F., & Monod, J. (1961). Genetic regulatory mechanisms in the synthesis of proteins. *Journal of Molecular Biology, 3*, 318–356.

Kant, I. (1790). *Kritik der Urteilskraft*.

Karpen, J. W. (2004). Ion channel structure and the promise of bacteria. *The Journal of General Physiology, 124*(3), 199–201.

Kauffman, S. A. (1986). Autocatalytic sets of proteins. *Journal of Theoretical Biology, 119*(1), 1–24.

Kauffman, S. A. (2000). *Investigations*. New York: Oxford University Press.

Kirschner, M. W., & Gerhart, J. C. (2005). *The plausibility of life: Resolving Darwin's dilemma*. New Haven: Yale University Press.

Kitano, H. (2004). Biological robustness. *Nature Reviews. Genetics, 5*(11), 826–837.

Kitano, H. (2007). Towards a theory of biological robustness. *Molecular Systems Biology, 3*(137), 1–7.

Kulasekara, H. D., & Miller, S. I. (2007). Threonine phosphorylation times bacterial secretion. *Nature Cell Biology, 9*(7), 734–736.

Lamelas, A., Gosalbes, M. J., Manzano-Marín, A., Peretó, J., Moya, A., & Latorre, A. (2011). Serratia symbiotica from the aphid Cinara cedri: A missing link from facultative to obligate insect endosymbiont. *PLoS Genetics, 7*(11), e1002357.

Letelier, J.-C., Soto-Andrade, J., Guíñez Abarzúa, F., Cornish-Bowden, A., & Luz Cárdenas, M. (2006). Organizational invariance and metabolic closure: Analysis in terms of (M,R) systems. *Journal of Theoretical Biology, 238*(4), 949–961.

Luisi, P. L. (2006). *The emergence of life. From chemical origins to synthetic biology.* Cambridge: Cambridge University Press.

Mansy, S. S., Schrum, J. P., Krishnamurthy, M., Tobé, S., Treco, D. a., & Szostak, J. W. (2008). Template-directed synthesis of a genetic polymer in a model protocell. *Nature, 454*(7200), 122–125.

Maturana, H., & Varela, F. J. (1980). *Autopoiesis and cognition. The realization of the living.* Dordrecht: Reidel Publishing.

Meyer, F. (1967). Situation épistémologique de la biologie. In J. Piaget (Ed.), *Logique et connaissance scientifique. Encyclopédie de la Pléyade* (pp. 781–821). Paris: Gallimard.

Mitchell, S. (2009). *Unsimple truths: Science, complexity, and policy.* Chicago: The University of Chicago Press.

Montévil, M., & Mossio, M. (2015). Biological organisation as closure of constraints. *Journal of Theoretical Biology, 372,* 179–191.

Montévil, M., Mossio, M., Pocheville, A., & Longo, G. (2016). Theoretical principles for biology: Variation. *Progress in Biophysics and Molecular Biology, 122,* 36–50. https://doi.org/10.1016/j.pbiomolbio.2016.08.005.

Moreno, A. (2016). Some conceptual issues in the transition from chemistry to biology. *History and Philosophy of the Life Sciences, 38*(4), 1–16.

Moreno, A., & Mossio, M. (2015). *Biological autonomy: A philosophical and theoretical enquiry.* Dordrecht: Springer.

Moreno, A., & Ruiz-Mirazo, K. (2009). The problem of the emergence of functional diversity in prebiotic evolution. *Biology and Philosophy, 24*(5), 585–605.

Moreno, A., Umerez, J., & Ibanez, J. (1997). Cognition and life. The autonomy of cognition. *Brain & Cognition, 34*(1), 107–129.

Moreno, A., Ruiz-Mirazo, K., & Barandiaran, X. (2011). The impact of the paradigm of complexity on the foundational framework of biology and cognitive science. In C. Hooker (Ed.), *Philosophy of complex systems* (pp. 311–333). Amsterdam: North Holland.

Mossio, M., & Bich, L. (2017). What makes biological organisation teleological? *Synthese, 194*(4), 1089–1114.

Mossio, M., & Moreno, A. (2010). Organisational closure in biological organisms. *History and Philosophy of the Life Sciences, 32,* 269–288.

Mossio, M., Saborido, C., & Moreno, A. (2009). An organizational account of biological functions. *The British Journal for the Philosophy of Science, 60*(4), 813–841.

Mossio, M., Bich, L., & Moreno, A. (2013). Emergence, closure and inter-level causation in biological systems. *Erkenntnis, 78*(2), 153–178.

Mossio, M., Montévil, M., & Longo, G. (2016). Theoretical principles for biology: Organization. *Progress in Biophysics and Molecular Biology., 122,* 24–35. https://doi.org/10.1016/j.pbiomolbio.2016.07.005.

Nghe, P., Hordijk, W., Kauffman, S. A., Walker, S. I., Schmidt, F. J., Kemble, H., et al. (2015). Prebiotic network evolution: Six key parameters. *Molecular Bio Systems., 11,* 3206–3217. https://doi.org/10.1039/C5MB00593K.

Nicolis, G., & Prigogine, I. (1977). *Self-organization in nonequilibrium systems: From dissipative structures to order through fluctuations.* New York: Wiley.

Pascal, R., & Pross, A. (2016). The logic of life. *Origins of Life and Evolution of Biospheres, 46*(4), 507–513.

Pattee, H. H. (1972). The nature of hierarchical controls in living matter. In R. Rosen (Ed.), *Foundations of mathematical biology*. Volume I, *Subcellular Systems* (pp. 1–22). New York: Academic Press.

Pattee, H. H. (Ed.) (1973). *Hierarchy theory*. New York: Braziller.

Piaget, J. (1967). *Biologie et Connaissance*. Paris: Gallimard.

Piedrafita, G., Montero, F., Morán, F., Cárdenas, M. L., & Cornish-Bowden, A. (2010). A simple self-maintaining metabolic system: Robustness, autocatalysis, bistability. *PLoS Computational Biology, 6*(8), e1000872.

Rosen, R. (1967). *Optimality principles in biology*. London: Butterworths.

Rosen, R. (1970). *Dynamical system theory in biology. Stability theory and its applications*. New York: Wiley.

Rosen, R. (1972). Some relational cell models: The metabolism-repair systems. In R. Rosen (Ed.), *Foundations of mathematical biology*. Volume II, *Cellular Systems* (pp. 217–253). New York: Academic Press.

Rosen, R. (1978). *Fundamentals of measurement and representation of natural systems*. New York: North Holland.

Rosen, R. (1985). *Anticipatory systems*. Oxford: Pergamon Press.

Rosen, R. (1991). *Life itself. A comprehensive inquiry into the nature, origin, and fabrication of life*. New York: Columbia University Press.

Ruiz-Mirazo, K., & Moreno, A. (2004). Basic autonomy as a fundamental step in the synthesis of life. *Artificial ife, 10*(3), 235–259.

Ruiz-Mirazo, K., Peretò, J., & Moreno, A. (2004). A universal definition of life: Autonomy and open-ended evolution. *Origins of Life and Evolution of the Biosphere, 34*(3), 323–346.

Schafer, J. R. A., Fell, D. A., Rothman, D. L., & Shulman, R. G. (2005). Phosphorylation of allosteric enzymes can serve homeostasis rather than control flux: The example of glycogen synthase. In R. G. Shulman & D. L. Rothman (Eds.), *Metabolomics by in vivo NMR* (pp. 59–71). Chichester: Wiley.

Semenov, S. N., Kraft, L. J., Ainla, A., Zhao, M., Baghbanzadeh, M., Campbell, V. E., et al. (2016). Autocatalytic, bistable, oscillatory networks of biologically relevant organic reactions. *Nature, 537*(7622), 656–660.

Shirt-Ediss, B. (2016). *Modelling early transitions towards autonomous protocells*. Ph.D. Dissertation. University of the Basque Country.

Stelling, J., Sauer, U., Szallasi, Z., Doyle, F. J., & Doyle, J. (2004). Robustness of cellular functions. *Cell, 118*(6), 675–685.

Stock, J. B., Ninfa, A. A., & Stock, A. M. (1989). Protein phosphorylation and regulation of adaptive responses in bacteria. *Microbiological Reviews, 53*(4), 450–490.

Umerez, J., & Mossio, M. (2013). Constraint. In W. Dubitzky, O. Wolkenhauer, K.-H. Cho, & H. Yokota (Eds.), *Encyclopedia of systems biology* (pp. 490–493). New York: Springer.

van Duijn, M., Keijzer, F., & Franken, D. (2006). Principles of minimal cognition: Casting cognition as sensorimotor coordination. *Adaptive Behavior, 14*(2), 157–170.

Van Segbroeck, S., Nowé, A., & Lenaerts, T. (2009). Stochastic simulation of the chemoton. *Artificial Life, 15*, 213–226.

Varela, F. J. (1979). *Principles of biological autonomy*. New York: North Holland.

Varela, F. J., Maturana, H. R., & Uribe, R. (1974). Autopoiesis: The organization of living systems, its characterization and a model. *Bio Systems, 5*(4), 187–196.

Varela, F. J., Thomson, E., & Rosch, E. (1991). *The embodied mind. Cognitive science and human experience*. Cambridge, MA: The MIT Press.

Veloso, F. (2017). On the developmental self-regulatory dynamics and evolution of individuated multicellular organisms. *Journal of Theoretical Biology, 417*, 84–99.

Wiley, H. R. (2013). Animal communication and noise. In H. Brum (Ed.), *Animal communication and noise* (pp. 7–30). Berlin/Heidelberg: Springer.

Winning, J., & Bechtel, B. (2018). Rethinking causality in biological and neural mechanisms: Constraints and control. *Minds and Machines, 28*, 287–310. https://doi.org/10.1007/s11023-018-9458-5.

Leonardo Bich is a 'Ramon y Cajal' Researcher at the IAS-Research Centre for Life, Mind, and Society of the University of the Basque Country (UPV/EHU), Spain. He obtained a PhD in Anthropology and Epistemology of Complex Systems from the University of Bergamo. He worked at the ImmunoConcept Lab of the CNRS & University of Bordeaux, at the Biology of Cognition Lab of the Universidad de Chile and, as a visiting fellow, at the Center for Philosophy of Science of the University of Pittsburgh. His research is focused on theoretical and epistemological issues related to biological organisation and autonomy, and on their implications for investigations in Origins of Life, Synthetic and Systems Biology, and Theoretical Biology.

Chapter 7
Robustness and Emergent Dynamics in Noisy Biological Systems

Christian Cherubini, Simonetta Filippi, and Alessandro Loppini

Abstract The concepts of robustness and stability play a central role in many natural phenomena ranging from Astrophysics up to Life. In this contribution we discuss these concepts by specifically focusing on a biological paradigmatic mathematical model for the nonlinear electrophysiology of clusters of animal beta-cells.

Keywords Robustness · Systems biology · Mathematical modeling · Computational electrophysiology · Noisy systems

7.1 Introduction: Robustness and Stability in Physics and Biology

Robustness represents a concept which belongs to many different disciplines ranging from medicine and psychology up to mathematical, physical, natural and technological sciences. Stability too is a notion commonly adopted in several contexts in order to define, for instance, behavioral characteristics of a person or social situations, although in the context of science such a term has a mathematically well-defined meaning. Stability and robustness are not synonymous but – depending on the context – they may be strictly linked. Here we will specifically focus on robustness and stability in biological systems from a biophysical point of view. To this aim, a brief introduction of these two concepts in physics must be given first.

C. Cherubini (✉) · S. Filippi
Unit of Nonlinear Physics and Mathematical Modeling, Departmental Faculty of Engineering, University Campus Bio-Medico of Rome, Rome, Italy

International Center for Relativistic Astrophysics – I.C.R.A, University Campus Bio-Medico of Rome, Rome, Italy
e-mail: c.cherubini@unicampus.it; s.filippi@unicampus.it

A. Loppini
Unit of Nonlinear Physics and Mathematical Modeling, Departmental Faculty of Engineering, University Campus Bio-Medico of Rome, Rome, Italy
e-mail: a.loppini@unicampus.it

© Springer Nature Switzerland AG 2018
M. Bertolaso et al. (eds.), *Biological Robustness*, History, Philosophy and Theory of the Life Sciences 23, https://doi.org/10.1007/978-3-030-01198-7_7

In Newtonian mechanics many systems are stable if they remain (almost) the same, deterministically or probabilistically, upon small (linear) or even large (nonlinear) perturbations while in classical thermodynamics (a smoothed outcome of statistical mechanics) the notion of quantities as entropy or other thermodynamic potentials is often adopted as a stability indicator (Kondepudi and Prigogine 1998).

In any case, the concept of stability is associated with a minimal set of configurations of the system so that the latter, once modified in energy or other physical indicators by external influences, tends to come back to its original specific configuration or remains very close to that (Strogatz 2014). However, energy arguments can be misleading when used as a general stability tool. In General Relativity for instance, the notion of energy in many cases is problematic, whether because the localization of the energy of dynamical relativistic gravitational systems in a general way is not possible, or because the notion of energy in the majority of cases is observer-dependent (Misner et al. 1973). Also in Newtonian dissipative systems described for instance by reaction-diffusion equations (Bini et al. 2006, 2010), the use of a classical field theory leading to a natural definition of energy is physically and mathematically meaningless due to the appearance of "backwards in time" diffusing species (Cherubini and Filippi 2009). Consequently, it would be preferable to avoid, when unnecessary, using the notion of energy as a prominent indicator for characterizing stability in a generalized sense, focusing instead on some other dynamical quantifier.

A commonly adopted definition of stability is that a certain mathematical distance of a configuration from a perturbed one remains small or even goes to zero at later times (Strogatz 2014). Stability has crucial relevance in many natural systems. In Quantum Physics, for instance, the dynamical stability of matter and radiation interactions is an essential ingredient for justifying the presence of the World around us. Before Niels Bohr's shocking proposal of non-radiating orbits for electrons in fact (quantization), Classical Electromagnetism applied to orbiting particles around a nucleus (the so-called Rutherford atom), led to the insane result that atoms would collapse in a period less than a nanosecond, so that an explanation for the Universe as observed at that time was not possible (Gasiorowicz 2003). Stability is a key ingredient in celestial mechanics too (Celletti 2010), ultimately leading to the possibility for Life to exist on the Earth. Stability is also a key component in General Relativity again, for instance in the case of black hole configurations, which represent a strong attractor for very complex dynamics involving stars, but also in Cosmology, where stability of masses up to a certain spatial scale (the so-called Jeans mass) plays a central role in creating diversified structures in the early Universe (Ohanian and Ruffini 1994), although close to the initial or final highly dynamical cosmological singularities a meaningful notion of stability becomes nontrivial (see for instance Hobill et al. 1994). Also complex systems as the Lorentz model (Jackson 1992), a cornerstone toy model in Climatology, show that dynamical stability, ultimately linked at mathematical level to delicate problems of geometry and topology in system's abstract phase space, is an essential ingredient for the persistence of admissible conditions which allow Earth to be as it is.

Many of the systems just discussed are characterized by a classical deterministic nature described by differential equations. They rely on an old point of view in Physics, dating back to Newton, in which the complexity of Nature, full of irregular shapes and motions, is smoothed out sufficiently in order to use the mathematics of Differential Calculus (functions, limits, derivatives, integrals, etc.) in a view to having acceptably good descriptions and predictions for physical systems. Many of these, however, can or must be described using nondeterministic laws where the presence of noise dramatically affects the underlying dynamics of the system, both at physical and mathematical levels. In this case, the problem must be framed in the realm of Stochastic Processes, where new appropriate mathematical tools, as the stochastic differential calculus describing continuous but non-differentiable (in the standard way at least) functions must be used in order to account for the effects of noisy inputs, which introduce uncertainty in the definition of any observable property of a system (Lemons 2002).

The change of paradigm is dramatic with respect to Newtonian physics here. While in order to go from point A to point B in classical mechanics we have one possible trajectory only, the stochastic description gives any possible path connecting A and B a certain probability in fact. The reader can readily recognize in the previous sentence a strong connection with Quantum Mechanics, and in fact, in both cases, the probabilistic description deals essentially with a tiny scale of the constituents of the system under exam which in some cases can generate higher size effects.

It is natural to investigate the problem whether stability could be an indicator for robustness of a physical system, but this is not a trivial point. In fact, while an almost well-defined notion of stability can be given in rigorous mathematical terms, the concept of robustness would require somewhat an interpretative external action to quantify, for instance, how rapidly the perturbed system comes back to the starting configuration or other.

Moreover, it could be the case that the perturbed system does not go back to the original configuration, but to a closer one, whose features are almost similar to the starting configuration's one. As an example for such a situation, one can imagine, for instance, a cheap metal hammer falling from a ladder. Because of the impact with the ground, the hammer's head shape gets slightly deformed. Both initial and final hammer's equilibrium states, although different, represent stable configurations to be used robustly for hammering nails. In mathematical terms, we would be tempted to say that the two tools belong to an equivalence class of hammers, both of them robust enough for homework uses. Another example we can discuss is a star as the Sun close to us. In this case, a fluid mass, heated up by nuclear reactions, generates enough pressure to balance the compression upon gravitational attraction, so that the configuration radiates in almost stationary regime energy in the outer space (Tassoul 1978). The configuration is stable and the energy emission is robust enough for Life on our planet for instance. These simple examples seem to suggest that in order to quantify robustness, one should specify a certain function of the system.

7.2 Robustness: The Point of View of Biophysics

The example just discussed are representative of phenomena which influence, but do not describe Life mechanistically. To this end instead, we need to apply several laws of Physics, of Classical and even Quantum types, to living systems. This task is performed by Biophysics, which can be seen as a relatively recent generalization of Physiology. What makes Biophysics more involved however is the introduction for studying Life of advanced mathematical tools, shared nowadays also with Bio-Mathematics and Bio-Informatics, in union with specific physical experimental techniques previously developed in other contexts as condensed matter and high energy physics. The payoff of such a methodology is that one can have impressively accurate descriptions of many biological systems with a predictive character well beyond standard Biology.

We have however to say, for the sake of completeness, that living systems present a very high level of complexity and the powerful outcomes of a biophysical approach as just described give their best when applied to selected subsystems participating to a living being, appropriately simplified in its key elements. We have to point out also that nowadays systemic approaches as the one of Systems Biology aim to bypass the Western World Occam's Razor's approach to scientific problems in favor of a multilevel democratic description which studies Life as a whole (Noble 2006). This concept is at the basis of the enormous success of complex networks theory in the field of biology (Giuliani et al. 2014), being such an approach intrinsically able to investigate a system taking into account all – or a large amount – of its components and specific interactions between them.

However, the Systems Biology program is only at the very beginning, so it will be more fruitful for us to approach a biological robustness problem using a standard biophysical point of view. To this aim, we introduce a paradigmatic model which represents many features of biological robustness, i.e. the endocrine pancreas (Hall 2015).

7.3 Modeling Robustness in Pancreatic β-Cells Populations

Pancreatic Langerhans islets are ellipsoidal aggregates of cells which are arranged in a complex architecture. These islets are characterized by the presence of β-cells, α-cells, δ-cells and pancreatic polypeptide cells. All of these communicate both via autocrine and paracrine signalling as wells as ultrastructural connections. Such communications have a leading role in controlling cells functionalities by smoothing cells heterogeneity and synchronizing electrochemical and biochemical cellular activities. In particular β-cells play a prominent role in Glucose-Stimulated Insulin Secretion, a key process in the regulation of blood glucose level. A complex electrical activity, triggered by glucose uptake, drives intracellular calcium oscillations leading to insulin secretion which is central to Life. Glucose here plays the role of a control parameter regulating the cellular dynamics. This system is a communicating network so that the structural architecture of β-cell aggregates is another essential aspect in regulating cells' electrical activity and function.

A peculiarity of β-cells is the following: electrically coupled cells in compact physiological configuration manifest a coordinated regular synchronous bursting, directly connected to pulsatile insulin secretion, while an isolated β-cell displays an entirely irregular spiking activity. The dynamic is intrinsically affected by glucose changes. Specifically, low glucose concentrations lead to noisy low-voltage fluctuations around a silent resting state in cell membrane potential and this pattern is not directly linkable to insulin secretion. On the other hand, higher levels of glucose initiate a coordinated and almost periodical bioelectrical pattern whose bursting activity results to be directly associated with an effective insulin release. Finally, higher glucose concentrations give rise to a continuous bursting response which is characterized by a sustained almost fixed noisy action potential which is not related to an oscillating insulin release. In summary, Nature has selected a particular glucose range for which coordinated and robust bioelectrical insulin-producing oscillations of the entire compact cluster are possible.

A question arises at this stage. What if the β-cells cluster is not compact anymore or if the biochemical and bioelectrical communications of the cells within it are compromised? This dramatic event is realized in Nature. In type-1 diabetes, an autoimmune attack determines important losses of β-cell mass (see Fig. 7.1 for a biological snapshot of such a dramatic dynamics) during several years, leading to severe communication defects and consequently to an impaired and insufficient pulsatile insulin secretion which compromises the whole organism's physiology (Hall 2015). In type-2 diabetes instead, an increase in peripheral insulin resistance or specific defects in β-cells cause impaired glucose homeostasis and sustained hyperglycemia. In this dynamics, inflammation plays a central role in causing cellular damage and death by affecting connexin proteins which play a central role in "cell to cell" communications.

We point out that diabetes represents today a major health problem worldwide. The tens of millions of cases already manifested and the much more expected for

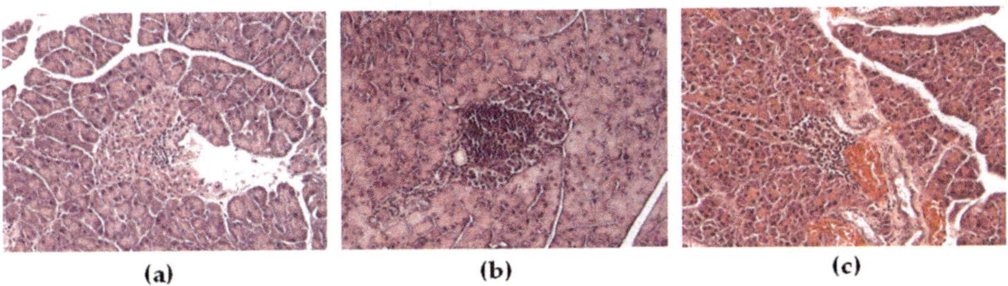

(a)   (b)   (c)

Fig. 7.1 Insulitis progression leading to type I diabetes in mouse pancreatic islets. (**a**) grade I insulitis: beta cells subject to a modest lymphocytes infiltration. (**b**) grade II insulitis: lymphocytes infiltration spreads almost over the whole islet. (**c**) grade III insulitis: the islet has almost been destroyed by lymphocytes infiltration. (Reprinted figure with permission from Portuesi et al. (2013). Copyright © 2012 John Wiley & Sons, Ltd)

the next decades not only in the US but also in the rest of the World have in fact induced many medical doctors to use the very impressive expression of a "Diabetes Pandemic" (Lancet Editorial 2011). Such a complicated biological phenomenology can be fruitfully analyzed by using an "in silico" model for the electrochemical activity of a cluster of beta cells in animals. Here we specifically focus on the Sherman-Rinzel-Keizer (SRK) stochastic model for the mouse (Sherman et al. 1988, Sherman and Rinzel 1991). Such relatively simple model essentially describes, via differential equations coupled to stochastically ruled parameters (in this case the fraction of open K-Ca channels at each time in each cell), the cells membrane action potential (V) as well as the intracellular calcium (Ca) dynamics (governing insulin secretion), in union with a third - so called - gating variable "n" which governs the dynamics of potassium. Each cell possesses in the model its set of (V, Ca, n) and interacts with the closest cells via a Von Neumann neighborhood communication pattern (Jackson 1992).

In this modeling, the presence of a stochastic noisy dynamical parameter for each cell plays a crucial role in determining a coordinated behavior by raising the dimensionality and compactness of the cluster. A lack of randomness in the equations (what is called in the biomathematical literature as the deterministic SRK model) in fact would determine a coordinated bioelectrical pattern which is almost insensitive to changes in cluster size, in striking contrast with experiments. The problem just described was numerically analysed in the case of a cubic cluster of 64 cells (Portuesi et al. 2013). Starting from an intact cluster with a normoglycemic state manifesting a well-coordinated electrical pattern associated to rhythmic intracellular calcium pulsations, gradual elimination of β-cells and a smooth increase of glycaemia levels, typical of the diabetic syndrome, were simulated determining crucial changes in the electrochemical patterns manifested by the beta-cells community. In particular, such a slow degradation of the cells' architecture evidences a synchronization loss which is reflected by an impaired intracellular calcium activity associated with a compromised insulin production (see Fig. 7.2).

What comes out from this analysis is that the robustness of the cluster is lost because of a cellular number decrease, although still when almost 70% of cells have been destroyed, proper glycaemia maintenance partially preserves the functionality of the system. Figure 7.2 gives quantitative information on the spatial robustness of the modeled cluster.

The emergence of coordinated patterns in this dynamics is paradigmatic and can also be interpreted by using tools typical of Quantum Field Theory (and of the companion theory of Statistical Mechanics), i.e. the algebraic approach of coherent state formalism (Loppini et al. 2014; Bertolaso et al. 2015). This point of view seems to suggest that coherent molecular domains appear in consequence of proper functional conditions. More in detail, by using a many-body physics point of view, one can notice that an increasing number of beta-cells in a cluster leads to a typical field theory "phase transition" situation triggered by changes in the glucose concentration (an order parameter) where the whole biological system is immersed.

Further progress can be done in this modelling activity by using the tools of Complex Networks (Cherubini et al. 2015), to analyse β-cells dynamics in the pres-

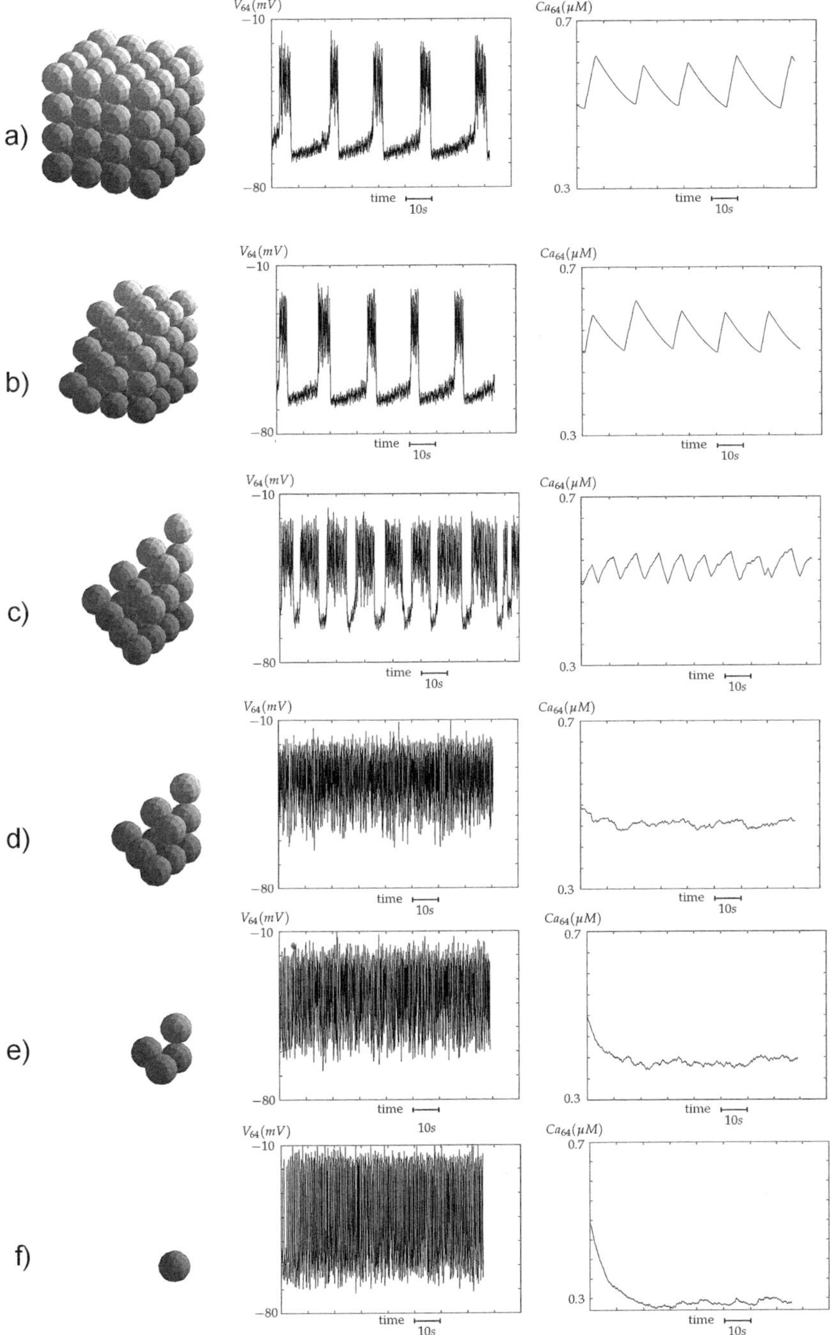

Fig. 7.2 Effects of autoimmune apoptotic process in a beta cells cluster (left column) on membrane potential (central column) and intracellular calcium (right column) dynamics of a representative cell. Stimulating glucose concentration increases with cluster damage from a normoglycemic condition up to severe hyperglycaemia. (Reprinted figure with permission from Portuesi et al. (2013) Copyright © 2012 John Wiley & Sons, Ltd)

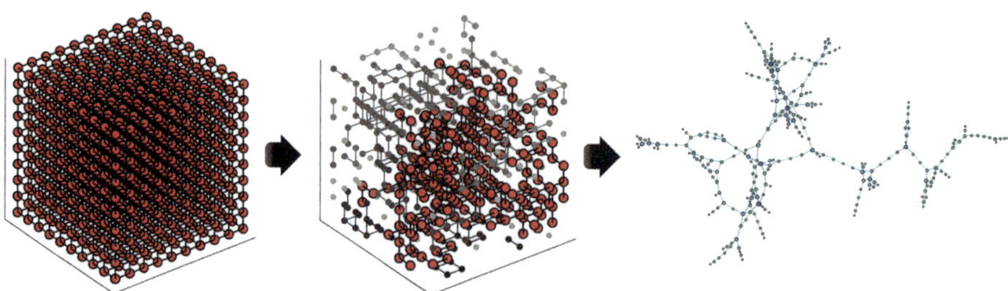

Fig. 7.3 Modeling of human-like beta cells network. Sphere and circles denote beta cells, while links denote coupling via gap-junctions. A site-bond percolation process is performed on a compact cubic cluster (left) based on beta cells percentage and spatial organization within pancreatic islets. This procedure results in a fragmentation of separated connected components (center). A two-dimensional projection of the largest component (right) can easily show the resulting complex topology. (Reprinted figure with permission from Cherubini et al. (2015). Copyright © 2015 by the American Physical Society)

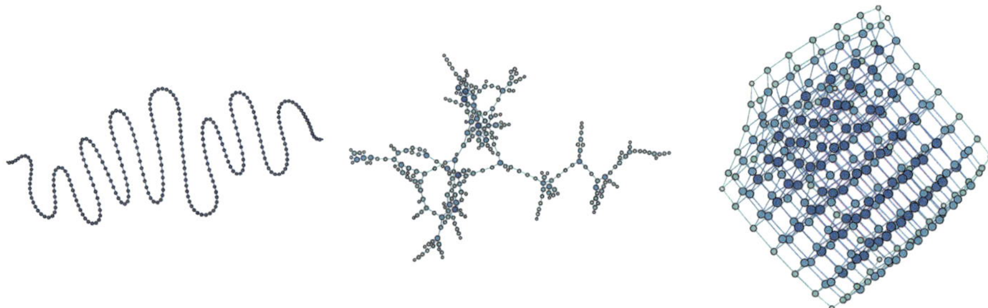

Fig. 7.4 Representation of different beta cells spatial assemblies including the same number of cells: (left) linear chain, (center) percolated cluster, (right) compact cluster. (Reprinted figure with permission from Cherubini et al. (2015). Copyright © 2015 by the American Physical Society)

ence of percolating phenomena on large clusters. In fact, percolation performed on compact configurations leads to β-cell aggregates which are characterised by non-trivial architectures that can be fruitfully visualised using complex networks theory (see Fig. 7.3).

Percolated systems can be studied in comparison with limiting cases in terms of network connectivity, such as chain configurations and compact three-dimensional lattice-like structure formed by the same number of cells (see Fig. 7.4), with the aim of investigating relevant differences in the emergent network bioelectrical activity induced solely by changes in the connectivity pattern. This procedure leads to different associated electrical oscillating patterns as shown in Fig. 7.5.

The presented study shows an interesting aspect of emergent dynamics with respect to the previously discussed ones, i.e. a cellular spatially fixed (we would say histological) organization, once "at work", shows a totally different organization in terms of synchronization patterns, i.e. it is possible to fruitfully contrast the concept of a spatial network to a functional one.

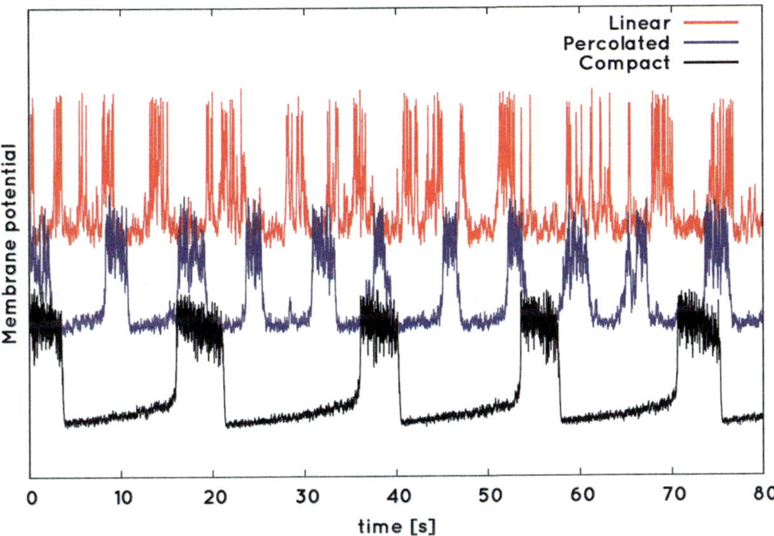

Fig. 7.5 Simulated membrane potential in a representative cell within spatial assemblies shown in Fig. 7.4: (top) linear chain, (middle) percolated cluster, (bottom) compact cluster. (Reprinted figure with permission from Cherubini et al. (2015). Copyright © 2015 by the American Physical Society)

In Fig. 7.6 the resulting dynamics is well summarized in a normoglycemic condition, considering clusters coupled by a physiological gap junction conductance. The most striking situation is the one of the percolated cluster shown in the central column, which induces (see central figure) a heterogeneous and modular functional complex network, remarkably different from the ones induced by the limiting cases of cluster connectivity presented in Fig. 7.4. Indeed, functional network modularity highlights an out-of-phase bursting across the β-cell population which can also be noticed in the space-time diagram showing the bio-electrical activity of entire cellular community (lowest row of Fig. 7.6). In linear chain and compact configurations, an almost unsynchronized (left bottom figure) and a completely synchronized (right bottom figure) bio-electrical pattern develop in time, respectively. Such features are grasped by homogenous functional networks with cells weakly or heavily connected. In comparison, percolated clusters (central bottom figure) promote the appearance of extremely coordinated sub-clusters, showing weak inter-modular correlations.

Biological robustness of this system manifests itself by varying the gap junction conductance staying at normoglycemic condition (see Fig. 7.7), as well as by changing glucose levels by keeping fixed at the physiological level instead of the gap junction conductance (see Fig. 7.8). In both cases, a functional networks analysis shows that only in the situation chosen by Nature (i.e. in physiological conditions) an efficient dynamics able to trigger a properly released insulin profile is present. Specifically, from an evolutionary perspective, were selected only the configuration

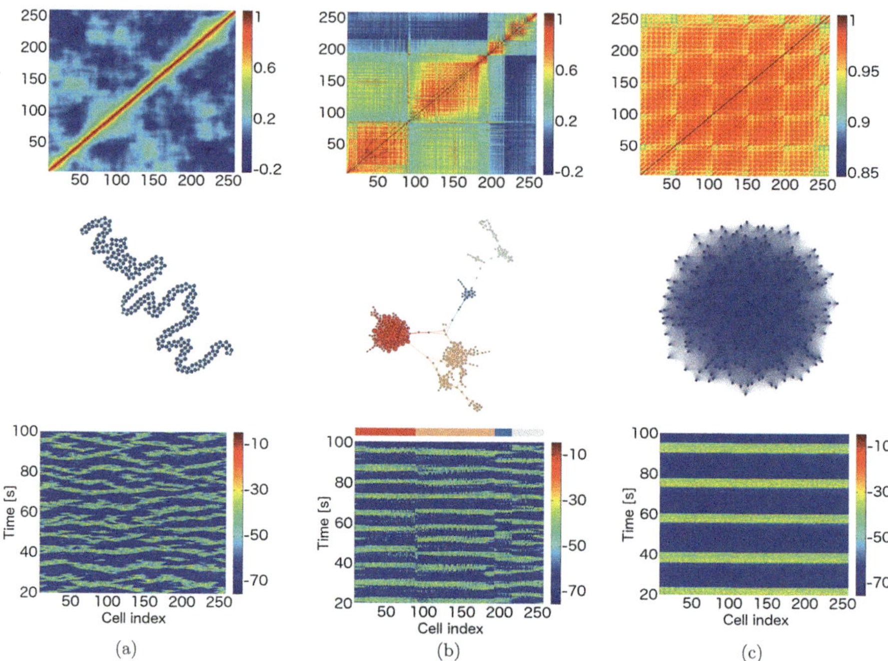

Fig. 7.6 Construction of functional networks from voltage time series. A correlation matrix (top row) is calculated from pair-wise correlation indices between simulated signals for different spatial assemblies shown in Fig. 7.4. A thresholding procedure is performed on the correlation matrix filtering only the most significant correlations, thus obtaining a binary matrix representing functional connections between cells. This matrix is used to build the corresponding functional network (central row). Analysis of functional organizations in comparison with the space-time plots of membrane voltage (bottom row) show that: (i) a low degree of functional coupling in the linear chain (**a**) underlies uncoherent membrane voltage oscillations across the cluster; (ii) intermediate levels of functional coupling and modular functional networks are linked to out of phase bursting across the percolated cluster (**b**); (iii) high degree of functional connectivity is linked to a strongly synchronized bursting in compact assemblies of cells (**c**). (Reprinted figure with permission from Cherubini et al. (2015). Copyright © 2015 by the American Physical Society)

of the system, in terms of controlling parameters, able to trigger effective patterns of bursting coordination, and related functional architectures, driving robust and pulsatile insulin release.

On the other hand, a slight difference in these parameters can induce significant modifications in the network's functional topology, i.e. in network activity, which impair the hormone production.

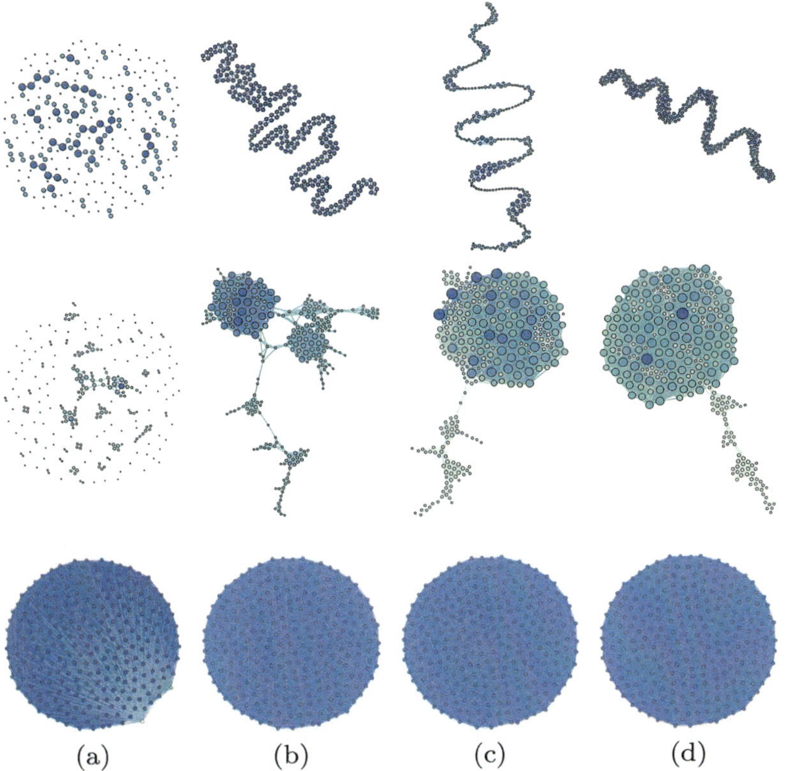

(a)   (b)   (c)   (d)

Fig. 7.7 Functional networks extracted from the linear chain (top row), percolated cluster (central row) and compact cluster (bottom row) at different values of the coupling conductance and fixed normoglycemic condition (glucose 7.1 mM). (**a**) 100 pS gap-junction conductance. (**b**) 215 pS gap-junction conductance (physiological mean value). (**c**) 300 pS gap-junction conductance. (**d**) 400 pS gap-junction conductance. A physiological value of coupling strength is required to obtain functional coupling. Moreover, in the case of percolated clusters, only around such value of coupling strength a modular organization of functional network arises. (Reprinted figure with permission from Cherubini et al. (2015). Copyright © 2015 by the American Physical Society)

7.4 Conclusion

Biological robustness is a very delicate outcome of a dynamical activity of specifically arranged cellular architectures. The model just presented shows that while noise is a fundamental ingredient for robust functional coordination of the network, directly reflected in appropriate insulin production, the change of biochemical parameters in the system, such as the glucose concentration or of the coupling strength between cells, dramatically affects the electrochemical pattern. In

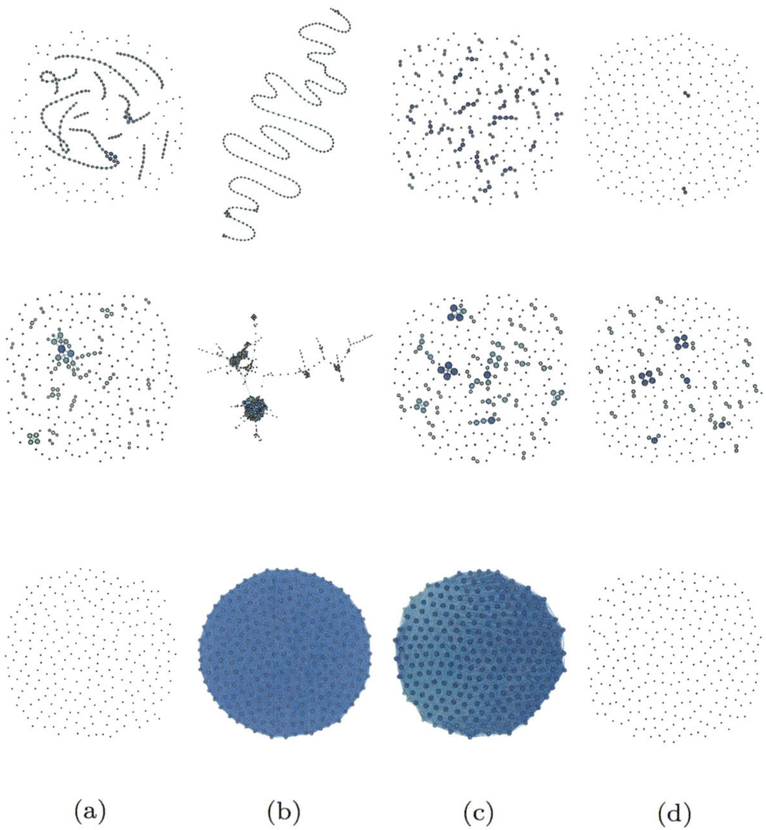

(a)   (b)   (c)   (d)

Fig. 7.8 Functional networks extracted from the linear chain (top row), percolated cluster (central row) and compact cluster (bottom row) at different values of the stimulating glucose concentration and fixed gap-junction conductance (215 pS). (**a**) 4.7 mM glucose (sub-threshold). (**b**) 8.7 mM glucose (typical postprandial level). (**c**) 12.7 mM glucose (mild hyperglycemia). (**d**) 16.6 mM glucose (severe hyperglycemia). At sub-threshold and hyperglycemic glucose levels all the beta cells assemblies show functional decoupling, while only in physiological postprandial glucose levels a proper functional coupling is obtained. (Reprinted figure with permission from Cherubini et al. (2015) Copyright © 2015 by the American Physical Society)

particular, the robustness of the function of the cluster, i.e. insulin production, is relevantly compromised by this action.

Summarizing,

- Deterministic analysis of such a biological system lacks many central features, especially in self-coordinating the cellular communities, a gap which instead is filled by adding the stochastic ingredient to the phenomenon.
- The noisy nature of the system generates pools of cells cooperating for the production of insulin and induces a robust emergent dynamics with respect to model parameters' changes.

– External control parameters as the glucose concentration must be confined in a strict range to ensure a coordinated and efficient insulin production. Glucose changes act relevantly on the robustness of this dynamics.

We must stress that the findings just shown would not have been possible without the use of specific tools typical of Complex Networks and Dynamics Systems. It is only by revealing the existence of very different histological and functional networks in the cluster that we realize that cellular communication towards robust biological activities - say insulin production - is a complicated matter which cannot be inferred by a simple histological description of the tissue. Life seems to be then a dynamical phenomenon where stochasticity plays a central role in synchronizing cells which are robustly in contact well beyond what we can see on a microscope. Indeed, a complex dynamics arise via communications of separate units, also when only trivial nearest-neighbors structural interactions are considered. Such interactions have the effect to shape deeply emergent behavior and functionality, and what emerges is a long-range coherent activity from underlying un-coherent dynamics. Further, one can ask why stochasticity is required. Wouldn't it be more reliable for an organism to adopt single deterministic cells as building blocks? The answer is not trivial and rely on the complex nonlinear behavior of the system, where heterogeneities induced by noise is beneficial both in enhancing the desired output and in ensuring metastable properties which are fundamental for robust functional plasticity. This point of view gives new hints both in how to mathematically model a biological system and in what to analyze in experiments.

Acknowledgments The authors acknowledge ICRANet and INdAM-GNFM for support.

References

Bertolaso, M., Capolupo, A., Cherubini, C., Filippi, S., Gizzi, A., Loppini, A., & Vitiello, G. (2015). The role of coherence in emergent behavior of biological systems. *Electromagnetic Biology and Medicine, 34*, 138–140.

Bini, D., Cherubini, C., & Filippi, S. (2006). Heat transfer in Fitzhugh-Nagumo models. *Physical Review E, 74*(4), 041905.

Bini, D., Cherubini, C., Filippi, S., Gizzi, A., & Ricci, P. E. (2010). On spiral waves arising in natural systems. *Communications in Computational Physics, 8*(3), 610.

Celletti, A. (2010). *Stability and Chaos in celestial mechanics*. Berlin: Springer.

Cherubini, C., & Filippi, S. (2009). Lagrangian field theory of reaction-diffusion. *Physical Review E, 80*, 046117-1–046117-14.

Cherubini, C., Filippi, S., Gizzi, A., & Loppini, A. (2015). Role of topology in complex functional networks of beta cells. *Physical Review E, 92*, 042702-1–042702-12.

Gasiorowicz, S. (2003). *Quantum physics* (3rd ed, Supplementary Material). Delhi/Massachusetts: Wiley.

Giuliani, A., Filippi, S., & Bertolaso, M. (2014). Why network approach can promote a new way of thinking in biology. *Frontiers in Genetics, 5*, 83.

Hall, J. E. (2015). *Guyton and Hall textbook of medical physiology* (13th ed.). Philadelphia: Saunders.

Hobill, D., Burd, A., & Coley, A. (1994). *Deterministic chaos in general relativity*. Boston: Springer.

Jackson, A. E. (1992). *Perspectives of nonlinear dynamics* (Vol. 2). Cambridge, MA: Cambridge University Press.

Kondepudi, D., & Prigogine, I. (1998). *Modern thermodynamics. From heat engines to dissipative structures*. Chichester: Wiley.

Lancet Editorial. (2011). The diabetes pandemic. *Lancet, 378*, 99.

Lemons, D. (2002). *An introduction to stochastic processes in physics*. Baltimore: Johns Hopkins University Press.

Loppini, A., Capolupo, A., Cherubini, C., Gizzi, A., Bertolaso, M., Filippi, S., & Vitiello, G. (2014). On the coherent behavior of pancreatic beta cell clusters. *Physics Letters A, 378*, 3210–3217.

Misner, C. W., Thorne, K. S., & Wheeler, J. A. (1973). *Gravitation*. San Francisco: W H Freeman & Co.

Noble, D. (2006). *The music of life*. Oxford: Oxford University Press.

Ohanian, H. C., & Ruffini, R. (1994). *Gravitation and spacetime* (2nd ed.). New York: W. W. Norton & Company.

Portuesi, R., Cherubini, C., Gizzi, A., Buzzetti, R., Pozzilli, P., & Filippi, S. (2013). A stochastic mathematical model to study the autoimmune progression towards type 1 diabetes. *Diabetes/Metabolism Research and Reviews, 29*, 194–203.

Sherman, A., & Rinzel, J. (1991). Model for synchronization of pancreatic beta cells by gap junction coupling. *Biophysical Journal, 59*, 547–559.

Sherman, A., Rinzel, J., & Keizer, J. (1988). Emergence of organized bursting in clusters of pancreatic beta-cells by channel sharing. *Biophysical Journal, 54*, 411–425.

Strogatz, S. H. (2014). *Nonlinear dynamics and chaos: With applications to physics, biology, chemistry, and engineering* (2nd ed.). Boulder: Westview Press.

Tassoul, J.-L. (1978). *Theory of rotating stars*. Princeton: Princeton University Press.

Christian Cherubini PhD is Associate Professor in Mathematical Physics at the Departmental Faculty of Engineering of Campus Bio-Medico University of Rome, Italy. He is also affiliated to the International Center for Relativistic astrophysics (I.C.R.A.). His research interests range from General Relativity, with particular attention to black hole physics, up to dynamical systems in biophysics with a specific focus on theoretical and computational aspects of electrophysiology.

Simonetta Filippi is Full Professor in Mathematical Physics at the Departmental Faculty of Engineering of Campus Bio-Medico University of Rome, Italy. She is also affiliated to the International Center for Relativistic astrophysics (I.C.R.A.). Her research interests range from astrophysics, with particular attention to classical and relativistic self-gravitating systems and to cosmology, up to complex systems with a specific focus on nonlinear cardiac dynamics and to the emergent phenomena of complex rhythms in biology.

Alessandro Loppini PhD holds a Postdoctoral position at the Departmental Faculty of Engineering of Campus Bio-Medico University of Rome, Italy. His research interests involve the study of nonlinear dynamics in complex systems, complex networks theory, and mathematical modeling of excitable cells and media.

Chapter 8
The Robustness/Sensitivity Paradox: An Essay on the Importance of Phase Separation

Alessandro Giuliani

Abstract Considering biological systems at different levels of organization as complex networks in which nodes (genes, proteins, metabolites…) are each other connected by (co-expression, physical interactions) is a very natural way of reasoning. Network approach allows scientist to make sense of the intricacies of biological regulation and, for the same mathematical nature of graphs, to obtain a multilevel description linking single node and whole network topological features. This paradigm allows for the detection of a clear signature of robustness: the ability of a system to keep separate different scales of response to environmental stimuli. A case study on the immune system allows for an immediate appreciation of this point.

Keywords Self-organization · Gene expression regulation · Networks · Systems biology · Innate immunity

8.1 Introduction

8.1.1 General Statements

In his book *The Origins of Order* (Kauffmann 1993) Stuart Kauffmann wrote "Living systems exist in the solid regime near the edge of chaos, and natural selection achieves and sustains such a poised state [..] Boolean systems, and by extension some large family of homologous nonlinear dynamical systems, with nearly melted frozen components can carry out the most complex, yet controllable behavior".

Kauffmann explains the simultaneous presence of strong resilience and rapid adaptation to changing environmental conditions by means of the location of biological systems 'near the edge of chaos', i.e., in a sort of liquid-crystal interface where the presence of islands of order of frozen elements (crystal side) stops the rapid spread of perturbations sensed by the periphery of the system. On the other

A. Giuliani (✉)
Environment and Health Department, Istituto Superiore di Sanità, Rome, Italy
e-mail: alessandro.giuliani@iss.it

© Springer Nature Switzerland AG 2018
M. Bertolaso et al. (eds.), *Biological Robustness*, History, Philosophy and Theory of the Life Sciences 23, https://doi.org/10.1007/978-3-030-01198-7_8

(liquid) side, the presence of an intermingled network of connections between elements allows the entire system to 'stay tuned' and thus to sense and to adapt to incoming environmental stimuli.

The impressive amount of simulations both Kauffman and other scholars (Kauffmann et al. 2003; Raeymakers 2002) performed on Boolean network models gave a satisfactory proof to the contemporary presence of crystal and liquid phases in the analyzed networks. In the following I will try to explain why the contemporary presence of a high sensitivity to environmental stimuli and the maintaining of an invariant structure is the 'biological way' to robustness.

In order to better clarify the above concepts, as a very first step we need an operational (albeit rough) basic definition of robustness. In order to get it, we let aside too sophisticated distinctions from cognate concepts like the (nowadays very fashionable) term resilience: in the following, the two terms will be interchangeable. Our definition of robustness can be synthetically expressed by the title of a famous film series: 'Die Hard'. The title evokes someone that is able to survive to very dangerous threats. This remarkable goal is achieved by means of rapid and effective reactions: it is the robustness of a warrior, not of a rock. The 'Die Hard' series follows the adventures of John McClane (portrayed by Bruce Willis), a police detective who continually finds himself in the middle of violent crises and intrigues where his reactivity represents the only hope against disaster.

Immune response is probably the biological phenomenon that best mirrors John McClane's character: immediate and efficient reactivity to potentially dangerous external stimuli and rapid recovery to the pristine state in order to be ready for the next adventure. Here I will try to show how Stuart Kauffmann's intuition of the contemporary presence of two distinct phases in the same system exactly corresponds to the functioning of the immune system and consequently to the adopted definition of robustness.

In order to do so I ask the reader a last imagination effort, namely to imagine the described system in terms of a network (graph) of interacting agents. This is the most direct representation style enabling to give the Kauffmann statement an operational basis.

8.2　Biological Networks

8.2.1　Yeast Metabolic Network

This is not so strange, after all the classical form in which biological systems are described (being they metabolic charts, regulation pathways, protein-protein interaction maps, food webs and so forth) corresponds to a set of nodes linked by edges, i.e. to a graph or a network. In these networks, the nodes are the basic elements of the described system (genes, proteins, metabolites and so forth) and the edges connecting them some rules of the kind 'is transformed into' or 'is increased by' (Huang 2009; Conti et al. 2007).

The development of high throughput techniques yielded larger and larger graphs and asked for some form of global analysis to get rid of their wild multiplicity.

This global analysis builds upon the so-called graph invariants: statistical descriptors catching some intuitive properties of the wiring architecture of the network. Examples of such invariants are the node degree (the number of edges relative to a particular node of the network) or, the average shortest path (the average of the length of the minimal paths connecting all node pairs). It is worth noting that such descriptors can be defined at both local (single node) and general (averaged across the entire network) levels so allowing for a straightforward change of scale of the obtained representations. The consideration of biological systems as graphs allows for a very peculiar multi-scale analysis: the values of the descriptors for a particular node depend upon the entire wiring of the network (this is what methodologists call a 'top-down' causation), while on the other hand the entire network properties descend on the node-level wiring (bottom-up causation). This implies network formalization allows for adopting a 'third way' with respect to classical 'bottom-up' and 'top-down' styles, we can define as 'middle-out' approach (Giuliani et al. 2014). This definition implies the focus of the analysis is neither on very general laws the system is supposed to obey (top-down) nor on the reconstruction starting from the regularities observed at the most microscopic scale (bottom-up), but rather on the consideration of the correlation structure of the system's constituting parts. The definition of the 'most favourable' viewpoint for system's partition neither stems from a peculiar hypothesis on the deep nature of physical world nor from the agency of a philosophical principle, but from purely stylistic considerations: the optimal layer for explanation is the one that allows to maximize the number of observed correlations. In the following I will discuss a practical example of this way of doing.

When a scientist accommodates a given corpus of knowledge at its most detailed (least aggregated) observation level in terms of variables (being they experimental or observational descriptors) whose pairwise correlations can be estimated by a suitable metrics (e.g. correlation coefficients, Euclidean distances...) he ends up into a situation like the one depicted in Fig. 8.1.

The figure reports a part of the metabolic network of the yeast: the nodes (variables) are the metabolites and the edges (arrows) point to the existence of a chemical reaction transforming the metabolite (an organic molecule) at the basis of the arrows into the chemical species (another metabolite) at the end. The metrics is in this case a distance between metabolites expressed in terms of the number of chemical reactions needed to transform a species i into another species j, the minimal distance being equal to 1 if the molecule i can be transformed into j by a single step (Palumbo et al. 2007).

Specific enzymes catalyse the different chemical reactions; if a mutation makes the reaction impossible by destroying the enzyme functionality, a part of the network can be detached from the rest of the system (e.g. this is the case of the interruption of the link between metabolite 921 and 415 in the bottom right part of

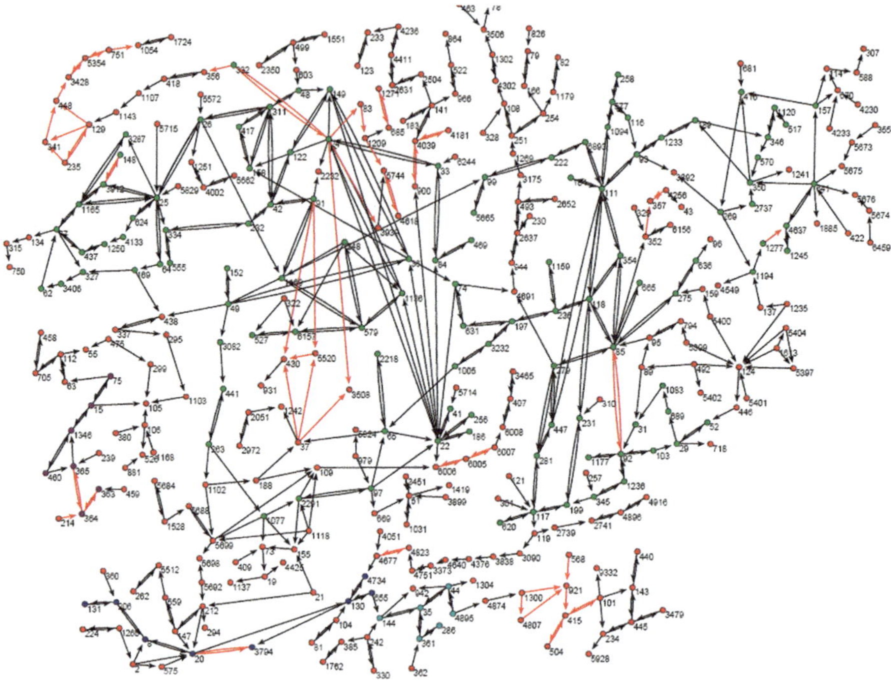

Fig. 8.1 The figure reports a part of the metabolic network of yeast. The nodes are the metabolites, the arrows (edges) point to the existence of a chemical reaction transforming one metabolite into another. Each chemical reaction is in principle reversible, but in the particular conditions of the cell a great part of the reactions have a specific favored direction. In the cases of reversible (in cell conditions) reactions, the arrows point on both directions. Each reaction is catalyzed by one (or more) specific enzymes; the elimination of the enzyme(s) catalyzing a given reaction corresponds to the cutting out of an edge from the network

Fig. 8.1). In this case, the yeast can survive if (and only if) the scientist supplements the culture medium with molecule 415 so to make the synthesis of metabolites 101, 9332,234, 5928 … (see Fig. 8.1) possible. On a macroscopic level, the elimination of that particular enzyme is lethal for the organism when kept in the minimal (not supplemented of 415 molecule) medium.

The essential character of the enzyme can neither be fully explained by going in depth into the enzyme structure and/or considering the specific role played by metabolite 415 in the molecule (bottom-up approach) nor by looking for a 'general law' of enzyme essentiality based on evolutionary theory (top-down approach). The only relevant explanation resides in the particular wiring architecture of the metabolic network and thus in the particular location of the enzyme in the graph. This appears as a 'top-down' principle (the node properties derive from the higher layer of network it is part of) but this is half of the story. On a different perspective we must recognize the wiring architecture of the network is constrained by the microscopic structural properties of the single enzymes (the enzymes engaged in subsequent reactions must physically interact), a classical 'bottom-up' principle (and this is the other 'half'). Taken in isolation neither 'top-down', nor 'bottom-up' strategies allow to predict the essential character of the enzymes: the heuristic must be of the 'middle-out' type focusing on network structure (i.e. on the among nodes correlations) and eventually expanding in the direction of higher or lower levels. Thus, if we want to study the robustness of the yeast and its adaptability to different environments (higher level) we will count the number of 'bottle-necks' in the entire network correspondent to crucial links that cannot be supplemented by alternative pathways. The higher the number of such crucial links the lower the robustness of the system as a whole: at this level of investigation, knowledge of the enzymes correspondent to the 'bottle-necks' is irrelevant.

In the opposite direction (lower level), we might want to test the hypothesis that network hubs (e.g. metabolite 22 in Fig. 8.1, an hub is a node involved in many edges) share some common physico-chemical features that could suggest the existence of still unknown organic reaction mechanisms of possible industrial application. In this case, the network wiring is only instrumental to identify the organic molecules of interest whose chemico-physical features will be investigated with no relevant relation to the biology of yeast.

While the study of yeast metabolic network is an example of static, purely structural, evaluation of the robustness of a system, we need to insert the time dimension in order to get a clear appreciation of the 'liquid' and 'crystal' phase separation advocated by Kauffmann.

Here we study the behaviour of a network having as nodes the around 25,000 genes active in macrophage (an important player of the immune response) cells and as edges, the mutual correlation between the gene levels of expression. We will not enter into the detailed wiring of this huge network, limiting ourselves to the description of its collective behaviour.

8.2.2   Gene Expression Regulation Networks

Microarray technology is the most widespread 'high-throughput' technology in biological research. This technology is based on a micro-electronic device that allows one to register at the same time the level of expression of around 30,000 genes in the same biological specimen, typically a population of cells growing on a test plate. The common experience of any experimentalist dealing with microarray data is the fact that any two independent samples of the same cell kind when correlated over the expression of 25,000 different gene products display a near to unity correlation (see Fig. 8.2).

This organization, spanning four order of magnitudes of gene expression levels and encompassing tens of thousands of gene products is a very remarkable fact of nature calling for an explanation and clearly supporting the crucial importance of a thorough investigation of its origin (Censi et al. 2010; Conti et al. 2007). Here it is sufficient to say that the particular wiring of the gene expression network gives rise

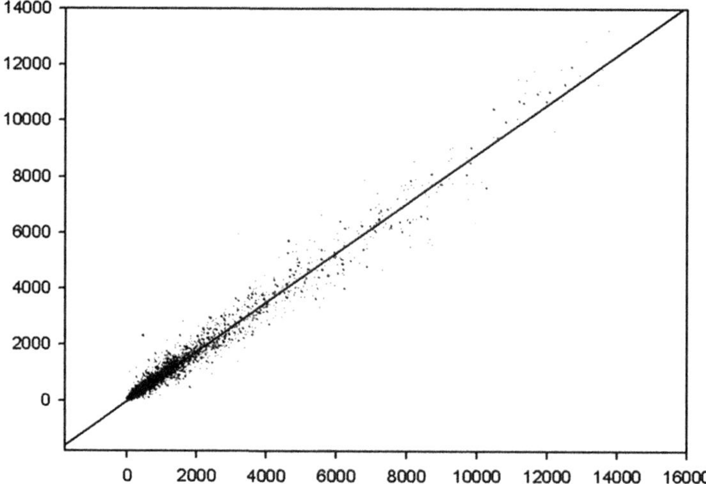

Fig. 8.2 Each point in the figure represents a specific couple of the same gene expression values relative to a sample of macrophages from a normal mouse (abscissa) and to a sample coming from an immune-deficient mouse (ordinate). Notwithstanding the huge phenotypic difference between the two animals (the immune-deficient animal has no response to LPS antigenic stimulus, while the normal one displays a complete immune response), the transcription profiles show a near to maximal correlation (Pearson r=0.99)

to this largely invariant profile as a resultant of its peculiar architecture (Lima de Faria 1988). The gene expression profile is the 'energy minimum' (stable attractor state) of the network. Figure 8.2 reports the transcriptome (the whole genome expression profile) relative to two mice, respectively normal (abscissa) and severely impaired in terms of innate immune response (ordinate) after an immune challenge constituted by an LPS (Lipo-Poly-Saccharide) injection (Tsuchiya et al. 2009a). The astonishing invariance of the transcriptome profile is a consequence of the existence of very few 'attractors' in the gene expression space.

This behaviour implies a strongly connected network of gene expression levels endowed with very few 'energy minima' or 'allowed states'. This almost-perfect and invariant order (strong resilience, crystal phase) goes hand-in-hand with a very striking phenotypic variability (liquid phase) when facing an external stimulus (Tsuchiya et al. 2009a) like in the case described here in which the mutants are not able to cope with the LPS antigen. Wild-type animals respond to the antigen with an impressive (and highly organized) cascade of events encompassing the sudden increase of growth factors, cytokines, and consequent cell proliferation. This cascade of events does not affect the 'bulk' of gene expression profile: the genome expression profile 'as-a-whole' is identical in wild-type (immune response present) and mutated animals (immune response abolished) (Fig. 8.1).

It is worth going into a more detailed characterization of the separation between 'local sensitivity' and 'global invariance' to external perturbation correspondent to the immunogenic stimulus (LPS). Figure 8.2 reports the autocorrelation (profile similarity) with t0 condition (system at rest) at different times focused on four different choices of genes and relative to different mutation status (Wild-Type mice, Myd88 and TRIF mice mutated in one of two essential macrophage receptors, while DKO corresponds to mice carrying both mutations). These choices correspond to: (1) entire genome, (2) only cytokine, (3) random extraction of genes, and (4) 'connector' genes respectively. These conditions correspond to different mesoscopic views of the co-expression network:

1. Entire genome corresponds to the analysis of the whole gene expression space.
2. Cytokine analysis focuses on 'specific effectors', i.e. on those peripherical nodes that are transiently decoupled from the global genome dynamics by the agency of LPS.
3. Random extraction of genes, in the hypothesis of a strongly connected network and of a sufficiently large sample (typically with number of genes >60, Felli et al. 2010), should give the same results as the whole genome (the gene expression response is scalable).
4. Connector genes are those genes that (weakly) link the effector genes (liquid phase) to the 'bulk' (crystal phase) of genome expression and thus are responsible of the connectivity of the network.

Figure 8.3 reports the autocorrelation distribution in time (Pearson r with the $t0$ sample) for the four choices of genes. In the case of the entire genome (top left panel) we observe a major departure from unit correlation correspondent to a greater response, as expected, in the case of wild type, but the system remains practically invariant. The maximum departure from total identity (autocorrelation = 1) is negligible, being the lower correlation equal to 0.98 (this can be appreciated by a look at the ordinate values).

Random choice of 100 genes (bottom left panel) displays the same pattern of the entire genome. We iterated many times these random choices observing a completely invariant picture starting from a minimal choice of around 60 genes (Tsuchiya et al. 2009b). This is a confirmation of the 'scalable' character of gene expression network that constitutes a strongly connected set whose global behaviour can be appreciated starting from a minimum sample of elements (let's keep in mind the total number of considered genes equals 25,000).

Top right panel refers to the cytokines local response: it is immediate to note the much bigger displacement from $t0$ (before LPS stimulus) with respect to the global response: here correlation coefficient goes down to 0.25 pointing to a drastic change in expression pattern.

As expected, the DKO genotype (correspondent to animals in which immune response is abolished) does not show any response in terms of cytokine expression, while the other three genotypes experience a marked departure from $t0$ state vector with a non-linear response reaching its maximum after 1 h since stimulus presentation.

Connector genes (bottom right panel) display an intermediate behaviour in between acute and collective motion. From collective motion, they inherit a minimal displacement from unit correlation of DKO, from their link with acute response they

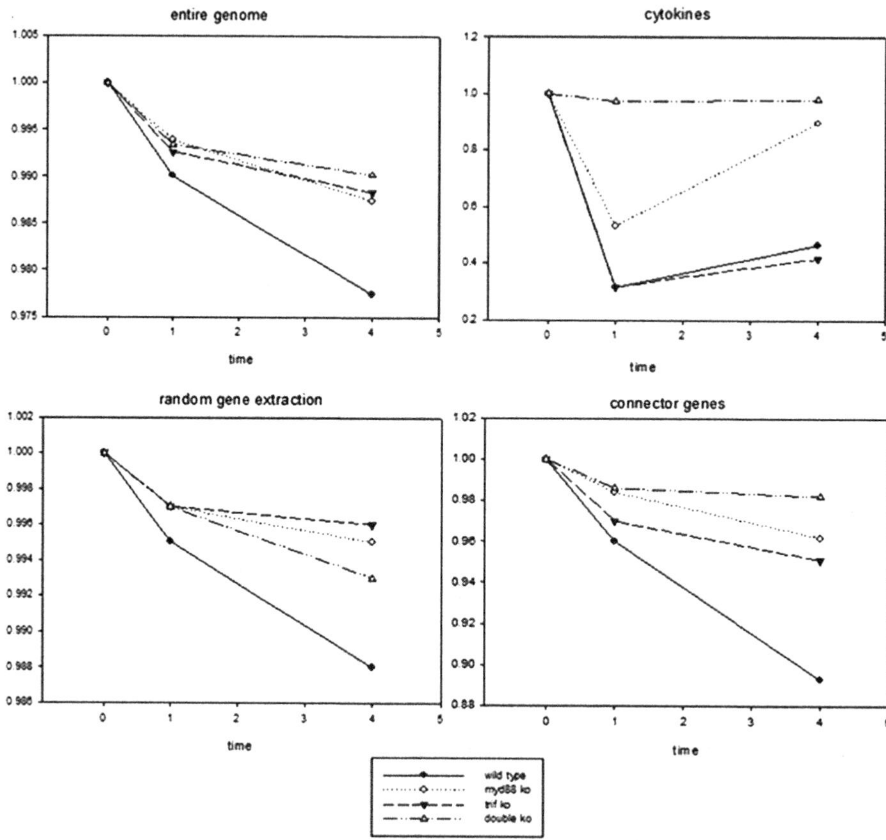

Fig. 8.3 The figure reports four different gene choices correspondent to different mesoscopic views on the gene regulation system (see text). The graphs have the Pearson correlation coefficient r with $t0$ state vector as ordinate and time (hours) as abscissa and report the dynamical response to LPS stimulus for each genotype, a lower value of r corresponds to a more marked response

inherit both the more marked displacement from unit with respect to global motion and the expected ranking of detrimental effect of mutations with Myd88 more affected than TRIF mutants.

The above results allow to delineate a solution to the resilience/sensitivity paradox: some nodes (genes) of the regulation network (in this case those corresponding to cytokines), thanks to their peculiar location in the network architecture, are responsible for the sensitivity. The elevated degree of interconnection is at the basis of the resilience properties of the system, here appearing as a global adjustment of the entire system following the stimulus. The spreading of the stimulus sensed by the cytokines (the Kauffmann 'liquid phase') to the entire network is mediated by connector genes and buffered by the strong 'attractive force' (the Kauffmann 'crystal phase') of the cell kind attractor. Our interpretation is consistent with the results obtained by Nykter and colleagues (Nykter et al. 2008) that analyzed the same biological system (global gene expression data from macrophage cell populations) demonstrating, by means of information theory based methods, the presence of criticality.

Criticality is another crucial landmark of self-organized dynamical systems and corresponds to a state of the system in which perturbations are propagated over long temporal or spatial scales. In the case of macrophages this corresponds to the spreading of the initial stimulus to the genome wide expression, which we observed to be mediated by the connector genes. Accordingly to Kauffmann hypothesis this spreading fails to colonize the entire gene expression profile thanks to the 'islands of order' (Kauffmann 1993) that do not permit the perturbation to become catastrophic.

Here we are able to identify, in analogy with the yeast metabolic network case, these order constraints as the presence of a cell-kind-ideal-profile energy minimum attracting the cell population back to its initial 'state' correspondent to the typical macrophage expression profile.

8.3 Conclusion

In a totally flat, 'ultra-democratic' society of genes in which all players are connected to each other with the same strength there is no possibility to 'go back' to the initial state after a perturbation. The consequences of the displacement of a part of the network from its equilibrium resting state should invade the entire network with no possibility of a rapid comeback to the equilibrium. We can equate this situation to pathologies like post-traumatic stress disorder (Yehuda 2002) where a punctual perturbation (the negative traumatic experience) continues to resonate in the patient mind without reaching a stop condition.

On the other hand the drastic separation of different modules, each in charge of a specific task, needs the presence of a super-controller external to the system, a dictator with a limitless power deciding which module to activate in any occasion. This 'magical thinking' is at the basis of the continuous discoveries of the 'gene-for-a-given-task' plaguing biological research since many years.

The only solution allowing for a 'self-organization' that does not need to invoke the action of intelligent agents while preserving the robustness (and thus the recovery after a perturbation) of the system is the presence of a network in which the elements assume (by the only effect of their location in the network) differentiated roles.

Peripheral nodes are in charge of the rapid response to the environment, their peripheral (in the literal sense of the need of trespassing many intermediate edges to reach the central core of the graph) location allows for increased flexibility (their lower degree corresponds to a lower number of constraints with the other players). On the other hand they can 'go back' to the resting phase thanks to the action of 'connector genes' that are connected to both the periphery and the core of the network, thus acting as mediators. The 'connector genes', thanks to their large number (around 200, to be compared to around 10 sensor genes and well above the 60 genes threshold correspondent to a reliable picture of the entire network behaviour) absorb the perturbation by 'fragmenting' the expression displacement into a drastically attenuated perturbation. They are the 'frozen island of order' that allow for the phase separation between the 'bulk of the network' and its sensor (tiny and liquid) module. This hierarchical organization descends from the wiring architecture of the gene co-expression network without invoking any 'external' device.

It is the natural way to robustness in an ever-changing environment.

References

Censi, F., Bartolini, P., Giuliani, A., & Calcagnini, G. (2010). A systems biology strategy on differential gene expression data discloses some biological features of atrial fibrillation. *PLoS One, 5*(10), e13668.

Conti, F., Valerio, M. C., Zbilut, J. P., & Giuliani, A. (2007). Will systems biology offer new holistic paradigms to life sciences? *Systems and Synthetic Biology, 1*(4), 161–165.

Felli, N., Cianetti, L., Pelosi, E., Carè, A., Gong Liu, C., Calin, G. A., Peschle, C., Rossi, S., Marziali, G., & Giuliani, A. (2010). Hematopoietic differentiation: A coordinated dynamical process towards attractor stable states. *BMC Systems Biology, 4*, 85.

Giuliani, A., Filippi, S., & Bertolaso, M. (2014). Why network approach can promote a new way of thinking in biology. *Frontiers in Genetics, 5*(83), 1–9.

Huang, S. (2009). Reprogramming cell fates: Reconciling rarity with robustness. *BioEssays, 31*, 546–560.

Kauffmann, S. A. (1993). *The origins of order. Self-organization and selection in evolution.* New York: Oxford University Press.

Kauffmann, S. A., Peterson, C., Samuelsson, B., & Troein, C. (2003). Random Boolean network and the yeast transcription network. *Proceedings of the National Academy of Sciences of the United States of America, 100*, 14796–14799.

Lima de Faria, A. (1988). *Evolution without selection. Form and function by autoevolution.* Amsterdam: Elsevier.

Nykter, M., Price, D. N., Aldana, M., Ramsey, S. A., Kauffmann, S. A., Hood, L. E., Yli-Harja, O., & Shmulevich, I. (2008). Gene expression dynamics in the macrophage exhibit criticality. *Proceedings of the National Academy of Sciences of the United States of America, 105*, 1897–1900.

Palumbo, M. C., Colosimo, A., Giuliani, A., & Farina, L. (2007). Essentiality is an emergent property of metabolic network wiring. *FEBS Letters, 581*(13), 2485–2489.

Raeymakers, L. (2002). Dynamics of boolean networks controlled by biologically meaningful functions. *Journal of Theoretical Biology, 218*, 331–341.

Tsuchiya, M., Piras, V., Choi, S., Akira, S., Tomita, M., Giuliani, A., & Selvarajoo, K. (2009a). Emergent genome-wide control in wildtype and genetically mutated lipopolysaccharides-stimulated macrophages. *PLoS One, 4*, e4905.

Tsuchiya, M., Selvarajoo, K., Piras, B., Tomita, M., & Giuliani, A. (2009b). Local and global responses in complex gene regulation networks. *Physica A, 388*, 738–1746.

Yehuda, R. (2002). Post-traumatic stress disorder. *New England Journal of Medicine, 346*(2), 108–114.

Dr. Alessandro Giuliani was born in 1959 in Rome where he lives, works and graduated at 'La Sapienza' University in Biological Sciences. He serves since 1997 as senior scientist at Istituto Superiore di Sanità (Italian NIH). Dr. Alessandro Giuliani is involved in the generation and testing of soft physical and statistical models for life sciences. His research deals with themes like protein sequence/structure prediction, complex network approaches, Systems Biology and, more in general, in developing a statistical mechanics approach for life sciences. He is the author of almost 300 publications on peer-reviewed journals.

Chapter 9
Can Engineering Principles Help Us Understand Nervous System Robustness?

Timothy O'Leary

Abstract Nervous systems are formidably complex networks of nonlinear interacting components that self organise and continually adapt to enable flexible behaviour. Robust and reliable function is therefore non-trivial to achieve and requires a number of dynamic mechanisms and design principles that are the subject of current research in neuroscience. A striking feature of these principles is that they resemble engineering solutions, albeit at a greater level of complexity and layered organisation than any artificial system. I will draw on these observations to argue that biological robustness in the nervous system remains a deep scientific puzzle, but not one that demands radically new concepts.

Keywords Neurons · Feedback loops · Degeneracy · Control theory · Stability · Regulation · Homeostasis

Nervous systems are robust. The very fact that they function at all is evidence of this. Consider the problem that biology solves as the human brain develops from an embryo to an adult. Some 86 billion neurons, each in principle capable of connecting to any other neuron, somehow establish appropriate connections as the brain forms. This process is repeatable in spite of substantial individual variability in genetics, environment and early life history. Remarkably, the end result is not a stereotyped pattern of connectivity; no two brains are internally configured in the same way. Yet to a greater or lesser extent brains reliably perform the multitude of functions required of them: they allow us to learn language, recognise danger, regulate body temperature, control appetite and sexual drive.

There is no strict blueprint for building a brain, but instead a web of interacting rules, some strict and some less so. I will argue that a key ingredient to this robust self-organization is feedback. I claim further that there is nothing special conceptually about biological mechanisms of robustness in the nervous system as opposed to

T. O'Leary (✉)
University of Cambridge, Cambridge, UK
e-mail: timothy.oleary@eng.cam.ac.uk

M. Bertolaso et al. (eds.), *Biological Robustness*, History, Philosophy and Theory of the Life Sciences 23, https://doi.org/10.1007/978-3-030-01198-7_9

robustness in other engineered systems. In practice, however, there is a tremendous gulf between the layers of complexity in the interacting mechanisms that give rise to nervous system robustness and the feedback systems that an engineer might design.

Biological mechanisms span many spatial scales, from molecules to ecosystems and many temporal scales, from the microsecond interactions in enzyme kinetics to the millennia required for evolution of robust living systems. Across all of these scales, feedback is at work, shaping the function and organisation of the nervous system (Davis 2006; Marder and Goaillard 2006). In addition to feedback, the architecture of nervous system exploits *degeneracy*, widespread functional overlap between multiple, distinct functional components. This overlap means that components that are lost or damaged can be immediately compensated for in certain circumstances. For these reasons, it is very difficult to reverse-engineer nervous systems (Csete and Doyle 2002). Nonetheless, the known biological principles underlying nervous system robustness, and feedback in particular, fit within an engineering framework.

I will introduce feedback in a general setting with familiar examples and attempt to illustrate some subtleties in understanding the behaviour of systems that employ feedback control. I will then describe several experimentally characterised examples of feedback control and our current understanding of how these mechanisms ensure robustness in the nervous system. Finally, I will discuss degenerate functional architectures that nervous systems exhibit at the cellular level and how these combine with feedback control to permit robustness and flexibility.

9.1 Feedback Control

To appreciate the importance of feedback, we need to define open and closed loop systems. Abstractly, a system is anything that transforms some time-varying input to an output. Many biological processes and components can be described using this input-output formalism; examples will be considered later but first we will set terminology and build a conceptual picture using familiar examples.

A coal fire is an open loop heating system: adding fuel (input) heats the room. The final temperature of the room (output) will inevitably depend on the ambient temperature of the surroundings, the size of the room, the degree of insulation, and so on. As the name suggests, closed loop systems have some form of output fed back into an input. For example, a modern thermostatic heating system takes a temperature set-point as the input, then modifies heat output according to the difference between the current temperature and the set point.

Now consider what would be necessary to make the open loop system (the coal fire) behave in the same way as the closed loop thermostat. We might imagine an fire-tending robot that knows just how much coal to add to the fire to achieve a specified room temperature. The robot could be very advanced and could allow for seasonal variations, the presence of extra bodies in the room or open windows that would diminish the insulating capacity of the room. We may now regard the robot as part of a rather expensive and technologically baroque heating system. With no knowledge of the details of fire-tending, we can simply request a temperature and the robot will get to work.

From the outside, the behaviour of this system is indistinguishable from a closed loop thermostat: a requested temperature is reached and maintained, possibly with some delay. Inside this system, there is a backward-going transformation from output (requested temperature) to input (amount of coal fed to the fire). This transformation is embodied in the feedback controller (the robot). Notice further, that the robot requires perfect knowledge of the *forward*–going transformation from coal to room temperature, that is, it needs an *internal model* of the system being controlled. This is the essence of the internal model principle, or, as Ashby called it, the good regulator (Conant and Ross Ashby 1970). The key point that Ashby appreciated is that for a feedback controller to reliably ensure a specified output in a controlled system, there must be an inversion of the controlled system's dynamics. The wider point to take from this is that a feedback regulation system *needs to be tuned to the system it is controlling*.

How can feedback control systems fail? We have discussed how the regulation system needs to be tuned to the system being controlled. Mistuning can result in a number of aberrant effects. If the gain of the feedback is too strong, the system could suffer from overshoot. In the coal fire analogy, this could equate to each shovelful of coal being too large, resulting in the room temperature exceeding the setpoint each time coal is added. Obvious pathologies can emerge due to sensor failure. If the temperature gauge on a thermostatic device fails or saturates, the heating system could be maximally activated or turned off altogether. Other, more subtle pathologies are possible due to temporal mismatch in the feedback loop. If additional delays are introduced into a system then the result of actions by the feedback controller might continually lag the feedback signal itself, leading to oscillations. It is worth noting that many of these kinds of pathologies are observed in medical conditions, such as motor tremors, cardiac arrhythmias and hypertension. The crucial role of feedback mechanisms in these phenomena are well known to physicians.

How can multiple feedback systems be integrated, as they must be in a biological system? There is a deceptively subtle connection between the open and closed loop systems and an operation that the influential control engineer, Jan Willems, called *zooming* (Willems 2007). Zooming, familiar to us now through graphical computer interfaces, loosely means adjusting the level of magnification in a control engineering description of a system. More precisely, as we zoom out, or decrease magnification, we lump together system components and view them as single entities that constitute components at the next level.

We performed a zooming operation in the preceding example by viewing a coal fire and a butler as being lumped together to form a 'heating system'. At this level of zoom, the transformation from requested temperature to room temperature is open loop in the most trivial sense: it is merely the identity mapping between requested temperature and room temperature.

What kind of system might we describe at this new level? We could imagine a large dinner party in the same room, with the imposed room temperature (input) having obvious effects on the guests' comfort (output). An open loop control would amount to the host directing the robot to maintain the room temperature at some specified value, resulting in complaints that the room is too hot or cold. On the other hand, the host might implement a closed loop control by asking the guests whether they are too hot or too cold at regular intervals, commanding the robot to decrease or increase the room temperature accordingly.

Zooming lets us understand how feedback control can make components modular and predictable when they are integrated into a system that performs higher level functions. In the preceding analogy it is clear that an open loop system at both levels would require the host to have a combined model of the guests' temperature preferences as well as the effects of adding a given amount of coal to the fire under a wide variety of circumstances. Minor perturbations or uncertainties in the coal-to-temperature or temperature-to-discomfort transformations would very easily cause the open loop system to fail. Feedback thus allows unreliable components operating in uncertain conditions to behave reliably, within limits. Assembling a massively interconnected system such as a nervous system would thus be impossibly fragile without these underlying feedback loops.

In summary, feedback control allows a system to operate in a reliable way in spite of perturbations to its internal dynamics or to the environment in which it exists. This reliability is not unconditional: there will be limits to the extent to which the feedback system can compensate for a perturbation. Examples of these include saturation of the sensing or actuating components of the system, or perturbations with temporal characteristics that prevent the feedback system from keeping pace. Finally, depending on the level that a system is analyzed, we can think of a feedback system as a feedforward system. That is, the internal working of the system ensures that a required output is delivered or maintained, but viewed externally, this feedback may not be apparent, unless the system fails in the ways outlined above.

9.2 Feedback Control in Nervous Systems

Now that we have a heuristic picture of feedback control, we can ask how this picture aids us in understanding aspects of nervous system robustness. The literature on this topic is vast, spanning levels of organisation from the molecules that make up nervous tissue to the level of the brain, periphery and their interactions with other organs and with the external world. I will focus on a level that connects the molecular make-up of neurons with their specialized function as electrically excitable elements in neuronal networks. In particular, I will describe findings in reduced biological preparations comprising neurons or small circuits that have been isolated from the rest of the nervous system. The lessons from this system, are, however, quite general.

For many decades it has been possible to remove living tissue from nervous systems and maintain this tissue in-vitro. Although this manifestly alters the context of the function of the neurons and neural circuits involved, it is a remarkable fact that we can observe key aspects of neurophysiological function in spite of having removed the rest of the animal. For example, the characterization of neuronal excitability itself was carried out in the 1950s in seminal work by Hodgkin and Huxley using an in-vitro preparation of a single axon, the squid giant axon (Hodgkin and Huxley 1952a, b). The giant axon, which is responsible for generating and transmitting action potentials or 'spikes' to muscle as part of a propulsion reflex continues to work even though it is removed from the animal (and the cell body of the nerve cell!) to which it belongs. This is a clear example of robustness: though the exact physiological properties of the axon are no doubt altered when it is excised from the animal, it continues to function and maintain the internal milleu that supports its function for several hours.

Similar 'reduced preparations' are possible using tissue from a variety of nervous systems. Mammalian central neurons can be removed from the brain in a brain slice preparation, or be mechanically dissociated and grown on an artificial substrate as a cell culture. Remarkably, neurons in dissociated cultures regrow elaborate axonal and dendritic trees and form functional connections over a period of several days (Banker and Cowan 1977, 1979). Again, this is clear evidence of robustness on many levels. The cells in a neuronal culture have been completely removed from the brain and are devoid of the normal biochemical cues and life support systems. Nonetheless, under a broad range of conditions, neurons can live for many weeks and recapitulate many of the physiological functions, such as intrinsic excitability and functioning synaptic connections that are essential to nervous system function in the intact brain.

Dissociated cultures are spontaneously active. Recording electrical activity in these preparations reveals continual barrages of action potentials that reverberate throughout the network. This activity appears several days after the neurons are plated in the culture dish and often appear to self-organize so that a large fraction of cells becomes synchronized, firing waves of action potentials every few seconds. By lowering the density of neurons in a culture dish it is possible to completely isolate individual neurons. Remarkably, these isolated neurons, devoid of any neighbours to form synaptic connections with, synapse onto themselves, forming 'autapses' that generate self-sustained activity in a closed loop (Bekkers and Stevens 1991).

Some of the effects seen in dissociated cultures – such as synchronous activity over large populations of cells or autaptic connections – would be pathological in the context of an intact nervous system. For this reason, some experimental physiologists dismiss the culture preparation as a valid system to study neurophysiology. But if we take a broader view, we see that the very fact that these unusual states arise is direct evidence of some kind of internal regulatory control, a kind that causes neurons to maintain a level of electrical activity in spite of the fact that all sensory input and all connections to the rest of the nervous system have been removed.

A plausible explanation for this emergence and maintenance of spontaneous activity is that individual neurons have some kind of feedback control mechanism that senses average levels of excitation and alters the physiological make-up of the cells and the wider network to maintain average excitation at a non-zero 'set-point' (Abbott and LeMasson 1993; Desai 2003; Turrigiano et al. 1994, 1995).

How can a neuron drive itself into a spontaneously active state? Neuronal excitability is a consequence of several physiological processes. First, all neurons (and, indeed, all cells) have a non-zero electrical potential across the cell membrane, maintained by various ion-selective transport proteins (ion pumps) that operate continuously. In addition to maintaining this steady 'resting' membrane potential, neurons have voltage-dependent ion channels that can open or close in response to fluctuations in membrane potential. In particular, neurons possess voltage-gated sodium channels that (i) open when the membrane potential is depolarized (made more positive with respect to the extracellular space) and (ii) selectively carry inward (depolarizing) ionic current. Thus the sodium channels create a positive feedback loop: a positive deflection in membrane potential causes an increase in positive current flowing into the cell, which further depolarizes the membrane potential. This process is fast, lasting less than a millisecond, and is terminated by 'inactivation' (delayed closing) of the sodium channels, and by the opening of a slower type of potassium channel that selectively carries negative ionic current, repolarizing the membrane potential. Thus, potassium channels act as a negative feedback loop, stabilizing membrane potential following a rapid excursion. The two types of channels thus cooperate to produce rapid, stereotyped action potential events, or spikes.

These two types of ion channel – the positive-feedback sodium channels and the negative-feedback potassium channels – must be maintained at appropriate densities in the membrane of a neurons. Too few sodium channels would be incapable of initiating a spike. Too few potassium channels would prevent the neuron regaining its resting potential following a spike. It is not difficult to see that this balance cannot arise for free. As a neuron grows, its electrical properties change and so too does the density and distribution of channels in the membrane. Moreover, each ion channel is a protein with a finite lifetime. While a neuron will live for a substantial fraction of the animal's lifetime, the ion channels and other proteins that constitute the building blocks of the neuron are continually broken down and replenished over a time-course of minutes to hours or days. Thus, there needs to be some kind of internal feedback mechanism to ensure that a dynamic balance of ion channels is maintained in the neuronal membrane.

How does a neuron maintain an appropriate balance of ion channels? One possibility is that neurons might somehow estimate the densities of ion channels of various types and regulate the synthesis and degradation of each type of channel to maintain a fixed density. Such proteostasis mechanisms are known to exist in many types of cells, for example many genes produce proteins that inhibit or terminate their own production. However, merely 'fixing' the balance of ion channels at particular density is akin to maintaining the heat output of the coal fire in our analogy above. And, as we saw in that analogy, a constant heat output does not guarantee a constant room temperature in typical situations.

In the case of a neuron, simply maintaining a fixed density of ion channels would not ensure reliable function in the face of the kinds of perturbations a neuron sees during development and throughout and organism's life. As we discussed, neurons grow and form new connections and lose existing connections as organisms develop and adapt. Therefore, to ensure robust function in the context of a dynamic circuit, neurons need to monitor ongoing activity and maintain their level of responsiveness accordingly.

This kind of activity-dependent feedback control has been observed in many neural systems (O'Leary and Wyllie 2011). In dissociated neural cultures, suppression of ongoing spontaneous activity results in an increase in the excitability of individual neurons in the culture. This phenomenon is widely known as *homeostatic plasticity* because the response counteracts the imposed perturbation. Theses observations have led to the hypothesis that neuronal activity is maintained at some internally monitored 'set-point' (Davis and Bezprozvanny 2001; Turrigiano 2007). Numerous experimental paradigms that artificially block spiking activity or excitatory synaptic drive for prolonged periods have shown that many ion channel types are affected by the suppression of activity, indicating a coordinated change. For example, Desai et al. (1999) reported a coordinated increase in the current carried by sodium channels and a decrease in potassium channel currents after blocking activity for several days. Similarly, other studies have shown that artificially elevating spontaneous activity using pharmacological means results in a decrease in excitability (O'Leary et al. 2010). These changes are mediated by alterations in the levels of several membrane channel types expressed in neurons, and in some cases by physical remodelling of the site of action potential initiation in axons (Grubb and Burrone 2010). The molecular signalling mechanisms underlying these phenomena are the subject of current research, and there is evidence that neurons sense ion flux and activation of specific voltage-gated calcium channels that could directly influence the synthesis of ion channels and receptors (Wheeler et al. 2012). Thus, the components that neurons use to sense ongoing electrical activity overlap functionally with the components that generate activity.

Changes in the intrinsic excitability of neurons in a network can regulate average spiking activity, but nervous systems must also be plastic in order to be adaptive. Current theories of how nervous systems adapt and store memories invoke synaptic plasticity as a key mechanism, and there is considerable experimental evidence to support such theories. A detailed discussion of the biological basis of synaptic plasticity and memory is beyond the scope of this chapter, though it suffices to say that alterations in synaptic strength need to be coordinated in some way to allow the nervous system to adapt while remaining functional. In particular, it is clear that any process that changes the coupling strength between millions of excitable cells has the capacity to destabilise a neuronal network. It is therefore remarkable that in addition to controlling intrinsic responsiveness in a homeostatic way, neurons also regulate the overall gain of the synaptic input they receive, in a process known as *synaptic scaling* (Turrigiano et al. 1998). Synaptic scaling is seen to occur in reduced neuronal preparations: artificially stimulating or suppressing neuronal activity leads to changes in the average strength of synaptic connections over several days.

Furthermore, the changes are homeostatic, in that they tend to counter the imposed manipulation: excitatory synapses strengthen on average when activity levels are suppressed and weaken when activity is artificially enhanced.

In summary, neurons possess an array of regulatory feedback mechanisms that tend to maintain average network activity with a physiologically appropriate range. These mechanisms act in parallel on multiple subcellular components on slow timescales of many hours or days. Recent work has shown, perhaps unsurprisingly, that these mechanisms are present and at work in the intact nervous system (Hengen et al. 2013; Maffei and Turrigiano 2008), especially during development. It is therefore reasonable to conclude that feedback control is an essential aspect of nervous system robustness.

In isolation, each of these kinds of feedback processes, whether they act on synapses or ion channels in axons, fits into the paradigm of a negative feedback control loop. In this sense these biological processes, which are key aspects of nervous system robustness, are no different from an engineered control loop such as a thermostatic heating system. However, considered together, the intricate layering of these mechanisms is extraordinarily complex. We do not yet have a satisfactory account of all the various mechanisms that exist, nor do we understand how they operate in concert to regulate nervous system function. Indeed, recent mathematical modelling efforts conclude that naive implementation of each of these mechanisms in idealised neuronal networks can easily result in pathological states (Harnack et al. 2015; O'Leary et al. 2014). Similarly, the abundance of autapses and widespread synchronous activity in dissociated neuronal cultures are consistent with nominally 'homeostatic' mechanisms operating outside their normal context, i.e. in an intact brain.

9.3 Robust Architectures: Degeneracy

Feedback control is one way to ensure robustness in an artificial or biological system. Another is to design the system, and the feedback control loops within it, in a way that allows overall function to be relatively invariant with respect to fluctuations or loss of the components themselves. A very simple example of a robust architecture is redundancy. Redundancy arises whenever there are spare copies of important components in a system that can operate in place of a failed or deleted component. There are very few genuine examples of redundant components in biological systems: it is rarely the case that any component, whether it is a gene, an ion channel or a whole neuron, is truly a 'spare' copy. Instead, biological systems, and nervous systems in particular, exhibit *degeneracy* (Edelman and Gally 2001).

Degeneracy arises whenever there is functional overlap between several components of a system, but removal of any one component fundamentally alters the operational relationship between the remaining components. For example, if an insect looses a leg, it may or may not lose the ability to walk. But the biomechanical relationships between the remaining legs will be fundamentally altered.

The components in nervous systems exhibit degeneracy at many levels. Deletion of neuron from animal's nervous system necessarily the dynamics and functional relationships in the remaining circuit, but in many cases the effects on overall nervous system function can be quite subtle. This 'graceful degradation' of the performance of a neuronal network is a well known property of artificial neural networks. For example, some of the earliest network models of memory formation and pattern recognition can continue to function following indiscriminate deletion of a fraction of the neurons (Hopfield 1982). Although this is not a property that everyday engineered artefacts such as cars and computers enjoy, engineers are increasingly using network architecture principles to understand robustness in human-engineered systems such as transport networks and power grids (Carlson and Doyle 2000).

Degeneracy is immediately evident at the level of single neurons. Earlier in this discussion, we considered the roles of two different kinds of ion channel, sodium and potassium channels, in generating action potentials. Strikingly, neurons express many tens, even hundreds of different ion channel types and many of these contribute to action potential generation and other aspects of neuronal excitability. Moreover, the protein subunits that make up a single ion channel are combined in may different ways, leading to thousands of possible subtypes of ion channels, each with different physiological properties. The reason for this bewildering combinatorial complexity is still the subject of current research. However, theoretical studies show that functional degeneracy that arises from multiple ion channel types can enhance the robustness and flexibility of neuronal signalling, because the different channel types overlap in their contribution to specific physiological functions (Drion et al. 2015).

It helps to illustrate these ideas pictorially. Figure. 9.1 abstractly depicts how two biological parameters combine to determine the function of a biological system. The blue shaded region indicates the region of parameter space where the system is functional. The two parameters, X and Y, represent *any biological parameters that the system can regulate using feedback control.* Examples could include sodium channel density and potassium channel density in a neuron, with the blue region indicating the possible combinations of channel densities for which the neuron can generate action potentials.

Panel A summarizes how feedback control can contribute to robustness. If a perturbation in the system causes the parameters to fall outside the functional region, the feedback control mechanism can return the system (indicated by the pink dot) to a functional point. From this picture it is also clear that the feedback system needs to be appropriately tuned, as discussed above: the direction and magnitude of the perturbation with respect to locus of the functional region needs to be inverted, or approximately inverted, by the feedback controller in order to restore the system to a functional state.

However, the shape of the functional space itself also determines the robustness of the system. In panel B there are two system configurations indicated by points 'F' and 'R'. Point F (for 'Fragile') is close to the boundary of the functional region. A minor perturbation in most directions will cause point F to fall outside the functional region. On the other hand, point R (for 'Robust') is well contained in the

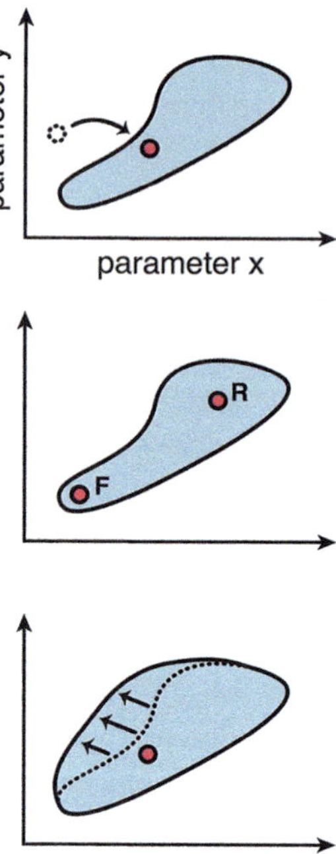

Fig. 9.1 Robustness in the functional parameter space of a biological system (see text)

functional region. Thus, the local relationships between system parameters and the global shape of functional parameter space combine to make certain system configurations more robust than others.

Degeneracy between the system parameters means that a change in one parameter can be counteracted by a change in the value of another. For example, a slight increase in the value of X at point F can be accompanied by an appropriate increase in Y to keep the system in the functional space. In this sense, degeneracy makes the task of regulating a single system property easier, by providing extra degrees of freedom in the possible functional configurations of system parameters.

Panel C shows another way that biological systems can achieve robustness. Suppose a third parameter exists that is not under feedback regulation within the neuron, but which can nonetheless modulate the relationships between the regulated parameters, X and Y. This third parameter might be, for example, an inherent biophysical property of the sodium channels in a neuron that is determined by the gene sequence of the channel. Changes in the gene sequence might increase the range of sodium and potassium channel densities that produce functional neurons, as indicated by the enlarged functional space in the figure.

In this example, the inherent robustness of the system and the context in which feedback control mechanisms operate are both altered. On evolutionary timescales we would expect this kind of component tuning to occur because there will be selection pressure for nervous systems that can survive acute perturbations and function in spite of biological variability. Biophysical and biochemical properties of enzymes, ion channels and receptors are thus tuned not only to produce novel functions, but to enhance the operational robustness of the nervous system. Although evolutionary tuning cannot be identified with feedback control within an organism, it can be viewed as feedback control of a sort: robust nervous systems enjoy a selection advantage, so the genes that promote robustness by tuning system properties will propagate more favourably.

9.4   Conclusion

I have provided a very selective glimpse of some of the biological mechanisms of robustness in the nervous system. By focussing on the molecular building blocks that give neurons their signalling properties I attempted to illustrate the importance of feedback control and degeneracy and how these operate on many levels and in parallel. In isolation, these mechanisms and principles are very similar to those employed by engineers to design robust artificial systems. In this sense, biological robustness is not conceptually unique.

On the other hand, biological systems contrast with engineered systems in their sheer complexity and in the fact that feedback control mechanisms operate at every spatial and temporal scale. Thus, if there is a true distinction between biological robustness and other forms of robustness, I claim it can be found in the degree of integration of many systems at many levels that can all be considered robust individually.

Acknowledgements   I acknowledge support from ERC-StG grant 716643 FLEXNEURO.

References

Abbott, L., & LeMasson, G. (1993). Analysis of neuron models with dynamically regulated conductances. *Neural Computation, 5*, 823–842.

Banker, G. A., & Cowan, W. M. (1977). Rat hippocampal neurons in dispersed cell culture. *Brain Research, 126*, 397–342.

Banker, G. A., & Cowan, W. M. (1979). Further observations on hippocampal neurons in dispersed cell culture. *The Journal of Comparative Neurology, 187*, 469–493.

Bekkers, J. M., & Stevens, C. F. (1991). Excitatory and inhibitory autaptic currents in isolated hippocampal neurons maintained in cell culture. *Proceedings of the National Academy of Sciences of the United States of America, 88*, 7834–7838.

Carlson, J. M., & Doyle, J. (2000). Highly optimized tolerance: Robustness and design in complex systems. *Physical Review Letters, 84*, 2529–2532.

Conant, R. C., & Ross Ashby, W. (1970). Every good regulator of a system must be a model of that system. *International Journal of Systems Science, 1*, 89–97.

Csete, M. E., & Doyle, J. C. (2002). Reverse engineering of biological complexity. *Science, 295*, 1664–1669.

Davis, G. W. (2006). Homeostatic control of neural activity: From phenomenology to molecular design. *Annual Review of Neuroscience, 29*, 307–323.

Davis, G. W., & Bezprozvanny, I. (2001). Maintaining the stability of neural function: A homeostatic hypothesis. *Annual Review of Physiology, 63*, 847–869.

Desai, N. S. (2003). Homeostatic plasticity in the CNS: Synaptic and intrinsic forms. *Journal of Physiology, Paris, 97*, 391–402.

Desai, N. S., Rutherford, L. C., & Turrigiano, G. G. (1999). Plasticity in the intrinsic excitability of cortical pyramidal neurons. *Nature Neuroscience, 2*, 515–520.

Drion, G., O'Leary, T., & Marder, E. (2015). Ion channel degeneracy enables robust and tunable neuronal firing rates. *Proceedings of the National Academy of Sciences of the United States of America, 112*, E5361–E5370.

Edelman, G. M., & Gally, J. A. (2001). Degeneracy and complexity in biological systems. *Proceedings of the National Academy of Sciences of the United States of America, 98*, 13763–13768.

Grubb, M. S., & Burrone, J. (2010). Activity-dependent relocation of the axon initial segment fine-tunes neuronal excitability. *Nature, 465*, 1070–U1131.

Harnack, D., Pelko, M., Chaillet, A., Chitour, Y., & van Rossum, M. C. (2015). Stability of neuronal networks with homeostatic regulation. *PLoS Computational Biology, 11*, e1004357.

Hengen, K. B., Lambo, M. E., Van Hooser, S. D., Katz, D. B., & Turrigiano, G. G. (2013). Firing rate homeostasis in visual cortex of freely behaving rodents. *Neuron, 80*, 335–342.

Hodgkin, A. L., & Huxley, A. F. (1952a). The components of membrane conductance in the giant axon of Loligo. *The Journal of Physiology, 116*, 473–496.

Hodgkin, A. L., & Huxley, A. F. (1952b). Currents carried by sodium and potassium ions through the membrane of the giant axon of Loligo. *The Journal of Physiology, 116*, 449–472.

Hopfield, J. J. (1982). Neural networks and physical systems with emergent collective computational abilities. In *Proceedings of the National Academy of Sciences* (Vol. 79, pp. 2554–2558).

Maffei, A., & Turrigiano, G. G. (2008). Multiple modes of network homeostasis in visual cortical layer 2/3. *The Journal of Neuroscience, 28*, 4377–4384.

Marder, E., & Goaillard, J. M. (2006). Variability, compensation and homeostasis in neuron and network function. *Nature Reviews, 7*, 563–574.

O'Leary, T., & Wyllie, D. J. A. (2011). Neuronal homeostasis: Time for a change? *The Journal of Physiology, 589*, 4811–4826.

O'Leary, T., van Rossum, M. C. W., & Wyllie, D. J. A. (2010). Homeostasis of intrinsic excitability in hippocampal neurones: Dynamics and mechanism of the response to chronic depolarization. *The Journal of Physiology, 588*, 157–170.

O'Leary, T., Williams, A. H., Franci, A., & Marder, E. (2014). Cell types, network homeostasis, and pathological compensation from a biologically plausible ion channel expression model. *Neuron, 82*, 809–821.

Turrigiano, G. (2007). Homeostatic signaling: The positive side of negative feedback. *Current Opinion in Neurobiology, 17*, 318–324.

Turrigiano, G., Abbott, L. F., & Marder, E. (1994). Activity-dependent changes in the intrinsic properties of cultured neurons. *Science, 264*, 974–977.

Turrigiano, G., LeMasson, G., & Marder, E. (1995). Selective regulation of current densities underlies spontaneous changes in the activity of cultured neurons. *The Journal of Neuroscience, 15*, 3640–3652.

Turrigiano, G. G., Leslie, K. R., Desai, N. S., Rutherford, L. C., & Nelson, S. B. (1998). Activity-dependent scaling of quantal amplitude in neocortical neurons. *Nature, 391*, 892–896.

Wheeler, D. G., Groth, R. D., Ma, H., Barrett, C. F., Owen, S. F., Safa, P., & Tsien, R. W. (2012). Ca(V)1 and Ca(V)2 channels engage distinct modes of Ca(2+) signaling to control CREB-dependent gene expression. *Cell, 149*, 1112–1124.

Willems, J. C. (2007). The behavioral approach to open and interconnected systems. *IEEE Control Systems, 27*, 46–99.

Timothy O'Leary   is an Assistant Professor in Information Engineering and Medical Neuroscience in the University of Cambridge. He trained as a mathematician at the University of Warwick and completed a PhD in experimental neurophysiology in the University of Edinburgh in 2009. His research combines experimental and theoretical approaches to understand how regulatory control and feedback operate in the nervous system. In 2014 he was awarded the Gruber International Prize for Neuroscience by the Society for Neuroscience for his contributions to understanding homeostatic plasticity in the nervous system and how these can lead to pathological behaviour as well as maintaining physiological function.

Chapter 10
Robustness vs. Control in Distributed Systems

Marta Menci and Gabriele Oliva

Abstract Understanding and controlling the behavior of dynamical distributed systems, especially biological ones, represents a challenging task. Such systems, in fact, are characterized by a complex web of interactions among their composing elements or subsystems. A typical pattern observed in these systems is the emergence of complex behaviors, in spite of the local nature of the interaction among elements in close spatial proximity. Yet, we point out that each element is a proper system, with its inputs, its outputs and its internal behavior. Moreover, such elements tend to implement feedback control or regulation strategies, where the outputs of a subsystem A are fed as inputs to another subsystem B and so on until, eventually, A itself is influenced. Such complex feedback loops are understood only by considering, at the same time, low- and high-level perspectives, i.e., by regarding such systems as a collection of systems and as a whole, emerging entity. In particular, dynamical distributed systems show nontrivial robustness properties, which are, from one side, inherent to the each subsystem and, from another, depend on the complex web of interactions. In this chapter, therefore, we aim at characterizing the robustness of dynamical distributed systems by using two coexisting levels of abstraction: first, we discuss and review the main concepts related to the robustness of systems, and the relation between robustness, model and control; then, we decline these concepts in the case of dynamical distributed systems as a whole, highlighting similarities and differences with standard systems. We conclude the chapter with a case study related to the chemotaxis of a colony of E. Coli bacteria. We point out that the very reason of existence of this chapter is to make accessible to a vast and not necessarily technical audience the main concepts related to control and robustness of dynamical systems, both traditional and distributed ones.

Keywords Dynamical systems · Control · Distributed systems · Biological systems · Robustness

M. Menci (✉) · G. Oliva
Departmental Faculty of Engineering, University Campus Bio-Medico of Rome, Rome, Italy
e-mail: m.menci@unicampus.it; g.oliva@unicampus.it

© Springer Nature Switzerland AG 2018
M. Bertolaso et al. (eds.), *Biological Robustness*, History, Philosophy
and Theory of the Life Sciences 23, https://doi.org/10.1007/978-3-030-01198-7_10

10.1 Introduction

According to the Merriam-Webster dictionary, a system can be defined as "a regularly interacting or interdependent group of items forming a unified whole" (Marriam-Webster Dictionary).

Control Theory is focused on *dynamical systems*, that is, systems characterized by specific quantities (e.g., position, velocity, humidity, etc.) that evolve as a function of time and that can be approached by means of mathematical models and methodologies for measuring and analyzing such a temporal evolution (Luenberger 1979).

From a Control Theory point of view, a system is not isolated from the rest of the world, but it may receive one or more *inputs* (e.g., stimuli or cues) and produce one or more *outputs* (i.e., stimuli to other systems). Systems are assumed as decomposable into a set of elements, each with its own inputs coming either from outside or from other elements within the same system and its own outputs, again, affecting other subsystems or crossing the boundaries of the system.

In particular, *distributed systems* are widespread in nature (e.g., the brain, colonia of bacteria, flocks of fishes, etc.) and are characterized by a large number of subsystems, interacting each with few neighbors on a local basis. As a result of such a sparse interaction, complex behaviors emerge that are not explainable by the knowledge of the single subsystems (e.g., synchronization, coordinated motion, conscience). Such systems, moreover, show non-trivial robustness and adaptation capabilities.

We point out that, unlike typical control problems in engineering, where the goal could be to attenuate the effect of a disturbance as much as possible, in biological systems, sensing changes in the input signals could be equally important for achieving proper function (Koshland et al. 1982). A consequence is that, as noted in Andrews et al. (2008) "the study of biology is essentially an analysis problem rather than a design problem".

The objective of this chapter is twofold. First, we aim at analyzing the concept of robustness in the framework of control theory. Then, we specialize and contextualize the concepts introduced for general systems in the case of large scale distributed systems, with particular reference to biological ones. Specifically, the outline of the chapter is as follows: in Sect. 10.2 we give an overview and some preliminary definitions that are related to modeling and control of dynamical systems; in Sect. 10.3 we discuss in detail the similarities and differences between the open- and closed-loop approaches to control; Sect. 10.4 is devoted to discuss the key concepts related to dynamical distributed systems; in Sect. 10.5 we analyze the main factors of robustness in feedback control, while in Sect. 10.6 we specialize such concepts in the case of dynamical distributed systems, discussing a case study concerning the chemotaxis of a colony of *E. coli bacteria*; Sect. 10.7, finally, contains some concluding remarks.

10.2 Control Theory Overview

The first step of Control Theory is to develop a *model* of the system. Modeling a phenomenon or a system means to replace it with something that is: (I) *easier to study* and (II) equivalent to the original phenomenon or system in its *important aspects* (Mitrani 1982). Making a model is thus the art of understanding which details matter and which do not. For instance, the Earth can be modeled as a surface and a sphere; but the surface proves a better model for car navigation purposes, and the sphere is more adequate while addressing gravitational phenomena. Thus, there is no strict rule of which model represents best our system; indeed, as noted by Box: "all models are wrong, but some are useful" (Box 1976).

A model of a system is not the system. In a mathematical model, for instance, the equations describing the system aim at reducing its complexity, by means of working assumptions (e.g., some parameters that vary slowly with time might be assumed to be constant, friction might be neglected, particles might be assumed to have perfectly elastic collisions, etc.). The goal of a simplified representation is to gain meaningful insights on the system and, possibly, develop successful ways to modify its behavior.

In the following, by "having a model" of a phenomenon, we will mean both having a representation of the phenomenon (e.g., by a set of dynamical equations) and having a way to reproduce the phenomenon. For instance, a model of a constant input to a system will be both a mathematical way to represent the input and the actual means required to produce an output having the same characteristics as the input (as discussed later, this might be required since, for instance, to be able to withstand a constant force, one must be able to produce a force of the same magnitude, but in the opposite direction).

By "control of a system", we mean the implementation of effective means to change the system so that "its behavior is in some way improved" (Luenberger 1979). Rather than changing the system once and for all (e.g., replacing some components with more robust or reliable ones), we exert a continuous effort throughout all the lifecycle of the system, in order to steer its behavior towards a desired one or to avoid dangerous configurations. The *control system* or *controller* is itself a system which influences the system being controlled. Control systems may be artificial (e.g., an artifact aimed at controlling a dike or a power plant) or biological (e.g., the complex regulatory systems a cell has to maintain its homeostasis); such controllers provide suitable inputs to the system, in order to change its behavior. Actually, it is the macro-system composed by the controller and the system being controlled that, as a whole, exhibits the desired behavior.

The mathematical model guides the design of an artifact for controlling the system, whereby its effectiveness depends on the extent to which the system behaves exactly as the mathematical model. Note that, although seemingly not applicable to the biological realm (as there is no actual and intentional controller design phase in this case), also biological systems might have a model of the inputs affecting them, i.e., they are able to reproduce stimuli of the same nature of the ones affecting them. This is a basic feature of any control strategy and is typically referred to as the *Internal Model Principle*; we will extensively discuss this feature in Sect. 10.5.2.

Robust control comes into play to face the challenge of achieving satisfactory control results when only rough and incomplete models are available to the controller of an artificial or biological system.

To this end, let us provide a definition of robust control. A control system is defined robust if it is *"insensitive to model uncertainty"* (Morari and Zafiriou 1989). In other words, a controller is robust if it operates properly also when some parameters, inputs or disturbances are slightly different from the nominal ones, even if it has been designed on the basis of rough models and simplistic assumptions.

10.3 Open-Loop vs. Closed-Loop Control

Let us now discuss the main antithetical paradigms in control, *open-loop* control and *closed-loop* control, shown in Figs. 10.1 and 10.2, respectively.

Within the open-loop paradigm, control is exerted on the system by pre-calculating or predicting the most adequate control action to be applied at each time instant, and then by enforcing the predicted control. Such a paradigm is, however, terribly fragile, as even slight errors in the mathematical model or the presence of perturbations would have unpredictable consequences.

Clearly, open-loop control, alone, is not the best way to control a dynamical system, since our models will be unable to fully capture the ongoing situation. This is especially evident when control needs to be exerted over a long period of time, or when the system has uncertainties or is subject to disturbances that are not known a priori. In other words, we need a better tool.

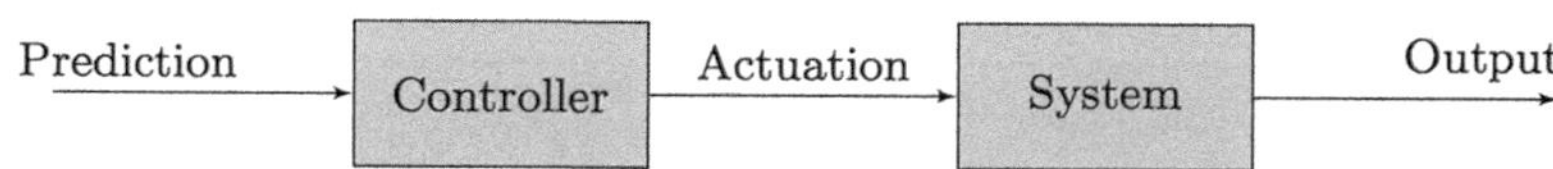

Fig. 10.1 Open-loop control scheme

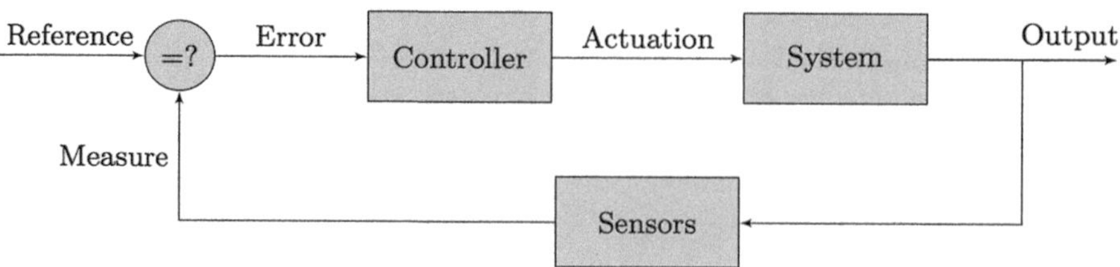

Fig. 10.2 Closed-loop control scheme

Biological systems, as well as engineered ones, have devised a better approach: closed-loop control. While in the former paradigm the control action is independent on the actual condition of the system, in closed-loop control we have the opposite.

The main ingredients of this latter control strategy are the ability to *measure* dynamically what is going on and, specifically, to quantify at each time instant the *error* between the desired and the ongoing situation. The goal of the control strategy is to reduce the error to zero. To achieve this, the control system must have a clear idea of the desired behavior for the system, i.e., it must have a model of the desired "reference" trajectory.

An essential feature of closed-loop control is the so-called *feedback*, which is the act of measuring the output of the system and "feeding it back" to the controller. This generates a control action that depends on the measured situation.[1] A simple example is adjusting the temperature of the water for taking a shower. You have a model of the desired temperature; you measure the actual temperature and move the handle from the red to the blue labelled side iteratively, aiming to approach the ideal temperature. A funny fact is that you never actually obtain the desired temperature; to succeed, you should keep moving the handle forever, oscillating in the two opposite directions, each time decreasing slightly the intensity of the change. What is worst, the strategy depicted in this example does not account for delays occurring between the time you move the handle and the time when the temperature changes, so although the control scheme is apparently simple, the controller might yet be quite complex and sophisticated in order for the system to behave as desired.

The paradigm that is coneptually opposite to closed-loop is sometimes referred to as *feedforward*. However, feedforward is also sometimes used as a technical term meaning the act of measuring directly a disturbance and compensating it without taking into account the state of the system; this is slightly different from open-loop control since the latter does not feature dynamical measurements. To avoid confusion, therefore, we stick to open-loop while describing the control strategy that is opposite to closed-loop. Note that the open- and closed-loop approaches might coexist within the same control system (see Fig. 10.3). Consider the case where we want to control a robotic arm on Earth; being the arm subject to gravity, in some applications (Egeland 1986; Park and Kim 1998) it is convenient to exert a control action to the arm that has two main components: a first open-loop component that compensates gravity, and a second closed-loop component that corrects the position or velocity of the robot's joints based on the position measures acquired by sensors such as encoders or cameras.

This example makes clear that open-loop is justified only when the assumptions made in the prediction have high likelihood to be correct (in the example, gravity is

---

[1] For the sake of simplicity, here we are assuming that one can control a system and steer its condition towards a desired one; this is not true in general and, for this to happen, the system must be *controllable*. Similarly, we are assuming that we can implement a successful feedback based on the available measures of the system's outputs; again, this is not true in general and the available measures must ensure that the system is *observable*. The interested reader can find detailed information about this issue on books about basic control engineering, such as (Luenberger 1979).

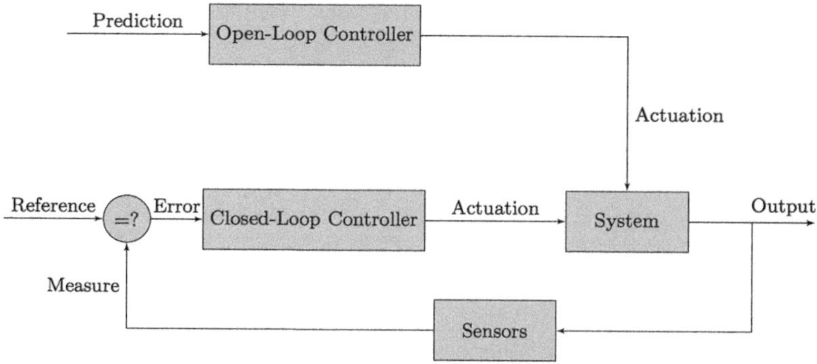

Fig. 10.3 Coexistence of open- and closed-loop control schemes

more or less constant on Earth, and the actual fluctuations have negligible effects) and is typically adopted as a complement of a feedback scheme.

10.4 Dynamical Distributed Systems

Dynamical distributed systems have a peculiar structure: the composing subsystems are typically geographically dispersed and interact directly only with neighboring subsystems. Moreover, distributed systems typically have a large scale (e.g., billions of neurons or bacteria).

A dynamical distributed system can be defined as "a continually running system in which an arbitrarily large number of processes (i.e., occurring at the subsystems, editor note) are part of the system during each interval of time and, at any time, any process can directly interact with only an arbitrary small part of the system" (Baldoni et al. 2007).

In this view, at any time, the interaction among the subsystems can be described by a *graph G*, which is a tool to represent a collection of elements, the *nodes* or *vertices* of the graph, and the existence of a connection or relation among the elements, the so-called *edges* or *arcs* (Fig. 10.4).

The "trademark" of a dynamical distributed system is the fact that the graph is not *complete*, meaning that the subsystems are not all connected with each other. Note that the structure and intensity of the interactions among subsystems may change over time, i.e., the graph may exhibit different links, or the links may model relations that remain valid, but having intensity that changes over time.

Another fundamental trait is that, although some subsystems may not directly influence each other, far away subsystems are able to influence other subsystems indirectly. In other words, provided that there is a *path* between two nodes over the graph, the nodes are able to influence each other.

As a result of such an interaction, the system, overall, exhibits "*emergent properties*" that are not an obvious consequence of the dynamics of the subsystems (the whole is more than the sum of its parts).

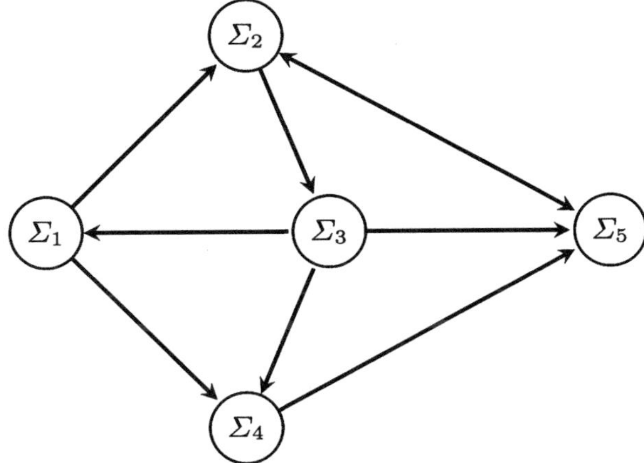

Fig. 10.4 Graph representing the relations among the subsystems $\Sigma_1,\ldots,\Sigma_5$ of a dynamical distributed system

Examples in this sense include the brain, which is made of neurons interacting each with a limited set of other neurons, and the synchronous flashing of fireflies (Winfree 2001; Strogatz 2001), which occurs even if not all fireflies are able to sense each other.

10.5 Robustness and Control

10.5.1 Feedback and Robustness

As discussed in the next subsection, a control system can exhibit robustness to specific issues if and only if specific circumstances occur. However, in this subsection, we argue that, per se, closed-loop control provides, to some extent, a degree of robustness with respect to modeling errors, uncertainties and disturbances (see a basic control engineering book like (Luenberger 1979) for a technical discussion on this topic).

Let us now discuss the robustness properties of feedback considering the most basic control scheme in the feedback control literature, that is, high-gain feedback control (Bode 1945). Within *high-gain* feedback controllers the magnitude of the control action is proportional to the error between the reference (e.g., the desired output for the system) and the actually measured output; specifically, it consists of the multiplication between the error and a multiplication constant, or *gain, K.*

Consider again the shower example and suppose you want to take a shower at a temperature of $T_{desired} = 30$ °C and that the measured temperature is $T_{measured} = 10$ °C. Hence, you have an error equal to $\Delta T = T_{desired} - T_{measured} = 20$ °C. The high-gain control strategy would be to rotate the handle of your shower of an angle $\alpha = K\Delta T$, where an angle is assumed as positive if the rotation goes towards the red labeled zone. Note that, as the name suggests, K should be quite a large number.[2] As a result of the control action, the temperature drops and we are expected to steer the handle in the opposite direction, making again a large number of turns. We keep doing this several times (actually, forever) but the rotation decreases each time; if the controller is successful, the rotation tends to become almost zero in a reasonable time.

Why does feedback always endow systems with some degree of robustness? We now give a very simplistic explanation that is not completely rigorous.[3]

In an open-loop perspective[4] let us suppose that, while affected by an input u, a system produces an output

$$y = Fu, \tag{10.1}$$

where F represents the dynamics of the system in response to u. Our goal is to achieve instead a situation where y = $y_{desired}$. The high-gain feedback approach consists in producing an input

$$u = K\left(y_{desired} - y\right), \tag{10.2}$$

so that by replacing it in Eq. (10.1), we obtain

$$y = KF\left(y_{desired} - y\right) = KF\,y_{desired} - KFy.$$

Rearranging the terms we get

$$y + KFy = KF\,y_{desired}$$

and

---

[2] When $K = 100$ you would need to rotate of about $\alpha = 2000$ degrees, i.e., make about 5.56 complete rotations! Interestingly, the more "cautious", yet successful approach we have in our everyday usage of the shower's handle suggests that we are unconsciously taking into account the delays occurring between the movement of the handle and the change of temperature, and we are limiting the magnitude of our control action accordingly.

[3] For simplicity, here we are presenting relations like $y = Fu$ as static, while in a rigorous discussion we should consider the relation between the output $y(t)$ and the input $u(t)$ at time t. Also, we are implicitly limiting our scope to linear and stationary systems, which, in a rough way, can be defined as systems where the ratio between the output and the input is constant over time. Considering linear and time-invariant systems, the response of a system is $y = Fu$ only if we are considering a frequency representation (e.g., Laplace transform), while if we are considering their temporal evolution, the signals and functions discussed in this example undergo the mathematical operation known as *convolution*. See (Luenberger 1979) for an exhaustive and formal discussion.

[4] The open-loop solution to this problem would be to set $u = y_{desired}/F$, so that $y = Fy_{desired}/F = y_{desired}$. This approach, however, is very fragile, and unless F has been calculated exactly, it is deemed to fail.

Table 10.1 Effectiveness of the high-gain feedback control strategy when $F = 10$ in Eq. (10.3) and different choices of K. As K grows, y becomes increasingly close to $y_{desired}$

K	y
1	$y = 0.5\ y_{desired}$
10	$y \approx 0.9090\ y_{desired}$
100	$y \approx 0.9909\ y_{desired}$
1000	$y \approx 0.9990\ y_{desired}$
$\vdots$	$\vdots$
∞	$y = y_{desired}$

$$\left(1 + KF\right) y = KF y_{desired}$$

hence

$$y = \left[KF / \left(1 + KF\right) \right] y_{desired}. \tag{10.3}$$

Table 10.1 shows the relation between $y_{desired}$ and y when $F = 10$ for different values of K; for sufficiently big gain[5] K, we get $y \approx y_{desired}$. Note that, in our example, this can be achieved for any F *greaterthan* 0, independently on F, hence this control scheme is robust to uncertainties and modeling errors. However, this result is true only for particular systems,[6] while for general systems we can only argue that feedback provides a given degree of robustness.

Based on the above example it should be clear that, by applying a feedback to the system, we are modifying its behavior and we are basically replacing it with another one (e.g. from the system in Eq. (10.1) to the one in Eq. (10.3)).

Note that feedback has other positive effects, which we cannot discuss here for reasons of space. Among others, feedback is known to have a linearizing effect on nonlinear systems, thus simplifying their dynamics.[7]

Even if we gave an intuitive idea of the robustness associated to closed-loop control systems, we should still discuss in which sense feedback systems are more robust, and what should be done to achieve robustness.

---

[5] Note that, although appealing, one can not select arbitrarily high values of K for several reasons: the actuation systems might not be able to produce stimuli that are too high in magnitude, or it might be onerous and thus impractical for both artificial and biological systems. Each system has typically an upper limit for K over which the system might experience saturations, instability or disruption of some of its composing elements.

[6] This result applies as presented only to the narrow class of linear and stationary systems under positive feedback; for general systems this is true only to some extent (see Morari and Zafiriou (1989) for more details).

[7] The interested reader is invited to verify that, when $y = u + 0.1u^3$ and u is chosen as in Eq. (10.2), then y becomes more and more similar to $y_{desired}$ as K grows.

10.5.2   Robust to What? The Internal Model Principle

What characterizes a system that is able to *track* a desired input or reject a disturbance? By tracking a desired input $y_{desired}$ we mean that we want to achieve (or get arbitrarily close to) the ideal situation where $y = y_{desired}$, while by disturbance rejection we mean that we want to achieve a situation where, in spite of the presence of a disturbance or unwanted input, the system behaves as if this disturbances were not present. Note that these goals typically require some time before being achieved, and there is typically an initial period, a *transient*, when the system is not able to replicate the desired input and is affected by the disturbances. However, as time passes, the output of the system gets closer and closer to the desired one and the disturbance has less and less effect.[8]

One of the central results of control theory is the so-called *Internal Model Principle* (see Francis and Wonham 1976, and Chapter 1 of *this volume*), stating that a closed-loop system is able to track a desired input or reject a disturbance if and only if such system has a model of the specific input/disturbance it is dealing with. Specifically, the controller must be able to generate outputs of the same typology of the inputs/disturbances of interest.

For instance if we want to be able to reproduce a constant input, our controller needs to be able to generate a constant output; if we want to reject a disturbance that grows linearly with time or oscillates at a given frequency, we need to produce an output that grows in the same way or oscillates at the same frequency. As previously discussed, to be able to resist to a constant force, we need to produce a force of the same type.

This principle has an important consequence: a control system is able to cope successfully only with those inputs and disturbances for which it has a model, while it has no possibility to handle inputs or disturbances of a different nature. If a controller can generate just a constant output, it could not react to a disturbance oscillating at a specific frequency. Robustness, therefore, is not general purpose, but each system is robust to specific cues while it might be prone to others.

Let us provide an example in the case of the bacterium *E. Coli*, which is known (Block et al. 1983; Alon et al. 1999) to be able to adapt to step changes in chemoattractant (i.e., instantaneous variations that remain stable for a given period of time, like changing all of a sudden the concentration of a compound diluted in water from 1 *mol/L* to 5 *mol/L* and then remaining to 5 *mol/L* for a given period of time).

Without going into details, it can in any case be shown that a (control) system is able to generate outputs changing as a step if and only if it has a mechanism of accumulation.[9]

---

[8]The objectives of having $y = y_{desired}$ or being unaffected by disturbances might never be completely achieved; in this case these goals are obtained only *asymptotically*, i.e., they are achieved completely only when time goes to infinity. In practice, however, there is a time after which such results are almost achieved, which is sufficient for most practical applications.

[9]In the case of linear and stationary systems it can be shown that the ability to generate a constant signal is mathematically equivalent to an integration operation (for more details see basic control books such as Luenberger 1979).

We point out that, to react to step cues, *E. Coli* has found a way to accumulate something to be used to counteract (or adapt to) the cue. To this end, *E. Coli* resorts to *methylation*, which is a biochemical process where a methyl group is added to the cytosine or adenine DNA nucleotides (Silverman and Simon 1977; Yi et al. 2000).

10.6   Control and Robustness in Distributed Systems

Distributed systems have a tendency to self-organization and to express complex emerging behaviors; this is evident in several contexts, ranging from economy to robotics and from social networks to biology (among other references, see (Jensen 1998; Bhalla and Iyengar 1999; Liu and Passino 2002; Peak et al. 2004) for rather classical results on the biological side and (Navlakha and Bar-Joseph 2015; Flemming et al. 2016; Singh et al. 2016; Wiggins and Stylianidou 2017) for some of the most recent developments).

Indeed, as remarkably noted by an anonymous poet in the seventeenth century while describing the flocking of fishes: "…and the thousands of fishes moved as a huge beast, piercing the water. They appeared united, inexorably bound to a common fate. How comes this unity?".

The above quote captures one of the main features of distributed systems, especially biological ones: the individual is not alone, but it is a subsystem of a large scale system, and its actions are influenced by the rest of the individuals. In other words, the behavior of an individual subsystem turns out to be remarkably different from the nominal behavior while isolated.

However, the large scale of the system, alone, does not justify its complexity. We point out that the main difference from a traditional system lies in the *sparsity* of distributed systems. By sparsity, we mean that each individual interacts with a limited amount of nearby subsystems. Yet, complex and global behaviors emerge as a result of the composition of a huge amount of elements, each interacting on a local or limited basis (see Babaoglu et al. 2006 for an interesting parallelism between sparsity in biological and technological systems).

Note that, as discussed in Sect. 10.4, each element of a distributed system is, itself, a proper system, with its inputs and outputs. We point out that, within a biological[10] distributed system (e.g., a colony of interacting bacteria), each individual typically implements all the ingredients of closed-loop control. Specifically, the individuals are able to measure their own internal condition (e.g., position, velocity, etc.) and the one of neighboring individuals (Miller and Bassler 2001; Thar and Kühl 2003; Waters and Bassler 2005). Moreover, each individual has actuators (e.g., means to change their behavior, such as flagella or cilia in bacteria) which can be exploited to adapt their state to the state of their neighbors.

---

[10] The same considerations hold true for artificial distributed systems such as swarms of autonomous mobile robots.

However we note that, with a high-level perspective, the outputs of a subsystem become the inputs for another one and vice versa. These mutual dependencies generate complex feedback loops, where the outputs of a subsystem A affect a subsystem B and, possibly, other subsystems in a chain up to a subsystem C, which generates outputs directly affecting A. Because of the interaction, from a high-level perspective, the inputs and outputs fade away in a complex web of inter-dependencies; the overall system's behavior emerges as a consequence of such complex and tangled paths.

At a first glance, distributed systems may seem to violate the internal model principle, due to their complexity. In fact, given the large number of elements or subsystems composing a distributed system, no individual subsystem might be able to have a model of all the perturbations affecting them, directly or indirectly, as a result of the dynamics of the other subsystems.

This aspect has been the object of an intense investigation in the scientific community, especially in the field of engineering, for what concerns *multi-agent systems* such as swarms of autonomous mobile robots. A convincing answer was given in (Wieland and Allgöwer 2009), where the authors provide a necessary condition for the elements of a distributed system to reach *consensus* or *synchronization*[11] (e.g., converge to the same position, move in formation, diffuse or relay information to all elements). Specifically, in order to achieve synchronization or coordinated behaviors, each element must be able to produce the consensus/synchronized trajectories. In 2011, this implication has been proved to hold in both directions (Wieland et al. 2011) (e.g., synchronization if and only if there is an internal model of the consensus trajectories).

In other words, although not having the perception of the behavior of all the subsystems, each element must be able to replicate the synchronized behavior (e.g., although initially each firefly blinks with different frequency, they either slow down or accelerate in order to match with the others; to be able to oscillate at the same frequency f, which might be a compromise among the fireflies, they must all be able to produce oscillations at that specific frequency).

Let us now discuss a key feature that characterizes the robustness of dynamical distributed systems which is the ability to maintain the overall functioning almost unchanged (or with limited degradation) even if a small fraction of the composing elements is destroyed or behaves erratically. This feature is typically known as *fault-tolerance* in the case of engineered systems (Cristian 1991), even though it finds applications also in the case of biological ones (e.g., the plasticity of the brain in response to a trauma, Johansson 2000; Bazan 2005). One of the reasons behind such a property lies in the large scale of the system at hand: a fault at the level of a single element (or a small fraction of elements) appears rather insignificant to the overall behavior. However, the large scale, alone, is not sufficient to yield such a

---

[11] By synchronization, we mean the fact that the different elements have all the same state or behavior, which may change over time in the same way over all systems. Examples in this sense include the internal clocks in a network for computers, the flashing of fireflies, the clapping of hands of people at a concert, etc.

robustness. As noted by Holme et al. in their seminal paper (Holme et al. 2002), disruptions affecting small portions of a graph have rather diverse effects depending on the topology of the graph, i.e., on the way links are formed. For instance, consider *scale-free* networks (Albert et al. 2000), which are characterized by the presence of a large imbalance on the *degree* of the nodes (i.e., the number of links that are incident at each node) and, in particular, by the presence of few strongly connected nodes (the *hubs*), while all other nodes have a limited number of links. Such networks are known to be quite robust with respect to accidental or random failures. In fact, if there are few hubs over billions of nodes, the probability that an hub is faulted is close to zero. Interestingly, the downside of this fact is that the effect of selective disruptions or attacks is quite severe on scale-free networks (e.g., the network connectivity might be disrupted by removing few important elements); this is from one side a source of vulnerability in engineered or technological networks (e.g., with respect to terroristic attacks) and from another side a possible way to implement targeted treatments or therapies (e.g., targeting few specific zones in anthrax-related proteins is known to cause the disruption of such structures, Di Paola et al. 2015).

Note that another source of robustness that is related to the topology of the network is the presence of redundant links that represent alternative paths that can be exploited in the case of failures at some of the subsystems.

10.6.1 Case Study: Chemotaxis of a Colony of E. coli *Bacteria*

To conclude this section, and in order to provide evidence of the robustness of distributed biological systems, we consider a case study which describes a chemotaxis processes for a colony of *E. coli* bacteria.

It is well known that the movement of living species is influenced by the environment they live in (Alon et al. 1999). In general, the reaction of an organism to an external stimulus is called taxis (from Greek *"to arrange"*). Different types of taxis can be mentioned, for instance phototaxis refers to the response to variation in light intensity, aerotaxis if the variation concerns oxygen concentration. Chemotaxis refers to several mechanisms through which cells can move in response to an external (usually chemical) signal, which is transduced into the cells by chemoreceptors. In particular, two cases are possible: negative chemotaxis if the organism is driven away from the source of chemical signal, and positive chemotaxis when the chemical factor tends to attract it.

In the context of chemotaxis, one of the most studied micro-organisms is the bacteria *E. coli*. A major reason behind this choice is that the protein network responsible for chemotaxis in such an organism has been characterized in detail (Barkai and Leibler 1997; Alon et al. 1999).

The motion of this small bacterium alternates between periods of smooth swimming and abrupt tumbles, which change the direction of the motion randomly. The presence of flagella on their membrane allows them to move and to seek optimum surroundings, directing their motion towards attractants or away from repellents. It has been observed that cells tumble less frequently in presence of an increasing concentrations of attractant: a chemical signal represents, in a sense, the control of the system.

Given the unfeasibility of comprehending all facets of such a complex process, it is convenient to resort to a mathematical model (Cucker and Smale 2007; Canale et al. 2015; Di Costanzo et al. 2018). Among others, the one introduced in Di Costanzo et al. (2018) is particularly effective, since it is able to take into account the spatio-temporal diffusion of the chemoattractant. In the following, for the sake of simplicity, we will allow collisions; specifically, the bacteria will be allowed to gather in the same location.

Within the model, the flocking/agreement state is achieved by means of a constant effort of each individual. Specifically, each individual measures, at each time, the relative difference of position and/or velocity with respect to neighboring individuals and the concentration of the chemoattractant, and uses such information to adapt its position and/or velocity, simultaneously leaving a trace of chemoattractant. The intuitive idea is that each individual adjusts its position and velocity depending on two complementary factors. From one side, each individual aims at the center of mass of the neighboring individuals; from another point of view, they tend to move in the direction that corresponds to an increase in the gradient of concentration of the chemoattractant. The two behaviors are mediated in the model, hence the resulting motion is a compromise between the two factors.

Note that, as discussed above, the model captures the ability of each bacterium to implement a closed-loop control strategy, where the state or output of a bacteria is fed back to itself, since it influences the other bacteria which, in turn, influence itself.

Moreover, the very fact that, within the model, bacteria reach a common location strongly supports the idea that, according to (Wieland and Allgöwer 2009; Wieland et al. 2011), each bacterium features an internal model of the overall consensus behavior.

Figure 10.5 shows the results of a numerical simulation featuring $n = 50$ bacteria. Each of them is represented by a disk and is initially placed at a random location. In Panels (a)–(c) we show different time instants of the nominal result of the simulation when all bacteria behave as intended; in panel (c), in particular, the bacteria have reached a common location. In Panels (d)–(f), instead, we analyze the effect of a failure affecting just one bacterium (the one shown in green in panel (a). Specifically, we assume the bacterium is not present. During the latter simulation, we observe that the bacteria reach a shared location, which differs only slightly from the expected one when all bacteria cooperate (the final location in the nominal case is shown by red ring).

Note that, although we considered few bacteria, the effect of the loss of a single one is negligible; this is a fortiori true for large colonies that feature billions of bacteria.

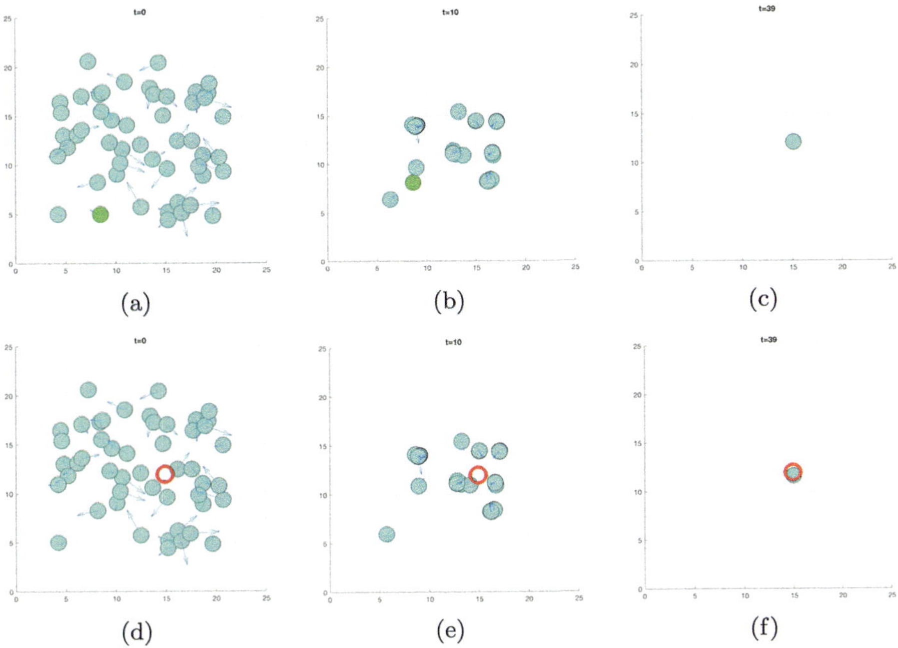

Fig. 10.5 Panels (**a**)–(**c**): numerical simulation of the behavior of a group of interacting E. Coli bacteria (colored disks) according to the mathematical model in (Di Costanzo et al. 2018). In panel (**c**) the bacteria reach a common location (the model does not account for collisions, hence the position of the bacteria coincides). Panels (**d**)–(**f**): same simulation as panels (**a**)–(**c**), assuming the bacterium shown with a green disc in panel (**a**) is not present. The nominal position of the bacteria when all of them are present is shown by a red ring

10.7   Conclusions

In this chapter we have characterized the robustness of dynamical distributed systems according to two coexisting levels of abstraction: the level of the single subsystem and the level of the system as a whole. Specifically, we have reviewed the main contents related to the robustness of systems, and the relation between robustness, model and control. Then, we have specialized these concepts in the case of dynamical distributed systems as a whole, providing an example in the case of distributed biological systems.

However, the path to mastering control in such complex scenerios, especially in biological ones, is still long, but worth the travel; indeed, as fascinatingly noted by Liu, Slotine and Barabási: "the ultimate proof of our understanding of natural or technological systems is reflected in our ability to control them" (Liu et al. 2011).

References

Albert, R., Jeong, H., & Barabási, A. L. (2000). Error and attack tolerance of complex networks. *Nature, 406*(6794), 378–382.

Alon, U., Surette, M. G., Barkai, N., & Leibler, S. (1999). Robustness in bacterial chemotaxis. *Nature, 397*(6715), 168–171.

Andrews, B. W., Sontag, E. D., & Iglesias, P. A. (2008). An approximate internal model principle: Applications to nonlinear models of biological systems. *IFAC Proceedings Volumes, 41*(2), 15873–15878.

Babaoglu, O., Canright, G., Deutsch, A., Caro, G. A. D., Ducatelle, F., Gambardella, L. M., Ganguly, N., Jelasity, M., Montemanni, R., Montresor, A., & Urnes, T. (2006). Design patterns from biology for distributed computing. *ACM Transactions on Autonomous and Adaptive Systems (TAAS), 1*(1), 26–66.

Baldoni, R., Bertier, M., Raynal, M., & Tucci-Piergiovanni, S. (2007, September). Looking for a definition of dynamic distributed systems. In *International conference on parallel computing technologies* (pp. 1–14). Berlin/Heidelberg: Springer.

Barkai, N., & Leibler, S. (1997). Robustness in simple biochemical networks. *Nature, 387*(6636), 913–917.

Bazan, N. G. (2005). Lipid signaling in neural plasticity, brain repair, and neuroprotection. *Molecular Neurobiology, 32*(1), 89–103.

Bhalla, U. S., & Iyengar, R. (1999). Emergent properties of networks of biological signaling pathways. *Science, 283*(5400), 381–387.

Block, S. M., Segall, J. E., & Berg, H. C. (1983). Adaptation kinetics in bacterial chemotaxis. *Journal of Bacteriology, 154*(1), 312–323.

Bode, H. W. (1945). *Network analysis and feedback amplifier design.* Huntington: R.E. Krieger Pub. Co.

Box, G. E. (1976). Science and statistics. *Journal of the American Statistical Association, 71*(356), 791–799.

Canale, E., Dalmao, F., Mordecki, E., & Souza, M. O. (2015). Robustness of Cucker–Smale flocking model. *IET Control Theory & Applications, 9*(3), 346–350.

Cristian, F. (1991). Understanding fault-tolerant distributed systems. *Communications of the ACM, 34*(2), 56–78.

Cucker, F., & Smale, S. (2007). Emergent behavior in flocks. *IEEE Transactions on Automatic Control, 52*(5), 852–862.

Di Costanzo, E., Menci, M., Messina, E., Natalini, R., & Vecchio, A. (2018). A hybrid mathematical model of collective motion under alignment and chemotaxis. *Discrete and continuous dynamical systems, Series B.* (Submitted).

Di Paola, L., Platania, C. B. M., Oliva, G., Setola, R., Pascucci, F., & Giuliani, A. (2015). Characterization of protein–protein interfaces through a protein contact network approach. *Frontiers in Bioengineering and Biotechnology, 3,* 170.

Egeland, O. (1986). On the robustness of the computed torque technique in manipulator control. In *Robotics and automation. Proceedings. 1986 IEEE international conference* (Vol. 3, pp. 1203–1208). San Francisco: IEEE.

Flemming, H. C., Wingender, J., Szewzyk, U., Steinberg, P., Rice, S. A., & Kjelleberg, S. (2016). Biofilms: An emergent form of bacterial life. *Nature Reviews Microbiology, 14*(9), 563–575.

Francis, B. A., & Wonham, W. M. (1976). The internal model principle of control theory. *Automatica, 12*(5), 457–465.

Holme, P., Kim, B. J., Yoon, C. N., & Han, S. K. (2002). Attack vulnerability of complex networks. *Physical Review E, 65*(5), 056109.

Jensen, H. J. (1998). *Self-organized criticality: Emergent complex behavior in physical and biological systems* (Vol. 10). Cambridge: Cambridge University Press.

Johansson, B. B. (2000). Brain plasticity and stroke rehabilitation. *Stroke, 31*(1), 223–230.

Koshland, D. E., Goldbeter, A., & Stock, J. B. (1982). Amplification and adaptation in regulatory and sensory systems. *Science, 217*(4556), 220–225.

Liu, Y., & Passino, K. M. (2002). Biomimicry of social foraging bacteria for distributed optimization: Models, principles, and emergent behaviors. *Journal of Optimization Theory and Applications, 115*(3), 603–628.

Liu, Y. Y., Slotine, J. J., & Barabási, A. L. (2011). Controllability of complex networks. *Nature, 473*(7346), 167–173.

Luenberger, D. (1979). *Introduction to dynamic systems: Theory, models, and applications.* New York: Wiley.

Miller, M. B., & Bassler, B. L. (2001). Quorum sensing in bacteria. *Annual Reviews in Microbiology, 55*(1), 165–199 ISO 690.

Mitrani, I. (1982). *Simulation techniques for discrete event systems* (No. 14). CUP Archive.

Morari, M., & Zafiriou, E. (1989). *Robust process control* (Vol. 488). Englewood Cliffs: Prentice hall.

Navlakha, S., & Bar-Joseph, Z. (2015). Distributed information processing in biological and computational systems. *Communications of the ACM, 58*(1), 94–102.

Park, J. H., & Kim, K. D. (1998). Biped robot walking using gravity-compensated inverted pendulum mode and computed torque control. In *Robotics and automation, 1998. Proceedings. 1998 IEEE international conference on* (Vol. 4, pp. 3528–3533). San Francisco: IEEE.

Peak, D., West, J. D., Messinger, S. M., & Mott, K. A. (2004). Evidence for complex, collective dynamics and emergent, distributed computation in plants. *Proceedings of the National Academy of Sciences of the United States of America, 101*(4), 918–922.

Silverman, M., & Simon, M. (1977). Chemotaxis in Escherichia coli: Methylation of che gene products. *Proceedings of the National Academy of Sciences, 74*(8), 3317–3321.

Singh, S., Rashid, S., Long, Z., Navlakha, S., Salman, H., Oltvai, Z. N., & Bar-Joseph, Z. (2016). Distributed gradient descent in bacterial food search. arXiv preprint arXiv:1604.03052.

Strogatz, S. H. (2001). Exploring complex networks. *Nature, 410*(6825), 268–276.

Thar, R., & Kühl, M. (2003). Bacteria are not too small for spatial sensing of chemical gradients: An experimental evidence. *Proceedings of the National Academy of Sciences, 100*(10), 5748–5753.

Waters, C. M., & Bassler, B. L. (2005). Quorum sensing: Cell-to-cell communication in bacteria. *Annual Review of Cell and Developmental Biology, 21*, 319–346.

Wieland, P., & Allgöwer, F. (2009). An internal model principle for consensus in heterogeneous linear multi-agent systems. *IFAC Proceedings Volumes, 42*(20), 7–12.

Wieland, P., Sepulchre, R., & Allgöwer, F. (2011). An internal model principle is necessary and sufficient for linear output synchronization. *Automatica, 47*(5), 1068–1074.

Wiggins, P. A., & Stylianidou, S. (2017). Emergent self-similarity in complex biological systems due to strong disorder. *Biophysical Journal, 112*(3), 240a.

Winfree, A. T. (2001). *The geometry of biological time* (Vol. 12). New York: Springer.

Yi, T. M., Huang, Y., Simon, M. I., & Doyle, J. (2000). Robust perfect adaptation in bacterial chemotaxis through integral feedback control. *Proceedings of the National Academy of Sciences, 97*(9), 4649–4653.

Marta Menci received her Laurea degree in Mathematics at University of Florence in 2016. During her university studies, she focused in applied mathematics, and her thesis was related to mathematical models for collective motions of agents, following a numerical approach. She is currently a Ph.D student in Bioengineering and Bioscience at Università Campus Bio-Medico of Rome, Italy. Her interests include mathematical models of biological phenomena involving cell motions, influenced by chemotaxis. In particular, she is interested in both analytical results concerning the existence of solutions and computer simulations, based on finite-differences schemes.

Gabriele Oliva received the Laurea degree and the Ph.D in Computer Science and Automation Engineering in 2008 and 2012, respectively, both at University Roma Tre of Rome, Italy. He is currently assistant professor in Automatic Control at the University Campus Bio-Medico of Rome, Italy. His main research interests include distributed systems, distributed optimization, applications of graph theory in technological and biological systems, and Critical Infrastructure Protection.

Chapter 11
The Robustness of Musical Language: A Perspective from Complex Systems Theory

Flavio Keller and Nicola Di Stefano

Abstract Within the field of systems theory, the term robustness has typically been applied to different contexts such as automatic control, genetic networks, metabolic pathways, morphogenesis, and ecosystems. All these systems involve either man-made machines, or living organisms. In this chapter, we will consider music as a peculiar complex system, involving both the realm of machines (the musical instrument) and the realm of biology (the player and the listeners). We will discuss some of the properties of music experience in terms of different attributes of robustness, focusing in particular on stability, the property enabling a complex system to maintain its function against a wide range of external and internal changes. We will provide examples of the human ability of isolating and maintaining stable information within the perceptual flow and despite changes in the external world that reach our perceptions, leading towards a characterization of robustness in music perception as referred both to the search for regularities and to the range of tolerance that perception admits to regularities. Finally, we will list four multiple interaction cycles that typically characterize music experience and that involve both internal properties of the organism and the environment.

Keywords Regulation · Functional integration · Cognition · Music perception · Auditory streaming · Missing fundamental

11.1 Introduction

The term robustness has typically been applied to complex systems such as automatic control, genetic networks, metabolic pathways, morphogenesis, and ecosystems (see chapter by S. Caianiello "Prolegomena to a History of Robustness" in this volume). One general characteristic of these systems is that they involve either

F. Keller · N. Di Stefano (✉)
Institute of Philosophy of Scientific and Technological Practice and Laboratory of Developmental Neuroscience, Università Campus Bio-Medico di Roma, Rome, Italy
e-mail: f.keller@unicampus.it; n.distefano@unicampus.it

© Springer Nature Switzerland AG 2018
M. Bertolaso et al. (eds.), *Biological Robustness*, History, Philosophy and Theory of the Life Sciences 23, https://doi.org/10.1007/978-3-030-01198-7_11

man-made machines or living organisms. In this perspective, music is a unique complex system in so far as it involves both the realm of machines (the musical instrument) and the realm of biology (the player). To obtain an optimal result, the two elements must interact optimally. It is precisely the fact that one realm must not prevaricate over the other that makes music unique. In other human endeavors characterized by instrument use, the aim of the instrument is to compensate for the "style" and deficiencies of the user, in order to guarantee safe, homogenous, and reliable results (this is the case of modern, computer-controlled airplanes or biomedical and surgical instruments). In music, the stylistic characteristics of the player are quintessential to the performance and when these characteristics are absent, the music produced lacks luster, color, artistry. For these reasons, playing a musical instrument is much more than using an instrument. Rather it is a complex activity fostered by an engaging interaction process, in which the instrument "mediates between human mind and physical energy" (Leman 2007, p. 138).

Furthermore, if we compare written music to written verbal language, we can easily see that music shows a much more complex organization: while written verbal texts have a strictly one-dimensional, sequential organization, most written music, except for simple melodies, has a multidimensional organization, as the multiple overlapping voices of a fugue or a chorale immediately show. Incidentally, this higher level of complexity explains why automatic music recognition software usually performs much more poorly than text recognition software.

Although robustness has been defined in many ways, one key characteristic is *stability*, a property that enables a complex system to maintain its function in the context of a wide range of external and internal changes. At the same time, a robust system also shows a peculiar *fragility*, i.e. the system tolerates losing some components, but is sensitive to the removal of some critical components (Kitano 2004). One example of robustness in living systems is embryogenesis: the organism goes through dramatic changes in morphology, adding new parts and mutual relations, while the overall developmental plan of the embryo is maintained. In a similar way, in the historical development of music, from its early forms to contemporary music, dramatic changes of forms and genres took place (for an interesting analysis of music history in terms of general system theory see Georgescu and Georgescu 1990; for a systemic perspective on musical consonance and dissonance see Di Stefano and Bertolaso 2014), and yet there appears to be a robust set of core properties that allows for music to be perceived as music, hence engendering in the listener music-conform experiences and behaviors, as opposed to other types of sounds (e.g. the ringing of the doorbell) or noises. In addition, music shows an intentional content, as evidenced by the fact that we listen in a different way to the same melody when it is used as a mere signal (e.g. in an alarm clock or as the ring tone on a cell phone) and when it is intended as 'real' music.

In this chapter, we will discuss some of the properties of music perception in terms of different attributes of robustness (i.e. *fragility* and *stability*), in particular the ability to maintain its fundamental properties in spite of a changing environment and the ability to incorporate novel elements without losing the essential properties of a system. We will then explore the impact of fragility on the emotional resonance of music perception. In the conclusion, we suggest that robustness in music experience can rely on multiple interaction cycles between the realm of biology (the player and the listener) and mediation technology (any kind of instrument used).

11.2   Stability and Fragility in Auditory Perception

11.2.1   Repetitions and Variations in Auditory Streaming

When related to music perception, the word 'robustness' can refer to the human ability to isolate and maintain stable information within the perceptual flow and despite changes in the external world that reach our perceptions. When we listen to a short piece of music, we are able to recognize relevant and stable features conveyed by changing qualities of acoustic stimuli (i.e. melody contour, timbre, major vs. minor modes). Similar musical materials are coherently organized into perceptual units, for example, by grouping events that have similar physical characteristics (rhythm, frequency or timbre) or that occur close in time. However, this ability to recognize similarities and variations as fundamental elements of musical language happens only under certain conditions. For example, everyone easily recognizes the simplest form of variation – i.e. the identical repetition of music materials at different points in time – but most complex forms of variation techniques (often adopted in the fugue or in contemporary music) are rarely recognized. A well-known example of simple repetitions and variations is Maurice Ravel's *Bolero*, in which two themes are continually alternated and repeated throughout the entire piece and each repetition presents a different orchestration. In this case, it is easy to isolate what is stable (i.e. melody and rhythm) and what is changing (i.e. timbre and orchestration). An opposite case is exemplified by Schoenberg's *Variations for Orchestra*, harshly criticized when first played by the Berliner Philharmoniker in 1928. In the *Variations*, Schoenberg adopted complex forms of imitation, such as the inversion canon that presents the imitation moving in contrary motion to the theme: where the theme steps up by a particular interval, the inversion steps down by that same interval. These kinds of imitation are rarely recognized by untrained listeners. With respect to the key features of robustness, *Bolero* offers an example of perceptual *stability*, while *Variations* offers one of perceptual *fragility*.

The perceptual process relies upon the ability to recognize regularities and repetitions. At the same time, it *forces* regularities in auditory streaming, by tending to ignore small differences in order to save patterns of regularities (see the pioneering writings of the *Gestalpsychologie*, e.g. von Ehrenfels 1988). Since processing is better for regular than for irregular sequences, we tend to hear as regular sequences that are not. This is because minimum variations in pitch or rhythm are not perceived as a break of regularities. These facts show that robustness in music perception is therefore linked to *stability*, i.e. the ability to ignore small differences in order to maintain stable patterns of repetitions and to 'save' regularities. For our perceptual system compares each newly arriving stimulus with preceding ones. This comparison can regard different musical qualities/quantities, such as pitch, tempo, or timbre. If the new stimulus is similar in duration (or frequency or spectrum) to preceding ones, it will be categorized as 'same'. Only a significantly different interval will be categorized as 'different'. Much of what is perceived as repetition or 'same' is far from being really the same. For example, a melody tuned at 440 *Hz* and the 'same' melody tuned at 444 *Hz* will be recognized as the *same* melody. As far as rhythm is concerned, though all musicians try to stick to the beat, none are *perfectly* on the beat. However, our perception 'adjusts' irregularities, giving us the impression of a regular beat, when the sequence of beats lies within an acceptable temporal window. Therefore, if a sequence is slightly irregular but all the intervals remain within the tolerance window, then we will perceive this sequence as the succession of 'same' intervals and so perceive a regular sequence (Drake and Bertrand 2003). These considerations lead us to reshape the definition of robustness when related to music perception as follows: robustness in music perception refers both to the search for regularities (*stability*) and to the range of tolerance that perception admits to regularities (*fragility*).

Abilities to detect similarities and variations in rhythm, melody and harmony are shown since infancy. For example, infants recognize a tone sequence when the tempo is altered so long as the relative durations remain unchanged (Trehub and Thorpe 1989). As for harmony, a fundamental role in the perception of harmony is played by consonance and dissonance. In studies of infant melody perception, in fact, the presence of consonant intervals, such as perfect fifth, has been associated with success in interval discrimination, and its absence with failure (Cohen et al. 1987; Trainor and Trehub 1993a, b). These findings have been related to the Pythagorean theory of consonance, as humans detect interval changes more easily in the context of small-integer ratios—the *octave* (2: 1), *perfect fifth* (3: 2), and *perfect fourth* (4: 3)—than in the context of large-integer ratios such as the *tritone* (45: 32 ratio) (Schellenberg and Trehub 1996a, b). The implication is that *perfect fifths* and *fourths* are inherently easier to encode than are tritons (Schellenberg and Trehub 1994, 1996a, b). All these findings support the idea that consonance and dissonance are structural elements within auditory streaming, thus representing an organizing principle of the perceptual process, rooted both in the listener and in the auditory signal (recent works support this hypothesis from different perspectives, see e.g. Pankovski and Pankovska 2017 and Trulla et al. 2018).

11.2.2 Ability to Select Salient Auditory Information Among Many Stimuli

Generally, when we listen to music we are not perceiving *only* music. A wide variety of noises – such as street traffic, police siren, air conditioning fan, rain – simultaneously reach our auditory system. However, we usually say that we listen to music, and not to noises *and* music. Why? How can we extract precise information among such a variegate multitude of noises and sounds?

This phenomenon is known as the 'cocktail party effect' (Kuyper 1972; Handel 1989; Arons 1992), which is the ability to focus one's auditory attention on a particular stimulus while filtering out a range of other stimuli, as people are asked to do when they carry on a conversation in a noisy room at a party. Something similar happens when we hear something relevant for us while conversing and our attention is immediately drawn by this auditory input.

This ability to select information from a whole is well exemplified in music perception. Albert Bregman (1990) uses the term 'auditory scene analysis' to indicate the ability to isolate and follow one particular group of notes as a different stream with respect to the perceived whole. For example, when listening to a symphony, we are led to integrate into the same auditory stream notes that are close together in timbre, rhythm, or frequency. This allows us to distinguish clearly a theme played by flute or oboe from ten or fifteen different overlapping voices. Clearly, when we are listening to a live orchestra, our hearing a specific voice is facilitated by the mechanisms that allow us to locate sound sources in space, but the ability to isolate a single voice is also preserved when we listen to recorded music. Incidentally, though this phenomenon occurs mostly in listening to polyphonic music, the same effect is produced when a single melodic voice aims at giving the impression of two voices (consider for example J. S. Bach's *Ciaccona* for violin solo).

There are different criteria for creating perceptual units within auditory streaming. Among harmonic criteria, for example, the similarity between tones is very common, i.e. it is possible to link different tones that represent multiples of the same fundamental frequency. Other criteria of similarity extend to timbre or rhythm. More abstract criteria for grouping notes and melodies regard the semantic level of music, i.e. the coherence of music materials. When perceived, a melody shows a tendency toward certain tones (dominant, subdominant, tonic), that play a peculiar role in the streaming and lead to other tones that are expected in the streaming (Stainsby and Cross 2016). The auditory scene analysis is linked to *stability*, and represents a case of robustness in music perception, occurring in particular when we reduce the information to better control parameters we are focusing on.

11.2.3 The Phenomenon of the Missing Fundamental

Though intrinsically related, pitch and frequency refer to different things. Pitch is the attribute of auditory sensations that allow us to order sounds along a scale extending from low to high, and is directly dependent on the frequency, sound pressure and waveform of the stimulus. Therefore, pitch is a perceptual correlate of physical properties of the signal, while frequency is a fundamental property of the waveform. If we consider that the range of audible frequencies spans from about 30 Hz to 20 kHz, we have to assume that only a small part of this range is composed by musical sounds. In fact, the lowest musical instruments can play tones at 30 Hz and the highest piccolo flute can reach about 5 KHz. Therefore, only sound signals from 30 Hz to 5 kHz are perceived as pitches, while tones exceeding 5 kHz lose their identity as musical pitches, even though they have similar physical features.

If perceived pitch usually corresponds to the fundamental frequency, we cannot assume that it is simply determined by the lowest frequency component. As research has proved, a pitch corresponding to the fundamental frequency of a complex tone was heard even when this frequency was removed from the spectrum. This pitch has been called "virtual pitch" or "missing fundamental" effect. The missing fundamental is a perceptual effect occurring with complex tones presented by Schouten (1938). The phenomenon can be reproduced depriving a complex tone of the fundamental frequency and leaving all the others the same. For example, a complex tone composed of 800, 1200, 1600, 2000 Hz waves will be heard as a sound with fundamental pitch at 400 Hz (the missing frequency).

The missing fundamental is a good example of "perceptual restoration of missing sounds" (Warren 2008) and supports the idea of robustness of perception as the tendency to maintain its fundamental properties in spite of a changing environment (*stability*). Incidentally, it is not restricted to sound perception, but extends to other senses. The phenomenon of "filling-in" the visual scene in the blind spot of the retina, using visual information from the surrounding regions, is another typical example of perceptual restoration (De Weerd 2006). The phenomenon of the missing fundamental can have paradoxical consequences for tones perception. If we listen to a pure tone at 700 Hz and a complex tone composed by waves at 800, 1200, 1600, 2000 Hz, we will hear the complex tone lower than the pure one, even though all the partials involved in complex tone are above it. This is because the complex tone is perceived with a missing fundamental at 400 Hz, and therefore the pure tone at 700 Hz is coherently perceived as higher. Finally, the missing fundamental phenomenon also has important technological consequences, in particular in the domain of audio processing applications. For example, the "missing fundamental" effect provides the possibility of giving the impression of lower bass in speakers that are not able to reproduce such a low bass, overtaking limitation and constrains in the design and production.

11.3 Fragility and Emotional Resonance of Musical Language

The relationship between music and emotions is a hotly debated issue that can be considered from two perspectives: i) the capacity of music to *generate emotions* in the listener; ii) the ability of music to *express emotions* (according to the intentions of the composer). In this chapter, we focus on the first perspective.

Music has an impressive ability to evoke emotions: even when we are not actively involved in producing it, our emotional state is altered by what we are listening to. Emotion is a key element in many theories on the origins and evolution of music.

For example, in the *Compendium Musicae*, Descartes refers to music as an act that pleases us and evokes different emotions (Descartes 1961). More recently, musicologist Deryck Cooke claimed that music is the language of emotions (1959, 32–33). Although emotions are one of the most preferred effects that people report as motive for listening to music, the emotional character of music remains particularly intriguing because – as Darwin foresaw – music has no immediate intrinsic biological or survival value.

The relationship between music and emotions has different levels (Juslin 2013). First, the emotional response of the listener, i.e. the emotions we immediately perceive while listening to music. Second, the emotional correlates of musicians or performers that can be very different from those perceived by listeners. For example, while the performer may feel anxious before playing a cadenza, the awaiting listener may be relaxed. Third, emotions expressed by musical contents (harmony, melody, timbre, pitch contour), are independent from the ones perceived by listeners. Here we refer to properties that characterize musical language and give rise to the specific emotional quality of music, explaining why a funeral march is perceived as sad and melancholy, while a nuptial march or waltz is generally perceived as happy.

The above classification rests on the distinction within the world of emotions between perceived and felt emotions (Gabrielsson 2002). Perceived emotions refer to emotion that we perceive or recognize from our surroundings and environments. Felt emotions refers to emotions we actually experience. Of course, in many cases perceived and felt emotions are identical. We may listen to joyful music and feel joy. However, this makes the relationship between music and emotions more complex: pleasant emotions may arise from listening to sad music due to the realization of our expectations or the alignment with our emotional state, and therefore sad music may evoke pleasant emotions (Kavakami et al. 2013).

Emotions represent a fundamental link between external stimuli and subjective experience, as they are both rooted in our sensorimotor system (the realm of biology) and evoked by an external input (the environment). As evidenced above, the *fragility* of robust systems is due to its sensitivity to the removal of some critical components. This aspect is manifest when we consider that minor changes in the musical stimulus affect the evoked emotions. For example, a theme played in minor mode and in a slow tempo differs only in a few tones from the major version of the same theme in a faster tempo, though the evoked emotions may be totally different (i.e. from sad and relaxed to happy and excited). On the one hand, considering emotions allows us to highlight the peculiar *fragility* of musical language. On the other, the role of emotion in music perception leads us to stress sensorimotor responses to music, as a form of expression that integrates and synchronizes different agents by translating the physical vibrations into regular body movements (Leman 2007; Windsor 2009; Leman et al. 2017), thus opening to the consideration of the impact of music on mental and physical health.

11.4 Conclusion

We started this chapter hypothesizing that robustness of music perception is rooted in the unique synthesis between machine (the musical instrument) and man (the player). We then considered a few examples of this idea of robustness in music perception, showing that information coming from the different sensory systems involved in music experience forms a whole in the corporeality of the subject. We emphasized that both key features of robustness - fragility and stability - characterize musical language. It is interesting to observe that recent results in the application of music to different diseases and disorders (e.g. Brotons and Koger 2000; Oldfield 2006; Geretsegger et al. 2014; Lagasse and Thaut 2013) are grounded on the highlighted properties of musical language, i.e. its ability to equally and actively involve different dimensions of human beings, from low-level sensorimotor synchronization to higher emotional and social communication and understanding. Through sounds, musical language activates a unique perception-action loop in which the subject is driven to react to acoustic perception through voluntary and involuntary movements (see e.g. Komeilipoor et al. 2015; Di Stefano et al. 2017). The effect of different strategies of music therapy can thus be seen as an indirect proof of the robustness of musical language, conceived as a unique complex system involving different but mutually interacting and communicating realms – the realm of external stimulation (the musical instrument or the musical medium) and the realm of biology (the player or the listener).

It remains to understand why this unique complex system is so surprisingly robust. We do not think that robustness of music experience lies in the music instrument *as instrument*. Tool use to enhance the agent's capabilities, and even to correct the agent's deficiencies of errors is a well-known phenomenon. Many animals can utilize tools to explore the environment and to grasp things that are out of reach. The main difference between music and other instrument-based activities resides precisely in the fact that music is biologically useless, if we consider it from the point of view of its purpose. Every tool has a specific purpose, while music has no specific purpose: this is precisely the reason why it can have so many different purposes.

In a recent review, Fernandez-Leon has argued that robustness comes through dynamically distributed, multiple interaction cycles that are not limited to the internal properties of an organism, but also include the environment (Fernandez-Leon 2014). We think that this conceptual framework is very well suited for understanding music. In music, we can observe at least four multiple interaction cycles in which both the biological organism and the surrounding environment are variously involved:

- Interaction cycles between the player's sensorimotor systems and his/her musical instrument, whereby multiple proprioceptive and exteroceptive sensorimotor loops are involved (think of a violinist who receives proprioceptive feedback from his musculoskeletalsystem, as well as acoustic feedback about the produced sound both through air conduction and bone conduction).

- Interaction cycles between the player and the listeners. Also in this case feedback comes through multiple sensory channels, e.g. the visual channel (facial expressions, gaze, gestures and movements), the acoustic channel (e.g. singing, hand clapping).
- Interaction cycles among different players. The interplay among musicians is one of the most characterizing aspect of any ensemble, from trio to big orchestra and chorus. In this perspective, interplay can be seen as the ability to react coherently to changes that occur in the ensemble during the performance and that can vary from accidental tempo variations to more complex forms of musical expressions (such as *legato*).
- Interaction cycles between the musicians and the conductor. This case probably sums up all the previous ones, as the conductor is at the same time a listener and a player (the instrument being the whole ensemble). A good interaction or 'feeling' between the conductor and the ensemble guarantees the opportunity to correct faults during the performance or to highlight certain musical nuances of the score.

These interaction cycles evidence the multimodal nature of musical language, in which lower-order sensorimotor regulatory loops and higher-order cognitive functions merge to produce one of the most characterizing, intriguing and complex forms of robust human behavior.

References

Arons, B. (1992). A review of the cocktail party effect. *Journal of the American Voice I/O Society, 12*(7), 35–50.

Bregman, A. S. (1990). *Auditory scene analysis*. Cambridge, MA: MIT Press.

Brotons, M., & Koger, S. M. (2000). The impact of music therapy on language functioning in dementia. *Journal of Music Therapy, 37*, 183–195.

Cohen, A. J., Thorpe, L. A., & Trehub, S. E. (1987). Infants' perception of musical relations in short transposed tone sequences. *Canadian Journal of Psychology, 41*, 33–47.

Cooke, D. (1959). *The language of music*. London: Oxford University Press.

De Weerd, P. (2006). Perceptual filling-in: More than the eye can see. *Progress in Brain Research, 154*, 227–245.

Descartes, R. (1961). *Compendium Musicae*. Rome: American Institute of Musicology.

Di Stefano, N., & Bertolaso, M. (2014). Understanding musical consonance and dissonance: Epistemological considerations from a systemic perspective. *System, 2*, 566–575.

Di Stefano, N., Focaroli, V., Giuliani, A., Formica, D., Taffoni, F., & Keller, F. (2017). A new research method to test auditory preferences in young listeners: Results from a consonance versus dissonance perception study. *Psychology of Music, 45*(5), 699–712.

Drake, C., & Bertrand, D. (2003). The quest for universals in temporal processing in music. In R. Zatorre (Ed.), *The cognitive neurosciences of music*. Oxford: Oxford University Press.

Fernandez-Leon. (2014). Robustness as a relational phenomenon. *Biological Reviews, 89*, 552–567.

Gabrielsson. (2002). Emotion perceived and emotion felt: Same or different? *Musicae Scientiae, 5*(1), 123–147.

Geretsegger, M., Elefant, C., Mössler, K. A., & Gold, C. (2014). Music therapy for people with autism spectrum disorder. *The Cochrane Library, 6*, CD004381.

Handel, S. (1989). *Listening: An introduction to the perception of auditory events*. Cambridge, MA: MIT Press.

Juslin, P. N. (2013). From every day emotions to aesthetic emotions: Towards a unified theory of musical emotions. *Physics of Life Reviews, 10*(3), 235–266.

Kavakami, et al. (2013). Sad music induces pleasant emotions. *Frontiers in Psychology, 4*, 311.

Kitano, H. (2004). Biological robustness. *Nature Reviews Genetics, 5*(11), 826–837.

Komeilipoor, N., Rodger, M. W. M., Craig, C. M., & Cesari, P. (2015). (Dis-) Harmony in movement: Effects of musical dissonance on movement timing and form. *Experimental Brain Research, 233*, 1585–1595.

Kuyper, P. (1972). The cocktail party effect. *Audiology, 11*(5), 277–282.

Lagasse, A. B., & Thaut, M. H. (2013). The neurobiological foundation of neurologic music therapy. *Music and Medicine, 5*, 228–233.

Leman, M. (2007). *Embodied music cognition and mediation technology*. Cambridge, MA: MIT Press.

Leman, M., Nijs, L., & Di Stefano, N. (2017). On the role of the hand in the expression of music. In M. Bertolaso & N. Di Stefano (Eds.), *The hand perception, cognition, and action* (pp. 175–192). Cham: Springer.

Oldfield, A. (2006). *Interactive music therapy – A positive approach*. London: Jessica Kingsley Publishers.

Pankovski, T., & Pankovska, E. (2017). Emergence of the consonance pattern within synaptic weights of a neural network featuring Hebbian neuroplasticity. *Biologically Inspired Cognitive Architectures, 22*, 82–94.

Schellenberg, E. G., & Trehub, S. E. (1994). Frequency ratios and the discrimination of pure tone sequences. *Perception & Psychophysics, 56*, 472–478.

Schellenberg, E. G., & Trehub, S. E. (1996a). Children's discrimination of melodic intervals. *Developmental Psychology, 32*, 1039–1050.

Schellenberg, E. G., & Trehub, S. E. (1996b). Natural musical intervals: Evidence from infant listeners. *Psychological Science, 7*, 272–277.

Schouten, J. F. (1938). *The perception of subjective tones* (Vol. 41, pp. 1086–1093). Amsterdam: K Akademie van Wetenshappen.

Stainsby, T., & Cross, I. (2016). The perception of pitch. In Hallam, Cross, & Taut (Eds.), *The Oxford handbook of music psychology*. Oxford: Oxford University Press.

Trainor, L. J., & Trehub, S. E. (1993a). Musical context effects in infants and adults: Key distance. *Journal of Experimental Psychology. Human Perception and Performance, 19*, 615–626.

Trainor, L. J., & Trehub, S. E. (1993b). What mediates infants' and adults' superior processing of the major over the augmented triad? *Music Perception, 11*, 185–196.

Trehub, S. E., & Thorpe, L. A. (1989). Infants' perception of rhythm. Categorization of auditory sequences by temporal structure. *Canadian Journal of Psychology, 43*, 217–229.

Trulla, L. L., Di Stefano, N., & Giuliani, A. (2018). Computational approach to musical consonance and dissonance. *Frontiers in Psychology, 9*, 381.

von Ehrenfels, C. (1988). On Gestalt qualities. In B. Smith (Ed.), *Foundations of Gestalt Theory* (pp. 82–117). Wien: Philosophia Verlag.

Warren, R. M. (2008). *Auditory perception. An analysis and synthesis*. Cambridge: Cambridge University Press.

Windsor, L. (2009). Measurement and models of performance. In Hallam, Cross, & Thaut (Eds.), *Oxford handbook of music psychology*. Oxford: Oxford University Press.

Flavio Keller is Professor of Human Physiology at the Università Campus Bio-Medico di Roma in Rome, Italy, where he is Director of the Neurosciences Lab. His training included studies with Dr. E.R. Kandel (Howard Hughes Medical Institute, Columbia University) and with Dr. M.E. Schwab (Brain Research Institute, University of Zurich). Since 2006, he has been an ordinary member in the Perception-Movement-Action Research Centre at the University of Edinburgh, and member of the Autism Research Consortium, Boston, U.S.A.

Nicola Di Stefano graduated in Philosophy from the University of Milan and in double-bass from "G. Verdi" Conservatory of Milan. Then he obtained his Ph.D. in Bioethics from Campus Bio-Medico University of Rome. He has epistemological and aesthetical interests and background. His current research activity focuses on music perception and aesthetics. He is member of Society for Music Perception and Cognition (SMPC) and of the European Society for The Cognitive Sciences of Music (ESCOM).

Chapter 12
Dynamical Rearrangement of Symmetry and Robustness in Physics and Biology

Giuseppe Vitiello

Abstract The mechanism of the dynamical rearrangement of symmetry in quantum field theory underlies the phenomenon of coherent boson condensation in the vacuum state. Coherent states appear to be related to fractal self-similarity. The dynamical paradigm of coherence opens the way to an integrated vision of natural phenomena and it may possibly rule morphogenetic processes. Robustness properties of physical systems, such as dynamical and functional robustness, topological robustness, multilevel and semantic robustness may find their root in coherence. Possible extension to biology and neuroscience is discussed.

Keywords Symmetry · Invariance · Conservation laws · Coherence · Quantum field theory · Fractals

12.1 Introduction

The symmetry properties of the mathematical formalism describing the system under study play a key role in physics. Symmetry is always related to the concept of transformation: if the motion equations describing the system evolution do not change their form when they undergo a certain group of transformations, then the dynamics described by those equations is said to be invariant or symmetric under such a group of transformations. Invariance under a group G of continuous transformations implies that the generators of G, i.e. the functionals describing the transformation process, are time independent (as prescribed by the Nöther theorem) (Itzykson and Zuber 1980). Knowing the constants of motion (the conserved charges) may be very helpful in finding some of the properties of the system under study. For example, invariance under time translation implies energy conservation,

G. Vitiello (✉)
Department of Physics "E.R. Caianiello" and Istituto Nazionale di Fisica Nucleare (INFN),
University of Salerno, Fisciano, Salerno, Italy
e-mail: vitiello@sa.infn.it

which means that the system is a closed, or isolated system: there is no exchange of energy between the system and other external systems (its environment).

The invariance under time evolution describes the robustness the system with respect to the flow of time, i.e. it is not affected by (it resists to) the flow of time. In the canonical (Lagrangian or Hamiltonian) formalism, such a robustness has the meaning of mechanical stability (conserved properties of the system). Things are, however, not always so simple.

In the following I discuss the dynamical rearrangement of symmetry and nonlinearity in open and far from the equilibrium systems (with possible extension to biological systems). The word robustness may then refer to a much wider spectrum of meanings than the one of mechanical stability.

Although it might appear unsatisfactory, I will adopt the physicist's attitude to avoid, as a rule, any philosophical or epistemological, in some wide sense, involvement in my discussion, and also any comparison or comment on different uses or meanings of words, which are technical terms in the present context, but might be of common use also in other disciplines. My only task is to illustrate in a qualitative way how robustness may be defined in the mathematical framework of quantum field theory (QFT) which is widely supported by laboratory observations. The comparative analysis of different definitions in different contexts is left to the interested reader.

12.2   A Two Level Description: Heisenberg Fields and Physical Fields

In the QFT of elementary particles and condensed matter physics, there are two levels of description: the dynamical level and the physical level (Umezawa 1993; Blasone et al. 2011). This last one is the level accessible to our observations, where observable properties (observables) of the system are detected by our measuring instruments. This is why we call it physical, or phenomenological level. At this level, we can observe ordered patterns and characterizing symmetries. We also assume that in regions far in space from the interaction region and much before and much after the interaction time (the asymptotic regions) the observed particles are non-interacting free particles. More generally, we assume that the elementary components of the system under study are characterized by specific individual properties, independent of the properties of the other components. In practice, however, it is never possible to switch off the interaction, and the assumption that particles (or other components of a system) can exist in a state of isolated object is only a convenient working hypothesis. The fields describing elementary components at this level are called physical fields. Their motion equations are linear equations since there are no interaction terms for physical fields.

The dynamical level is the one describing the interactions. We have no access to such a level with our observations, since then we would spoil the interaction with our interferences. For example, the photons or other quantum particles, used as probes to observe, e.g., an atom, may carry enough energy and momentum to perturb the state of the atom and the interactions among its elementary constituents. The fields describing interacting components are called Heisenberg fields. The dynamics is described in terms of nonlinear equations for the Heisenberg fields.

The objective of QFT is finding how it happens that the interaction of the Heisenberg fields generate the physical properties of the system described in terms of physical fields. We indeed remark that the linearity of the free field equations is far from being trivial since it describes how the nonlinear interaction manifests itself to us in the asymptotic region.

12.3 Spontaneous Breakdown of Symmetry and Dynamical Rearrangement of Symmetry

Suppose that the Heisenberg field equations are invariant under a group G of continuous transformations. It can happen that under specific boundary conditions the state of minimum energy of the system (the ground state or vacuum |0>) is not invariant under some or all the transformations of the group G. In such a case we say that *symmetry is spontaneously broken* (Itzykson and Zuber 1980; Umezawa 1993; Blasone et al. 2011).

One may then show that the equations for the free fields are invariant under the group of transformations G', with G' $\neq$ G. We then say that the *dynamical rearrangement of symmetry* (DRS) G $\rightarrow$ G' occurs (Umezawa 1993; Blasone et al. 2011).

For example this is the case of ferromagnets: below the critical temperature T_c (the Curie temperature), the vacuum state, characterized by a specific value of the magnetization M pointing in a certain direction, say z, is invariant under the group of the rotations around the z direction, but not under full group G of rotations around the axis x, y and z. QFT offers indeed infinitely many possible spaces of the system states and the vacuum associated to each of them may be characterized by different values of the magnetization, M, M', M'', etc., also including the vanishing magnetization $M = 0$ for $T > T_c$. In such a last case ($M = 0$, unbroken symmetry) the system behaves no more as a magnet, its vacuum state being invariant under the rotation group G around x, y and z axis.

We thus see the usefulness of distinguishing between invariance and symmetry: the fact that the dynamics is invariant under G means that the generators of G are time independent. This strongly characterizes the Heisenberg field equations at the dynamical level. At the physical level, however, it may happen that the original invariance G of the dynamics manifests itself in different physically observable realizations (ordered patterns, see below for the definition of order), each one characterized by a different group, say G', G'' $\neq$ G', G''' $\neq$ G' $\neq$ G'', and so on, all different from G, depending on different boundary conditions.

The dynamical process leading from the group G to G', G'', etc. is the morphogenesis process by which the observable patterns in physical systems are generated. Here the word morphogenesis is defined only within the strict mathematical definition of DRS in condensed matter physics and particle physics; it refers to crystal patterns, ferromagnetic domains, superconductor ordering, vortices, dislocations, coherent vacuum structure, Higgs induced mass generation, quark confinement, etc.[1]

The many possible realizations of the basic invariance of the theory may be thought of as a possibility of dynamical polymorphism (in the flat mathematical sense, namely with the limitations and within the above said strict definition for the word morphogenesis. See footnote 1.). These formal mechanisms do not depend on the specific physical system under study; they are common to elementary particle physics and to condensed matter physics and appear as characterizing properties of the QFT formalism. In such a sense, they represent a robustness feature of the QFT formal structure. A sort of meaningful aspect of the QFT mathematics (syntax), thus a semantic robustness of QFT. Also in this case, I stress that the words syntax and semantic are here used in the sense just specified, i.e. syntactic is here defined to refer to the elements of the mathematical apparatus of QFT, e.g. the equations, the fields, the functions and the distributions, the set of parameters entering the formalism, etc.; semantic is defined to refer to the construction of the QFT as a strategy to model the physical system of interest, its mathematical two level structure, the general mathematical mechanisms of invariance and symmetry above described. Footnote 1 applies to these words too. The reader may find the mathematical formalism clarifying their use in this context in QFT textbooks such as Bogoliubov et al. 1975; Schweber 1961; Umezawa et al. 1982; Blasone et al. 2011.

So far, few aspects of the QFT mathematical structure have been summarized, putting special emphasis on the invariance and symmetry properties of the Heisenberg field equations and of the physical fields equations. In the following sections we will see how the robustness of the QFT formalism provides a powerful tool to uncover robustness at various levels of the physical systems.

12.4 Boson Condensation, Ordered Patterns and Low Energy Theorem

Let me go back to the ferromagnet example. The existence of the privileged direction along which the magnetization points signals the loss of isotropy in the behavior of the magnet, like for the magnetic needle of a compass which points toward the North pole. The continuous rotational symmetry, corresponding to zero magnetization, is broken. The magnetization M is called the order parameter. In the presence of an external magnetic field, the magnet (the needle of the compass) gets aligned to it, in

---

[1]The word morphogenesis is thus defined by specific mathematical QFT quantities. Its relation with other definitions or uses, in a qualitative or quantitative way, in other disciplines is not object of discussion here. The comparison among different definitions in different context does not belong to the task of this paper and therefore I will not discuss it.

one specific direction. Ordering means the possibility to distinguish among (infinitely) many directions, all of them equivalent in the case of isotropy. Thus we see that order arises as lack of symmetry, it is the manifestation of symmetry breakdown.

The states which exhibit non-zero magnetization M are called non-symmetric, or ordered, ground states and are denoted by $|0>_M$. They are the states where a prevalent number of elementary constituent magnets point in a given direction (thus the states of ordered elementary components). The state with zero magnetization $|0>=|0>_{M=0}$ is called the symmetric, or disordered, or normal state.

Note that the symmetry breakdown considered here is not due to the addition of a symmetry breaking term to the dynamical equations (which is called explicit symmetry breakdown). The spontaneous breakdown of symmetry (SBS) mechanism occurs, instead, when the ground state is not invariant under the continuous group G under which the Heisenberg field equations are invariant. The system, driven by its own dynamics, spontaneously sets into one of the non-symmetric states compatible with the boundary conditions. This is why the breakdown of the symmetry is said to be spontaneous. The agent triggering the symmetry breakdown may be a weak stimulus, which can be switched off after symmetry breakdown has occurred.

The magnetization M acts as a label for the vacuum state: $|0>_M$, $|0>_{M'}$, etc. A change of a relevant boundary condition parameter, e.g. temperature, $T \to T'$, may induce a change in the magnetization value, $M \to M'$: in order to adjust itself to the changes of the boundary conditions the system moves from ground state to ground state, $|0>_M \to |0>_{M'}$. We say that the system undergoes a phase transition process. It moves from a dynamical regime to another different dynamical regime, with different values of the observables, e.g., $M \neq M'$. Systems with different magnetization have in fact different physical behaviors. The basic invariance at the dynamical level may thus manifest itself into a variety of stable ordered patterns at the physical level of the observables.

Since the generators of the invariance group G are time-independent, mathematical consistency requires that invariance cannot be lost and thus the equations for the physical fields must be invariant under the "rearranged" symmetry group G'. One can show that G' includes transformations that are translations of physical fields, say $B(x,t)$: $B(x,t) \to B(x,t) + c$, with c denoting a constant, and that the mass of these $B(x,t)$ fields, called the Nambu-Goldstone (NG) boson fields, or particles, must be zero (the NG theorem).

This result is of extraordinary importance. Ordered patterns are formed in the ground state of the system when symmetry is spontaneously broken. Correlation modes among the system elementary components, creating macroscopic patterns, must be then present in the system. In the case of ordered patterns extending over the whole system, the correlation modes or quanta should be able to propagate over distances of the order of the system size (they are thus called long-range correlation modes). One can show that they are the NG modes, indeed. Since they do not carry energy (have zero mass), at their lowest (zero) momentum, their condensation in the ground state does not modify its (zero) energy value. Different condensation densities correspond to different values of the order parameter and thus to different ground states or phases for the system. Changes in the condensation density produce phase transitions. It is known that these changes imply singularities and criticality (Hilborn 1994; Vitiello 2004a; De Lauro et al. 2009), also with appearance of avalanche-like and self-similarity phenomena. On the other hand, ordered patterns become observable when the NG modes are excited out of the ground state. One can show that the system is not affected by the excitation of low energy, i.e. long wavelength fields B(x,t). This is called "the low energy theorem" and it is actually a robustness theorem: the system is robust against actions exciting long wavelength perturbations in it. These perturbations are the ones that do not break the system in parts or pieces (the system is not fragile under such perturbations). For instance, one may apply a pulsed force to a crystal such that, below a certain threshold of the force intensity, the crystal does not get broken, but an elastic wave is excited in it, or, stated in different words, a number of low energy phonons (the NG quanta associated to the elastic wave) is excited (see any solid state physics textbook).

Summarizing, the transformations B(x,t) $\rightarrow$ B(x,t) + c acts as regulating the condensation of the NG modes in the ground state. It can be shown that this happens in such a way to recover the original invariance and to preserve the conservation of the constants of motion through the DRS G $\rightarrow$ G'. The regulation of boson condensation acquires thus the physical meaning of a dynamical stability principle and the DRS expresses the dynamical robustness of the system, namely its robust response to the symmetry breakdown aimed to recover the conservation laws of the basic dynamics.[2]

12.5   Coherence, Structure and Function

The generator (see Sect. 12.1 for its definition) of the boson condensation transformation B(x,t) $\rightarrow$ B(x,t) + c is actually the generator of coherent states. These states have an infinite (i.e. as large as one wants) number of elementary components oscillating in phase, i.e. with a constant phase difference (phase locking); they are thus related by a constant phase correlation spanning over the whole volume of the coherent pattern (Perelomov 1986; Umezawa 1993; Blasone et al. 2011, Auletta et al. 2009). Among the characterizing properties of coherent states, there is the one of being self-regenerative: by destroying or extracting from a coherent state one or more of their components, the state reconstructs itself so that coherence is preserved.

A large class of phenomena and systems is identified by the coherent boson condensation, from crystals to ferromagnets, superfluids, superconductors, in elementary particle physics and in a wide range of temperatures, ranging from thousands

---

[2]These conclusions hold also when a gauge field (e.g. in electrodynamics, in the electro-weak interaction of the standard model and in the quantum chromodynamics) is present. In such a case the condensation of NG bosons still characterizes the ground state and its robustness properties. The NG bosons do not belong to the spectrum of observable particles and the gauge field, which acquires a non-vanishing mass, is confined to self-focusing channels (the so called Anderson-Higgs-Kibble mechanism) (Itzykson and Zuber 1980; Umezawa 1993; Blasone et al. 2011). Such a self-focusing of the gauge field propagation is another manifestation of the robustness of the system in recovering the original dynamical invariance.

of degrees (diamond crystal melting point is 3545 °C, the cobalt magnetization is lost at 1075 °C) to hundreds of degrees (magnetization loss is at 770 °C for iron, the sodium chloride crystal (kitchen salt) melts at 804 °C), to very low temperature (below − 252 °C in superconductors containing compounds of niobium, −153 °C for copper and bismuth compounds). Such a class of phenomena and systems exhibit observable ordered patterns and physical behaviours which are strictly dependent on the specific coherent dynamics ruling their large multitude of elementary components (Umezawa 1993; Blasone et al. 2011).

Note that the NG bosons are real quanta of correlation waves and they interact with other components of the system. Therefore, they enter in the list of the system basic components, which specifies the system structure; although of dynamical origin, they do belong to the system structure. In fact, for example, the crystal of NaCl is not made solely by the atoms of sodium Na and chlorine Cl, but also by the phonon quanta, which are the NG quanta condensed in the crystal ground state and are responsible of the crystal coherent ordered pattern. Without the phonons there is no NaCl crystal. The phonons belong thus to the system structure. They are also responsible of the crystal physical behaviour, e.g. the crystal stiffness, the crystal heat and electrical conductibility, etc. There are many ways one can use a crystal in practical applications. A crystal may function as an insulator, or may function as an optical device, etc., with different performances depending on the density of condensed phonons (a pure crystal or a non-homogeneous one, etc.), on the working temperature, etc. The phonons thus determine also the system function. These considerations are quite general and with the necessary changes (instead of phonons, we might consider the magnons, which are NG quanta of the spin wave in a magnet, etc.) they might be extended to the whole class of systems above identified. I stress that the definitions of structure and the one of function are the ones referred above within the limitations and the context of the present discussion. To them applies what stated in footnote 1.

In conclusion, in QFT it is not possible to separate the notions of structure and function; the dichotomy and the antinomy between them vanishes (Vitiello 1998). Therefore, the dynamical stability principle which guaranties the robustness of structure also guaranties the robustness of functions and vice versa. This is a mathematical QFT result, experimentally fully confirmed, by which we can compute the values of physical quantities determining the system functional properties (Vitiello 1998; Blasone et al. 2011).

12.6 Coherence and Change of Scale, from Micro to Macro

Another consequence of the fact that SBS produces coherent patterns in the ground state is the transition from the microscopic dynamical level to the level of the macroscopic behavior of the system. This is the transition from the behavior of a collection of a multitude of individual components to their behavior as a system (Vitiello 2001). The question of how such a transition may happen is of great scientific interest. In QFT one may derive the dynamical formation of large scale ordered patterns as the macroscopic manifestations of the microscopic quantum dynamics of the elementary components (Blasone et al. 2011). The multitude of elementary components defines what in the present case is termed the syntactic level. The macroscopic manifestations of their coherent dynamics defines the semantic level. The usual limitations and the strict mathematical definitions of these terms apply also in this case (cf. first footnote of this chapter).

The linear size over which the correlations extend, much greater than the size of the elementary components, determines the macroscopic linear size of the system as a whole. The system macroscopic behaviour cannot be derived as the sum of the behaviours and properties of the elementary components, by adding small contributions in order to obtain a finite result out of many perturbing effects, as one would proceed in perturbative physics. It is a nonlinear phenomenon not derivable by means of the perturbative formalism. For example, system properties of magnets and crystals, such as the magnetization, the electric conductivity, the elasticity, etc., are not properties of the individual atomic or molecular components. The physical behaviour of these systems can be mathematically derived only by recurring to the microscopic quantum dynamics. In this sense, they are called macroscopic quantum systems.

The nonlinearity of the basic dynamics thus leads to the formation of coherent domains or tissues and other macroscopic patterns (forms) in which the components are organized or, more specifically, self-organized, as a result of the non-perturbative dynamics. The final output, as we have seen in the previous sections, are the functional properties of the system (its being a crystal, a magnet, a metal with electric conductivity, etc.). These functional properties are then protected by the basic invariance of the theory and their stability or persistence (life-time) may extend on a much longer time scale than the ones of the microscopic interaction of the components. This might be of particular relevance in far from the equilibrium physical systems, where one has to cope with several levels of organization, with different time-scales and space-scales. All these features have suggested to apply the paradigm of QFT here discussed to biology (Del Giudice et al. 1983–2006; Vitiello 1995, 2001; Kurian et al. 2016) and to neuroscience (Freeman 1975, 1991, 2014; Freeman and Quian Quiroga 2013; Freeman and Vitiello 2006, 2016; Pessa and Vitiello 2003, 2004; Vitiello 2015, 2016).

In both these sectors, many results have been obtained by using the tools provided by statistical and stochastic analysis, also using numerical simulations. Statistical and stochastic mechanics are of course vital tools in the physicist's tool-box, together with commutative and non-commutative algebraic and group theoretical methods, functional integration methods, quantum mechanics and QFT. One has to use, case by case, the proper tool or the proper set of tools, depending on the system under study. Some of these tools focus mostly on the kinematical and probabilistic properties of the system, other ones on the formal properties of the set of operator functionals describing the system. Quantum mechanics and QFT focus on the dynamical properties. Thus, it may happen that in complex systems the analysis pursued by use of some tools, although necessary in order to treat some specific aspects of the system, may result to be not sufficient for a comprehensive description of the overall cooperative features of the various organization levels and domains of the system.

The QFT attempts describing the dynamical phenomenon of coherence, may lead to results hard to be reached by purely statistical, stochastic and algebraic approaches. The QFT approach is not aimed of course to substitute or negate these approaches. On the contrary, it is aimed to provide the dynamical scenario underlying the great amount of results obtained in other approaches, the great richness of the phenomenology and the structured heterogeneity of the systems. These cooperative efforts coming from different approaches may be much fruitful for example in the study of the living phase of matter (Caianiello and Bertolaso 2016).

In such a perspective it is extremely appealing the proposal put forward by Walter J. Freeman in neuroscience (Freeman 1975, 1991; Freeman and Quian Quiroga 2013; Kozma and Freeman 2016), in postulating meaningfulness as a key ingredient in the action-perception cycle in brain activity. The concept of meaning is associated by Freeman to the amplitude modulated (AM) neuronal patterns, commonly observed by EEG, ECoG, fMRI, and other techniques. These AM patterns are assemblies of myriads of coherently oscillating neurons, which form under the incoming stimuli (inputs) to the brain. It is remarkable that many formal connections, also supported by laboratory observations, can be made with the mathematics of QFT illustrated up to now (Freeman and Vitiello 2006, 2016; Freeman et al. 2012). For instance, by translating in mathematical terms the relation proposed by Freeman in neurosciences between coherent neuronal patterns and meaning, one may define mathematically in the QFT framework the association between ordering and meaning. The dynamical formation of collective modes in ordered states then becomes the dynamical formation of meanings. The functional robustness discussed in previous sections becomes then semantic robustness (again the observation of footnote 1 applies to the case of the present definitions) and the self-regenerative property of the coherent state against destruction of part of the population of its elementary components appears then as robustness of the meanings against variations, or corruption, or deletion of some of the syntactic elementary components.[3] The mathematical details at the basis of these statements can be found by the interested reader in the above cited publications.

12.7 Topological Robustness and Robustness Against Noisy Quantum Fluctuations

So far we have considered the field translation $B(x,t) \rightarrow B(x,t) + c$, with c being a complex constant number, independent from space and time. In such a case there is homogeneous condensation in the ground state and the modulus square of c, $|c|^2 = c^*c$, represents the number of condensate modes in the ground state. A proper mathematical definition of the generator of such a $B(x,t)$ translation requires that

---

[3] Such an observation has led to a possible mathematical formulation, within the Chomsky Generative Grammar approach, of the formation of concepts out of syntactic elements in linguistics (Piattelli-Palmarini and Vitiello 2015).

one must consider first the transformation $B(x,t) \rightarrow B(x,t) + c \cdot f(x,t)$. Here $f(x,t)$ is a solution of the motion equation for $B(x,t)$, thus preserving the invariance of the equations under such a transformation. After the computation is done, one considers the limit $f(x,t) \rightarrow 1$, so that the homogeneous condensation is recovered. It is however interesting to consider also cases in which $f(x,t)$ is not reduced to 1. In such a case, the condensate density depends on x and t through $|f(x,t)|^2$, and we have non-homogeneous boson condensation, namely we have domains where condensation occurs whose form and distribution is controlled by $f(x,t)$ (acting as a form factor), and regions where there is no boson condensation (where $f(x,t) = 0$). A very interesting scenario, indeed, which may also be characterized by non-trivial topology. The function $f(x,t)$ may in fact carry divergent or topological singularities which reflect on the topological properties of coherent domains. In this way a number of topologically non-trivial solutions, such as vortices, domain walls, etc., may be obtained. These solutions are characterized by conserved topological charges.[4] They cannot decay in the vacuum since the topological charge of the vacuum is zero (it is topologically trivial): we have topological robustness.

Finally, I remark that the average of the number N of condensed modes in the coherent state, say $|\alpha\rangle$, where α denotes the coherence strength, is given by $|\alpha|^2$: $\langle N \rangle = |\alpha|^2$. The average of the quantum fluctuations ΔN on the number N in the coherent state is given by $|\alpha|$. Therefore, $\langle \Delta N \rangle / \langle N \rangle = 1/|\alpha|$. This means that for large coherence strength $|\alpha|$ quantum fluctuation are negligible with respect to average condensate number, namely the system behavior is not affected by quantum fluctuation, it behaves as a classical system. This is why the order parameter (such as the magnetization M) is a classical field, its values do not depend on quantum fluctuations. Coherence of the boson condensation thus guaranties the classical behavior of the macroscopic quantum system, or, in other words, its robustness against noisy quantum fluctuations. Examples of such kind of systems are magnets, crystals, superconductors, laser beams, etc. Their macroscopic physical properties may also be described as averages in terms of statistical or stochastic methods, but these provide only an approximate description, a phenomenological one, without delving into the depth of the microscopic dynamic origin of the systems macroscopic properties. Descriptions in terms of statistical and stochastic averages, if not supplemented by QFT, might be necessary, in view of difficulties due to the system complexity, heterogeneity, etc., but are often not sufficient to explain the dynamic origin of the high stability of the macroscopic functional properties of the mentioned systems, as above quantitatively defined (conductivity, electrical current flows, optical properties, reactivity to changes in temperature, to radiations or other inputs from the environment, etc.) (Anderson 1984; Umezawa et al. 1982). These conclusions are drawn for the physical systems so far considered (condensed matter and elemen-

---

[4]An illustrative example of topological charge is the one of the kink solution in one-dimensional nonlinear dynamics. The kink is characterized by the hyperbolic tangent tgh x, whose limits at $x \rightarrow \pm\infty$ are ± 1, respectively, so that the topological charge Q is in this case defined by $Q = \frac{1}{2}(\text{tgh } x\,|_{+\infty} - \text{tgh } x|_{-\infty}) = 1$. In the case of a vortex solution the topological charge is given by the number of "turns" around the vortex core and is called the *winding number*.

tary particle systems with ordered patterns) and are fully supported by the experimental observations. Framed in such a contemporary scenario, Schrödinger (1944) quest of a basic quantum dynamics for living systems, in his *What is life?*, still preserves its full legitimacy.

12.8 Dissipation, Chaos and Fractal Self-Similarity

One further feature of boson condensation is the fractal-like self-similarity of coherent states. Self-similarity is one of the characterizing properties of fractals (the "most important property" of fractals (p. 150 in Peitgen et al. (1986)). Such a property has been framed in the theory of entire analytical functions and their functional realization expressed in terms of (squeezed) coherent states (Vitiello 2009, 2012, 2014). The analysis is supported by many experimental observations (Capolupo et al. 2014; Chen et al. 2010); for instance, fractal patterns are observed in crystals submitted to deforming stress actions at low temperature; they appear as the result of non-homogeneous coherent phonon (boson) condensation providing an example of emergence of fractal dislocation structures (Chen et al. 2010) in non-equilibrium dissipative systems (Vitiello 2009, 2012, 2014). Fractals provide a further example of macroscopic quantum systems, in the specific sense that their scale-free macroscopic properties cannot be derived without recurring to the underlying quantum dynamics. The quantum dynamics thus controls the morphogenesis processes of the fractal macroscopic growth. Interesting specific examples, which I do not report here for brevity, are provided by applications in condensed matter physics, neuroscience and biology. The interested reader can find details in Capolupo et al. (2013, 2014), Kurian et al. (2016), Freeman, et al. (2012), Loppini et al. (2014), Atmanspacher (2015).

Scale-free behavior denotes invariance of the dynamics under scale transformations. The conclusion is that coherence is then related to robustness of the dynamics against scale transformations. At each scale the dynamics is the same. In some sense, we have a multilevel robustness within the unifying, i.e. scale-free, coherent dynamics. Perhaps from here it could emerge an understanding of the reasons of the ubiquitous diffusion of fractals in nature. In addition to the experimental observations of self-similarity in elementary particle, condensed matter physics and fractal-like systems, further investigation is however needed in order to fully substantiate such a unified ecological vision.

I further observe that in dissipative systems, mathematical consistency requires the balance of the energy fluxes between the system and its environment. Formally this is achieved by doubling the operator algebra A describing the system: $A \rightarrow A \times A$ (Celeghini et al. 1992). The environment is then described as the system Double, whose dynamics is the time-reversed of the system dynamics, since the energy flux outgoing from the system is incoming into the environment, and vice versa.

The same structure of QFT is based on such a doubled algebraic structure, which is called Hopf algebra. I do not insist in discussing the mathematical details of the formalism, which can be found in Celeghini et al. (1998) (see also Blasone et al. 2011). I only remark that the open system is permanently coupled to its environment (its Double). The dynamics of the system and the one of the environment are each other entangled in such a way that one is the reference of the other. One is open and actually depends on the other one, reciprocally. From a computational point of view, such a dynamical self-consistency contains in itself a truth-evaluation function, which is continuously updated in the realization (optimization) of the fluxes balances. We thus realize one further form of robustness which crucially characterizes physical open systems: the truth (reliability) or the fallaciousness (falsity) of the system behavioral processes is not established by comparisons with predetermined, external catalogs of standards (protocols, control procedures, etc.). On the contrary, it is internal to the system-environment dynamical relationship. Far from being self-referential (typical of closed systems), the open system is forced by its same openness to a continuous updating of its dynamics in the unavoidable interaction with its Double (Vitiello 2004b, 2001; Piattelli-Palmarini and Vitiello 2015). The identification of the system, in quantitative mathematical terms, is not solely in its structural/functional persistence; it incorporates the differences implied by the evolutionary continual redefinition of the truth/fallacy in the always renewed relation with the environment. For an analysis in theoretical computer science see Basti et al. (2017).

12.9　Conclusions

We have seen that in QFT there might be several types of robustness (r.), all of them derived from mathematical properties of symmetry, its spontaneous breakdown and its dynamical rearrangement. The list of considered types includes: mechanical, semantic, dynamical, functional, topological r., r. against noisy quantum fluctuations, r. against scale transformations, multilevel r., r. in the evolutionary continual re-definition of the truth, with the mathematical specifications and limitations declared in several points of the text and in footnote 1.

The coherent structure of the QFT vacuum manifests itself in the auto-similarity properties of fractal structures for a large number of physical systems, in experimental observations in particle physics and condensed matter physics. Within such a class of systems and phenomena, and within the limitations of the definitions introduced above in our discussion, the vision of nature sectioned in separated domains is replaced by the vision of nature unified by the dynamical paradigm of coherence ruling morphogenetic processes. The appearance of forms through coherence becomes the formation of meanings. We have seen in Sect. 12.6 that in neuroscience the formation of meanings out of perceptual experience has been proposed to play a central role in the action-perception cycle in brain activity (Freeman 1975, 1991; Freeman and Quian Quiroga 2013; Kozma and Freeman 2016). In particular, the formation of meaning is associated by Freeman to the dynamical formation of collective modes in ordered amplitude modulated neuronal patterns, observed e.g. by EEG, ECoG, fMRI, etc.

The duplication of the DNA macromolecular chain through PCR (polymerase chain reaction) seems also to confirm the key role played by coherence in the formation of *codes*. The coherent structure of the electrical dipole waves mediating the interaction between DNA and enzyme in conventional PCR processes suggests that DNA is the vehicle through which coherence propagates in living matter (Kurian et al. 2016; Saiki et al. 1988; Montagnier et al. 2011–2017).

Of course, I am considering here the familiar DNA duplication by the widely used PCR as a physical and chemical many-body experiment, not in its biological implications. The consolidated physicist attitude is always inclusive, in the direction to add a novel perspective, on the basis of the mathematical apparatus supported by experimental evidence, not to substitute methods and results obtained by use of different tools. As stressed in previous sections, statistical and stochastic tools, for example, are absolutely necessary, even when they seem not sufficient in filling the gap between the microscopic components behavior and the global macroscopic properties of physical systems. A very dangerous mistake would be indeed the one of reciprocal exclusion and competition among different approaches and investigation tools. In such a respect, it is much instructive to report a comment by Freeman (2014) on the present state of brain studies: "… Cognitive neuroscience is in disarray. […] I cannot emphasize the depth and intensity of this conflict strongly enough. The battle has been on for half a century between those defending the neural pulse paradigm, also commonly called 'the Neuron Doctrine'[5] in a move to take the high ground, versus those conceiving neural activity as alternating between great clouds of pulses on axons and matching waves of dendritic current. The stakes are high, because there is growing awareness among adherents to the neural pulse paradigm that neurocognitive operations are conducted by millions and even billions of neurons, simultaneously. Their temptation is to perseverate by up-scaling existing techniques for recording pulses to match their sampling needs. As a practical matter the requisite hardware would be incompatible with brain survival, as any neurosurgeon will attest." In order to stress how deep and how general, not limited to neurosciences, is Freeman lesson inviting to consider not only "great clouds of pulses", but also "matching waves", it is then worth to recall the famous Gramsci passage: …in "a scientific discussion…the most 'advanced'[6] thinker is he who understands that his adversary may express a truth which should be incorporated in his own ideas, even if in a minor way. To understand and evaluate realistically the position and the reasons of one's adversary (and sometimes the adversary is the entire thought of the past) means to have freed oneself from the prison of ideologies, in the sense of blind fanaticism." (Gramsci 1932).

---

[5] Quotation marks in the original text.

[6] Quotation marks in the original text.

Acknowledgments I thank the anonymous referee and the Editors of this volume for their suggestions and help. Of course, the responsibility of statements and positions presented in this paper is only mine. I am glad to dedicate this work to the memory of Walter J. Freeman in the occasion of one year since his departure and to celebrate the wisdom of Antonio Gramsci on the 80th year since his tragic death.

References

Anderson, P. W. (1984). *Notions of condensed matter physics*. Menlo Park: Benjamin.

Atmanspacher, H. (2015). Quantum approaches to consciousness. *Stanford Encyclopedia of Philosophy*. http://plato.stanford.edu/entries/qt-consciousness/

Auletta, G., Fortunato, M., & Parisi, G. (2009). *Quantum mechanics*. Cambridge: Cambridge Univ. Press.

Basti, G., Capolupo, A., & Vitiello, G. (2017). Quantum field theory and coalgebraic logic in theoretical computer science. *Progress in Biophysics and Molecular Biology, 130 A*, 39–52 arXiv:1701.00527v1 [quant-ph].

Blasone, M., Jizba, P., & Vitiello, G. (2011). *Quantum field theory and its macroscopic manifestations*. London: Imperial College Press.

Bogoliubov, N., Logunov, A. A., & Todorov, I. T. (1975). *Introduction to axiomatic quantum field theory*. Reading: W. A. Benjamin, Advanced Book Program.

Caianiello, S., & Bertolaso, M. (2016). Robustness as organized heterogeneity. *Rivista di filosofia Neo-scolastica, 2*, 293–303.

Capolupo, A., Freeman, W. J., & Vitiello, G. (2013). Dissipation of 'dark energy' by cortex in knowledge retrieval. *Physics of Life Reviews, 10*, 85–94.

Capolupo, A., Del Giudice, E., Elia, V., Germano, R., Napoli, E., Niccoli, M., Tedeschi, A., & Vitiello, G. (2014). Self-similarity properties of nafionized and filtered water and deformed coherent states. *International Journal of Modern Physics B, 28*, 1450007 20 pages.

Celeghini, E., Rasetti, M., & Vitiello, G. (1992). Quantum dissipation. *Annals of Physics, 215*, 156–170.

Celeghini, E., De Martino, S., De Siena, S., Iorio, A., Rasetti, M., & Vitiello, G. (1998). Thermo field dynamics and quantum algebras. *Physics Letters A, 244*, 455–461.

Chen, Y. S., Choi, W., Papanikolaou, S., & Sethna, J. P. (2010). Bending crystals: Emergence of fractal dislocation structures. *Physical Review Letters, 105*, 105501.

De Lauro, E., De Martino, S., Falanga, M., & Ixaru, L. G. (2009). Limit cycles in nonlinear excitation of clusters of classical oscillators. *Computer Phys. Comm., 180*, 1832–1838.

Del Giudice, E., & Vitiello, G. (2006). The role of the electromagnetic field in the formation of domains in the process of symmetry breaking phase transitions. *Physical Review A, 74*, 022105.

Del Giudice, E., Doglia, S., Milani, M., & Vitiello, G. (1983). Spontaneous symmetry breakdown and boson condensation in biology. *Physics Letters A, 95*, 508–510.

Del Giudice, E., Doglia, S., Milani, M., & Vitiello, G. (1985). A quantum field theoretical approach to the collective behavior of biological systems. *Nuclear Physics B, 251*(FS 13), 375–400.

Del Giudice, E., Doglia, S., Milani, M., & Vitiello, G. (1986). Electromagnetic field and spontaneous symmetry breaking in biological matter. *Nuclear Physics B, 275*(FS 17), 185–199.

Del Giudice, E., Preparata, G., & Vitiello, G. (1988a). Water as a free electric dipole laser. *Physical Review Letters, 61*, 1085–1088.

Del Giudice, E., Manka, R., Milani, M., & Vitiello, G. (1988b). Non-constant order parameter and vacuum evolution. *Physics Letters, 206B*, 661–664.

Freeman, W.J. (1975/2004). *Mass action in the nervous system*. New York: Academic.

Freeman, W. J. (1991). The physiology of perception. *Scientific American, 264*, 78–85.

Freeman, W.J. (2014). Review of the book by K. H. Pribram, *The form within: My point of view.* Westport: Prospecta Press. 2013.

Freeman, W. J., & Quian Quiroga, R. (2013). *Imaging brain function with EEG.* New York: Springer.

Freeman, W. J., & Vitiello, G. (2006). Nonlinear brain dynamics as macroscopic manifestation of underlying many-body dynamics. *Physics of Life Reviews, 3,* 93–118.

Freeman, W. J., & Vitiello, G. (2016). Matter and mind are entangled in two streams of images guiding behavior and informing the subject through awareness. *Mind & Matter, 14*(1), 7–24.

Freeman, W. J., Livi, R., Obinata, M., & Vitiello, G. (2012). Cortical phase transitions, non-equilibrium thermodynamics and the time-dependent Ginzburg-landau equation. *International Journal of Modern Physics B, 26,* 1250035.

Gramsci, A. (1932). *Quaderni del carcere.* Quad. N. 10, 1932–35, V. Gerratana (Ed.), Einaudi, Torino 1977, p. 1263 (*The open Marxism of Antonio Gramsci.* Translated and annotated by C. Marzani, Cameron Associates, New York 1957).

Hilborn, R. (1994). *Chaos and nonlinear dynamics.* Oxford: Oxford University Press.

Itzykson, C., & Zuber, J. (1980). *Quantum field theory.* New York: McGraw-Hill.

Kozma, R., & Freeman, W. J. (2016). *Cognitive phase transitions in the cerebral cortex – Enhancing the neuron doctrine by modeling neural fields.* Cham: Springer.

Kurian, P., Capolupo, A., Craddock, T. J. A., & Vitiello, G. (2016). Water-mediated correlations in DNA-enzyme interactios. *Physics Letters A, 382*(1), 33–43 arXiv:1608.08097.

Loppini, A., Capolupo, A., Cherubini, C., Gizzi, A., Bertolaso, M., Filippi, S., & Vitiello, G. (2014). On the coherent behavior of pancreatic beta cell clusters. *Physics Letters A, 378,* 3210–3217.

Montagnier, L., Aissa, J., Del Giudice, E., Lavallee, C., Tedeschi, A., & Vitiello, G. (2011). DNA waves and water. *Journal of Physics: Conference Series, 306,* 012007.

Montagnier, L., Del Giudice, E., Aïssa, J., Lavallee, C., Motschwiller, S., Capolupo, A., Polcari, A., Romano, P., Tedeschi, A., & Vitiello, G. (2015). Transduction of DNA information through water and electromagnetic waves. *Electromagnetic Biology and Medicine, 34,* 106–112.

Montagnier, L., Aïssa, J., Capolupo, A., Craddock, T. J. A., Kurian, P., Lavallee, C., Polcari, A., Romano, P., Tedeschi, A., & Vitiello, G. (2017). Water bridging dynamics of polymerase chain reaction in the gauge theory paradigm of quantum fields. *Water, 9*(5), 339 Addendum, *Water, 9*(6), 436.

Peitgen, H. O., Jürgens, H., & Saupe, D. (1986). *Chaos and fractals. New frontiers of science.* Berlin: Springer.

Perelomov, A. (1986). *Generalized coherent states and their applications.* Berlin: Springer.

Pessa, E., & Vitiello, G. (2003). Quantum noise, entanglement and chaos in the quantum field theory of mind/brain states. *Mind and Matter, 1,* 59–79 arXiv:q-bio.OT/0309009.

Pessa, E., & Vitiello, G. (2004). Quantum noise induced entanglement and chaos in the dissipative quantum model of brain. *International Journal of Modern Physics B, 18,* 841–858.

Piattelli-Palmarini, M., & Vitiello, G. (2015). Linguistics and some aspects of its underlying dynamics. *Biolinguistics, 9,* 96–115.

Saiki, R. K., Gelfand, D. H., Stoffel, S., Scharf, S. J., Higuchi, R., Horn, G. T., Mullis, K. B., & Erlich, H. A. (1988). Primer-directed enzymatic amplification of DNA with a thermostable DNA polymerase. *Science, 239,* 487–491.

Schrödinger, E. (1944). *What is life?* Cambridge: Cambridge University Press (1967 reprint).

Schweber, S. S. (1961). *An introduction to relativistic quantum field theory.* New York: Harper and Row.

Umezawa, H. (1993). *Advanced field theory: Micro, macro and thermal concepts.* New York: AIP.

Umezawa, H., Matsumoto, H., & Tachiki, M. (1982). *Thermo field dynamics and condensed states.* Amsterdam: North-Holland Pub. Co.

Vitiello, G. (1995). Dissipation and memory capacity in the quantum brain model. *International Journal of Modern Physics B, 9,* 973–989.

Vitiello, G. (1998). Structure and function. An open letter to Patricia Churchland. In S. R. Hameroff, A. W. Kaszniak, & A. C. Scott (Eds.), *Toward a science of consciousness II* (pp. 231–236). Cambridge: MIT Press.

Vitiello, G. (2001). *My double unveiled*. Amsterdam: John Benjamins Publ. Co.

Vitiello, G. (2004a). Classical chaotic trajectories in quantum field theory. *International Journal of Modern Physics B, 18,* 785–792.

Vitiello, G. (2004b). The dissipative brain. In G. Globus, K. H. Pribram, & G. Vitiello (Eds.), *Brain and Being. At the boundary between science, philosophy, language and arts* (pp. 315–334). Amsterdam: John Benjamins Publ..

Vitiello, G. (2009). Coherent states, fractals and brain waves. *New Mathematics and Natural Computation, 5,* 245–264.

Vitiello, G. (2012). Fractals, coherent states and self-similarity induced noncommutative geometry. *Physics Letters A, 376,* 2527–2532.

Vitiello, G. (2014). On the isomorphism between dissipative systems, fractal self-similarity and electrodynamics. Toward an integrated vision of nature. *Systems, 2,* 203–216.

Vitiello, G. (2015). The use of many-body physics and thermodynamics to describe the dynamics of rhythmic generators in sensory cortices engaged in memory and learning. *Current Opinion in Neurobiology, 31,* 7–12.

Vitiello, G. (2016). (2016). Filling the gap between neuronal activity and macroscopic functional brain behavior. In R. Kozma & W. J. Freeman (Eds.), *Cognitive phase transitions in the cerebral cortex – Enhancing the neuron doctrine by modeling neural fields* (pp. 239–249). Cham: Springer.

Giuseppe Vitiello is Professor of Theoretical Physics at the University of Salerno, Department of Physics and Istituto Nazionale di Fisica Nucleare (INFN). His research activity is focused on elementary particle physics and the physics of living matter and brain. He is author of about 140 research papers, many reports to International Conferences and of the books: Quantum Mechanics, coauthored with H. Umezawa (Bibliopolis, Napoli 1985 and, translated in Japanese, Nippon Hyoron Sha. Co.Ltd., Tokyo, Japan 2005), My Double unveiled (John Benjamins Publ. Co., Amsterdam 2001). Quantum field theory and its macroscopic manifestations, co-authored with Massimo Blasone and Petr Jizba (Imperial College Press, London 2011). Together with Gordon Globus e a Karl Pribram, he is editor of the book Brain and Being. At the boundary between science, philosophy, language and arts. (John Benjamins Publ. Co., Amsterdam, 2004).

Chapter 13
Difference and Robustness: An Aristotelian Approach

Alfredo Marcos

Abstract The paper starts by recalling the ordinary and etymological sense of the word "robustness", for placing it then in the context of the current systemic view. Then I focus discussion on systems robustness in the paradigmatic case of living organisms. We discover that the notion of robustness is closely linked, in the case of organism, with the notion of difference, given that organisms arise precisely by a differentiation process (Sect. 13.1). So, in order to understand the ontology of robustness, we need to explore the ontology of difference; and in order to do so, we must distinguish between constitutive and comparative difference (Sect. 13.2.1). Then I deal with the problem of unity of the constitutive differences: I wonder if it is possible to unify the many differences that an organism exhibits in a single difference. It is an important question, given that unity is one of most essential characteristics of the organism (Sect. 13.2.2). And the answer to this question raises immediately a query for the intelligibility of this final and unique constitutive difference (Sect. 13.2.3). Such intellibility is possible thanks to the formal nature of the final difference, but it requires also a pluralistic approach. Besides that, we have to sketch the ontological and epistemological relationships between difference, identity and similarity (Sect. 13.3), which will be crucial for intelligibility of the difference, since according to a certain tradition, intelligibility depends on identity.

Keywords Robustness · Constitutive difference · Comparative difference · Identity · Similarity

A. Marcos (✉)
University of Valladolid, Valladolid, Spain
e-mail: amarcos@fyl.uva.es

© Springer Nature Switzerland AG 2018
M. Bertolaso et al. (eds.), *Biological Robustness*, History, Philosophy
and Theory of the Life Sciences 23, https://doi.org/10.1007/978-3-030-01198-7_13

13.1 Introduction: Robustness and Difference

"Robustness" is usually defined as the condition of being robust. The word "robust" itself derives from the Latin "robur", meaning "oak". The oak tree can be said to be paradigmatically robust. In English, we have the common expression "strong as an oak", and in other languages, such as Spanish, similar expressions exist. The contrary of "robust", in ordinary language, is expressed by words like "weak", "fragile", "vulnerable" or even "sickly". In more abstract terms, we say that something is robust if it is able to maintain its identity, unity and correct functioning in the face of internal and external stressors. Something is fragile or sickly if it breaks down or comes apart as a result of minor perturbations.

The term "robustness" has recently entered into the language of science, as a result of the influence of systems theory. In this context, "robustness" refers to a condition proper to systems. Not that there are no fragile systems, yet every system has one degree or another of robustness. Indeed, robustness is a condition that is by definition present in all systems, to a greater or lesser degree. A system with zero robustness would be, in reality, not even a system, since it would completely lack identity, permanence over time and functional stability. It would be a mere random and ephemeral aggregate of elements.[1] In contemporary science, the notion of robustness is completely linked to that of system. In fact, both notions apply to the same domains. Wherever we speak of systems (economic, social, conceptual, mathematical, engineering, informational, chemical, biological, etc.) we also speak of robustness. Clearly, the nuances that this term acquires in each of these domains are in function of the characteristics possessed by the systems in question.

An economic system is not the same as a computational or biological system. Even within each class of system we find distinct subtypes. Within the range of biological systems, for instance, an ecosystem is not the same as an organism. As a result, robustness is not the same across all of the systems in question. To be more specific, they all possess analogous formal characteristics, simply by the fact that all are systems and are viewed from the perspective of general systems theory. But there are important ontological distinctions among them. A computational system, as well as, for instance, a building and a city, ontologically depends on the actions and intentions of human beings, while a biological system can maintain its identity even if no human being is observing it. A conceptual system is, obviously, more abstract than an ecosystem, and we can similarly enumerate other ontological differences between system types. In Aristotelian terms, we can say that a computational system, for instance, is only *ousia* in an accidental sense, as well as a building, while a conceptual system is so in a secondary sense, while a biological system (and above all an organism) is *ousia* in a full and paradigmatic sense.

What I propose here, in view of the great diversity in possible types of system, is to begin—from the ontological point of view—with the study of the paradigmatic

---

[1] On the other hand, the degree of robustness of a system is always finite, since otherwise the system would be eternal, which cannot be said even of the entirety of the universe as a whole.

case of a robust system, i.e. that of a living organism (Nicholson 2014, 347–359), such as an oak. On that basis, and proceeding via analogy, we may perhaps come to be able to study in an organized fashion the characteristic of robustness in other systems of highly diverse types. In other words, the general principles that determine the robustness of a system, which are paradigmatically present in organisms, will require an analogical interpretation and application in other types of systems. In the Aristotelian tradition, the most relevant ontological concepts have an analogical application. To begin with, this happens with the concept of "being", which according to Aristotle is said in many ways. It can be said with regards to an animal, a color or a number (an animal *is*, a color *is*, a number *is*), but it is said differently in each case. That is, the term is not univocal. But neither is it equivocal, for all these entities, each in its own way, in effect, *are*. What mediates between univocity and equivocality is analogy. An analogy establishes a certain proportionality. Thus, between the lens of the human eye and the lens of a photographic camera there is a relation of analogy, since they fulfill the same function in different systems, despite all the differences that may exist between these two entities, by their origin, their material composition, durability, size... Similarly, when we apply the term "robustness" to both an oak and a computer we do so analogically.

Organisms, then, constitute themselves through a process of differentiation, also called development or ontogenesis. In this context, the term "canalization" is sometimes used as a synonym of robustness. Waddington popularized the use of this term in order to speak of developmental stability in the presence of certain perturbations. During ontogenesis, the organism maintains its identity and functionality despite internal and external perturbations. It also maintains its unity despite the fact that numerous differences or distinct characteristics arise out of what was originally an undifferentiated cell. Organisms, then, exhibit a notable robustness during their development. If we define robustness - as we have done - in terms of the ability of a system to maintain its identity, then it seems clear that any attempt to study the robustness of organisms ontologically will require us to think about the notion of difference, since the constitution of an organism is achieved precisely through a process of differentiation, and since its identity is maintained—or is even attained—in and through its internal differences. It is also reasonable to conjecture that the concept of difference will also play a crucial role in the confluence zone between evolutionary biology and developmental biology, currently called evo-devo (Amundson 2005).

13.2 Aristotle and the Ontology of Difference

Some contemporary authors, such as James Lennox (Lennox 2017), think that the rescue of certain Aristotelian concepts (difference, identity, form, function...) may be advantageous for contemporary science, and especially for biology. So, the concept of difference might play an important role in current debates. Actually, the emphasis on the notion of difference places us into the atmosphere that is typical of postmodern philosophy. The Heideggerian critique of the forgetting of difference, as well as his own emphasis on the importance of that notion (Heidegger 2002), was immediately picked up by other contemporary philosophers, such as Deleuze (1968), Lyotard (1983) and Derrida (1967). And with this revitalization of the topic, we also encounter an important difficulty, which is that of the intelligibility of difference. There is an older tradition according to which only identity is truly intelligible. If we seek to grasp the constitution of an organism by way of the notion of difference, we may run the risk of not understanding what an organism is.

Allow me, then, to discuss the notion of difference at this point. Perhaps, once we have reconsidered it, we can employ it to understand the ontology and characteristics of organisms, including their robustness, and will be able to employ this case as a paradigm for comprehending it in other systems. We may even be able to do so without putting at risk the identity of the organism, which is seemingly threatened by the process of differentiation. And it may further be possible that our investigation of the notion of difference will show that, at base, and in its own way, not only identity is intelligible, but difference as well.

13.2.1 Constitutive Difference and Comparative Difference

The notion of difference does not appear with postmodern philosophy, but rather has a long history. It was a key notion as far back as the biology of Aristotle. Thus, when we read Aristotle from the perspective of his biology, as has become more common in recent decades, his thought connects in a very natural way with that of contemporary philosophers of difference, and especially with the thought of Deleuze. Aristotle's biological texts call each characteristic or trait of a living being *difference* (*diaphorá*). Being viviparous or herbivorous, or having wings or a biliary vesicle, are differences. In fact, the biology of the Greek thinker is structured according to differences, and not according to species. For example, it is possible to encounter closely together references to the dolphin and to the horse, due to the fact that they are both viviparous. Also: Aristotle is interested in the mole because it embodies two differences, namely blindness and viviparity, which rarely appear together. The biology of Aristotle deals, in sum, with differences, such as viviparity, blindness or herbivority, and not with this or that species (Pellegrin 1982; Balme 1987a).

But in Aristotle the notion of difference has, at least, two meanings that it is important to distinguish. Difference can be understood in a *logical sense (logikos)*, as a trait that differentiates, distinguishes or separates one class from another, or else in its *physical sense (physikos)*.[2] In this second sense it is a matter of a trait qua something constitutive of a concrete living being. We can find analogues for both senses in

---

[2] Here "physical" is read as being opposed to "logical". The physical is that which has reality apart from thought. In this sense, the physical is not opposed to the biological. In fact, all living beings have their own reality whether or not they are thought of. Nor is there any assertion of reductionism here: it is not held that biology can be reduced to physics, but only that living beings have their own reality.

today's terminology. According to the first sense, we say that two entities are different from each other in virtue of this or that characteristic. Here, the difference compares and classifies. According to the second sense, we speak of the process of differentiation of an organism, which is its ontogenesis, the genesis of the heterogeneous starting with the homogenous, and, as a result, the constitution of the entity itself. Here the difference constitutes the being as such. Of the two kinds of difference, the latter is intuitively more relevant to the question of robustness than the former.

Aristotle inherits the first sense of difference from Plato, and maintains it. He adds, however, the second, which is properly biological. The concept of differentiation is used in this second sense today in embryology, in order to indicate the process by which more differentiated tissues arise out of others that are less differentiated. The first sense is more classificatory, comparative and static, while the second sense is more dynamic and constitutive. The first is principally logical and the second physical and, in the case of living beings, biological. One could also say, therefore, that comparative differences are in reality a subproduct of the constitutive. That is, as a living being gradually differentiates itself, and a second living being does the same, both become increasingly different from each other. This is why living beings, even of different species, resemble more each other in the earliest embryonic stages.

In the contemporary philosophy of difference we can also encounter an idea of constitutive difference that is very close to that of Aristotle, which is what Deleuze (1968) has termed *internal difference*. If we wish to come to know the characteristics of organisms, including their capacity for canalization, we must go first to what is primary and constitutive, to difference in the physical sense. Difference in its physical sense directs us towards the organization of the organism itself, towards its constitution, i.e. towards its individual form, towards its essence.

The allusion to an individual form—which is a *form of life* in the case of living beings—might appear strange, and it might seem even stranger that we would identify the concrete and individual organism with its essence. But, in my opinion, there are good arguments in favor of this identification. Aristotle himself, if we take into account recent research into his biology, can be perfectly well interpreted as being a defender of a form that is quantitatively and qualitatively individual, which is identifiable, in each being, with its essence.[3] As David Balme (1987b, 306) wrote: "In *GA* [*De Generatione Animalium*] as in *Metaph. Z*, neither essence nor form correspond to 'species' ". In this regard, Theodore Scaltsas (1994, 3) has written: "The essence cannot belong to the subject, it must be the subject itself". Quoting Balme (1987a, 19) again: "In *HA* [*Historia Animalium*] the aim is evidently to collect and analyze differentiae so that animal form can be defined, and such definition will be able to be individual". And Aristotle himself wonders: "What is there that impedes certain realities from being identified with their essence, given that essence is substance?"[4]

---

[3] In the interpretation I present here, I follow contemporary authors such as Pierre Pellegrin (1982) and David Balme (1987, 1987a). I have argued in favor of this interpretation in various works (Marcos 1996, 2012).

[4] *Metaphysics*, 1031b 31 and ff.

In sum, we have access to knowledge of an organism by way of knowledge of its constitutive differences; and ontogenesis—with its capacity for canalization— would consist, precisely, in a process of differentiation. Nevertheless, the question of the unity of the organism is still a concern, since we are still speaking of differences, in plural. Is it possible that all of them could be integrated into a single difference?

13.2.2 The Unity of Differences in the Final Difference

The constitutive difference can only be one and unique, since it constitutes a being that is one and unique. Even more, the constitutive difference, in reality, is identified with the very substance it constitutes. In the case at hand, the constitutive difference would be the organism itself. Various texts from the treatise *On the Soul* and from the rest of the biological works of Aristotle point in this direction. Allow me to consider one of them, perhaps the most significant. There is a passage from the treatise *On the Parts of Animals* in which Aristotle affirms that "the difference is the form in the matter".[5] For some, this affirmation might appear strange.[6] And it is if we read it from the logical point of view, but not if we do so from the physical point of view.

From the logical point of view, the species is the result of adding the difference to the genus. That is to say, one gets the impression that the species is closer to matter, and that matter is the principle of individuation of the species. But from the physical point of view things change, and it is the difference that is closer to the matter. The difference is the form in matter, the form in the concrete individual entity, and, in reality, it is that entity itself, since there are many texts in which Aristotle affirms the unity of matter and form.[7] It is the form, understood as difference, which here plays the role of principle of individuation of an undifferentiated or generic matter. As a result, the constitutive difference is not a form in the abstract, but rather the form in the matter, the concrete organism, if we are speaking of living beings. This is not an unusual reading, but rather it links in a very natural way to other texts found in Book II of the treatise *On the Soul* and in the *Metaphysics*, Books VII and VIII.

It is clear, then, that from the physical point of view the constitutive difference is unique and is identified with the thing itself; in the case of living beings, this is the concrete organism. However, we are now faced with the problem of the relationship between the physical and the logical. That is to say, can we somehow capture this difference by means of our concepts? Can we define each organism, come to know its individuality? Can we integrate the distinct traits or differences proper to the

---

[5] *On the Parts of Animals*, 643a 24.

[6] There have even been cases of editors and translators of Aristotle's texts that have sought to amend this reading. Nevertheless, it figures precisely in this way in all manuscripts save one (Inciarte 1974, 276).

[7] *Metaphysics* VIII 6; On the Soul II 1; *On the Parts of Animals* I.

organism in one single and final logical difference that corresponds with the constitutive physical difference? In this case, the final difference would be at once the substance and the definition of the thing.[8] We are in the presence of the problem of the intelligibility of the difference, or, if one wishes, of the concrete individual, which in the case at hand is the same as speaking of the intelligibility of the organism. I will now proceed to a consideration of this question.

13.2.3 *The Intelligibility of the Final Difference*

If the separation between *logos* and *physis* were unbridgeable, we would have to renounce any knowledge of concrete organisms. Aristotle seeks in two ways to close the gap between logical differences and physical difference. He does so, first of all, via a reworking of the theory of definition. And he fails. He then makes a new attempt, seeking a new form of knowledge (*alle gnosis*[9]), this time by way of practical philosophy and a new constellation of ideas, including the ideas of analogy, metaphor, similarity, prudence and practical truth. The contemporary philosophy of difference has taken note, and rightly so, of the failure of the first attempt, that of the definition and the univocal logos, but has not been able to give sufficient value to the potential of the second.

The reform of the logical apparatus of the definition is carried out by the Greek thinker via the following steps: First, he reduces to one all the genera that can figure in a definition: "In the definition nothing else enters except the genus called first and the differences".[10] Next, he reduces the genus to the species: "The genus does not exist at all apart from the species of that genus, or if it exists it is as matter".[11] In a third step he reduces the species to the differences: "It is clear that the definition is the statement constituted on the basis of the differences".[12] And the fourth step reduces all the differences to the last: "It is clear that the final difference will be the substance and the definition of the thing".[13] Later, if the whole process was well performed, that is, always dividing by the difference of the difference,[14] then, the entire definition—and even more, the substance itself—would be contained in the final difference, which would be both logical and physical at the same time.

But in the treatise *On the Parts of Animals*[15] there already appear numerous problems that arise from attempting to establish connections between the *physis* and the *logos* via the path of the definition. The definition that leads to the final difference,

---

[8] *Metaphysics*, 1038a 19–20.

[9] *On the Generation of Animals*, 742b 32.

[10] *Metaphysics*, 1037b 30 – 1038a 4.

[11] *Metaphysics*, 1038a 5–8.

[12] *Metaphysics*, 1038a 8–9.

[13] *Metaphysics*, 1038a 19–20.

[14] *Metaphysics*, 1038a 9; *On the Parts of Animals*, 642b 5 – 644a 12.

[15] *On the Parts of Animals*, 643b 10 and ff.

dividing by the difference of the difference (animal→ legged animals→ quadruped→ soliped…), does not appear workable. As a result, when investigating animals one must proceed via various series of differences that are, so to speak, parallel, and which do not resolve into one single such series. Thus, the most recommendable thing is to proceed as common knowledge does, putting those individuals together that share a certain constellation of differences, which we cannot yet reduce to one single difference. For instance, we understand birds as being blood-bearing, oviparous, winged, feathered animals with hollow bones, beaks and no teeth. Each of these differences can be the extreme limit of a distinct series and none of them need necessarily include all of the others.

It appears that the idyllic communion between *physis* and *logos* ends here. The closest we find them to one another is in the context of formulas such as "Socrates is rational" (his specific difference), which expresses more about Socrates than does "Socrates is human" (his species).[16] Even so, it does not correctly express the integration of the various differences in the concrete individual. We discover here— pace Parmenides and Hegel—that the relationship between thought and reality cannot be one of identity. It should come as no surprise that Aristotle himself would demonstrate in certain passages a clear lack of confidence towards a univocal logos and the virtualities of the definition.[17]

What would this other form of knowledge (*alle gnosis*) consist in, this knowledge that would bring us closer to the concrete organism? In the first place, in order that this knowledge be possible at all, it is important to note the formal nature of the difference qua principle of individuation. Only if we recognize this formal aspect in individuals will they turn out to be intelligible. Aristotelian philosophy can be interpreted in many ways, and historically commentators have done precisely that. However, in our times, and on the basis of a close attention to the biological texts, a reading is emerging according to which the form is individual. It is individual in a quantitative sense in each and every being, and it is individual in a qualitative sense in a graded manner. That is, the individual qualitative differentiation admits of degrees; thus, an ant, whose behavioral flexibility is very limited, since its behavior is genetically regulated in a very rigid way, shows fewer differences with respect to other ants than does a dolphin, whose ability to learn is considerably greater, with respect to other dolphins. Therefore, in certain cases, what we learn about the species can nearly exhaust what we can learn about each individual. But in other cases, once the traits of the species have been learned, there still remains much to be learned about each organism.

For this task, the integration of multiple scientific methods will be necessary, as will a certain analogical interpretation of scientific concepts, but attention will also have to be paid to other forms and sources of legitimate knowledge that will get us closer to the plane of the concrete individual. These will include, for instance, the arts, with their capacity for metaphor creation, as well as philosophy, praxis and the

---

[16] *On the Parts of Animals*, 645b 13–22; *On the Soul*, 402b 10–16, 415a 16–20.
[17] *Sophistical Refutations*, 165a 5–14.

daily experience of life interpreted through the modulation of prudence. As Sandra Mitchell (2004, 81) suggests, "both the ontology and the representation of complex systems recommend adopting a stance of integrative pluralism, not only in biology, but in general".

In my opinion, if we wish to achieve a kind of knowledge that is reasonable and close to the concrete individual, this pluralism of methods will have to include those offered by the natural sciences, and will probably have to go beyond even those. Even more, the natural sciences themselves will have to be reoriented towards methodological forms that are not exclusively reductionist, towards analogical interpretations of their conceptual systems and, as Thomas Nagel (2012) has recently suggested, towards a reconsideration and broadening of our models of causality. It is possible to conjecture, in this regard, that the deep explanation of the robustness of certain systems resides in the capacity of the system itself to act on its own parts, in a kind of top-down causality. This functional or teleological perspective seems to demand an ontology that recognizes the substantial existence of the system as such and not only of its parts. This top-down causality, in my view, is found in a paradigmatic manner in the development and functioning of living organisms. It is no accident that in Aristotle it is organisms that are substances in a paradigmatic sense. Nor is it an accident that developmental canalization is the paradigmatic instance of robustness.

It appears that through this pluralist reorientation of the sciences, and a sensible integration of diverse sources of knowledge, we can attain—or nearly so—to knowledge of the individual, which is important, because we construct the scaffolding of science on the basis of a certain knowledge of concrete individuals, through similarity among differences. This is how we elaborate and apply concepts, laws, classifications and models.

13.3 Difference, Identity and Similarity

The intelligibility of difference can be cast into doubt, however, since according to a certain philosophical tradition, intelligibility depends on the identity of the object. But I have characterized the organism here as the process and result of differentiation. The organism is seen, in this way, as a difference. However, has the organism, seen as difference, thereby lost its identity? Is it, then, knowable at all? We must, as a result, investigate the relationship between difference and identity.

One of the most profound and influential studies of identity and difference is that undertaken by Martin Heidegger. A lecture he gave in 1957 has been published, together with another text from the same period, under the title *Identity and Difference*. This text became especially fashionable in the circles of postmodern philosophy, and has been seen as marking the beginning of the so-called *philosophy of difference*. We can say that, together with difference, we also receive identity: Heidegger (2002, 21) spoke about "the close relation [*zusammengehörigkeit*] of identity and difference". He argues that "what the principle of identity [...] states is exactly what the whole of Western European thinking has in mind: [...] If science could not be sure in advance of the identity of its object in each case, it could not be what it is. By this assurance, research makes certain that its work is possible. Still, the leading idea of the identity of the object is never of any palpable use to the sciences" (Heidegger, 2002, 26).

I suggest interpreting Heidegger's text in the following terms. The *physical identity* of the beings that the sciences deal with is a condition of possibility of those very sciences. If each thing were not one and the same with itself, it would be difficult to think scientifically. The world – and this term is already exaggerated – would be a chaos that is totally refractory to reason. On the other hand, however, the physical identity of each being with itself is not a very useful notion in doing science. Science needs also another form of identity that connects beings, that takes them out of their individuality, puts them in contact and joins them together. This kind of identity would be identity in concept: any two horses or any two drops of water are just that, and can be respectively bundled together under the same concept, thanks to their conceptual identity; they are subsumed under one and the same concept. We can speak here of *logical identity*, as opposed to the *physical identity* of any concrete being with itself.

But this type of logical identity, or identity according to concept, has come in for fierce criticism from some postmodern thinkers (from Heidegger onwards, including thinkers such as Deleuze, Derrida and Lyotard). The basis of their criticism lies in the fact that identity, understood in this way, leaves differences out; instead, the peculiarities of each being and each process, never exactly the same as any other, pass to a second level of reality. Difference is thus forgotten.

According to postmodern thinkers, this forgetfulness goes hand-in-hand with an attempt to impose identity over difference. Postmodernity can be seen, in fact, as being the fruit of the cultural and vital malaise of a reason that has forgotten difference. Postmodern thought has been based on this critique ever since its ultimate origins in Nietzsche and Heidegger.

I think that the postmodern attempt to rescue difference may be valued positively, along with its emphasis on the dynamic and vital aspects of reality and its denouncing of the excesses of a reason that focuses exclusively on identity. We hear, for instance, Deleuze's voice against a background of Bergsonian resonances leading us to the mobile, the fluid, the concrete, the diverse, the living. Nevertheless, we should ask ourselves whether difference alone will ever allow us to regain identity. The question is important. Without a certain minimum stability, without identifiable objects, the work of science would become impossible.

However, forgetting differences distances us from the real world, from things themselves. If reason aligns itself exclusively with identity, then it becomes separate from life and experience, from development, from time, from the diverse, from the plural, from the concrete and real. But the unilateral favoring of difference does not auger well for the results of science, bringing instead fragmentation, deconstruction, relativism and, finally, nihilism. Heidegger was right when he invited us to think of identity and difference *together.* How can this be done? Would the mediation of similarity be useful here?

First, let us return to the Aristotelian distinction between the logical (*logikos*) and physical (*physikos*) points of view.[18] The Spanish philosopher Xavier Zubiri (1980, 22) clarifies the meaning that the *physical* has here: "'Physical' is the original and ancient expression for designating something which is not merely conceptual, but real". This distinction would be bereft of meaning if being and thinking were indeed one and the same. Affirming the identity of being and thinking means forgetting or denying difference. However, the everyday experience of the search for knowledge is the experience of effort, of the making of mistakes, of inaccuracy. The fallible, contrived and unpredictable nature of human knowledge makes us see that there is a distance, a *difference*, between being and thinking.

In a complementary way, achievements and acquisitions, moments of insight and even our very survival clearly indicates that the gap between being and thinking is not unbridgeable. Reality is not concept. Nevertheless, the two are not totally refractory to each other; they may be linked through the work of a subject. Reality is *not identical* with the concept, but it is intelligible, in a contrived, unpredictable, not algorithmic, fallible but revisable and critical way. It is therefore probable that the very relationship between being and thinking may be better described through the concept of *similarity*.

Secondly, the distinction between the *logikos* and *physikos* points of view will have to be applied to the very notions of identity and difference. Recall that identity, from the physical point of view, is the relationship that each entity has with itself. It is true that the beings around us are subject to change. Change, however, need not always mean the loss of identity. Beings can change some of their properties over time without losing their physical identity. In part this is the very basis of their capacity for robustness and resilience. In turn, when we consider different entities under a single concept, we are thinking of identity in logical or conceptual terms, apart from time, change and physical processes. Both types of identity are indispensable in human knowledge, the former as a condition of possibility for knowledge and the latter as a result of conceptual construction and as a tool for explanation and application.

On the other hand, we can also discern a logical and a physical way of looking at difference, as we have seen above. Both prove necessary, as was the case with the two meanings of identity. Without difference in the physical sense, there would be no identifiable objects, only an undifferentiated magma–or nothing, pure and simple. For its part, difference in the second sense, the logical sense, is the key to establishing comparisons and drawing up classifications.

This observation allows us to clarify the relationship between identity and difference in the physical sense. Neither has priority: the self-identical is constituted by differentiation, and difference is always the constitutive difference of an entity which is identical with itself. In Heideggerian terms, they belong to each other. We have to think them together.

However, and thirdly, we know that neither the physical identity of each substance with itself, nor their differences alone, serve to construct concepts, laws,

---

[18] *Physics* 204b 1–12; *Metaphysics* Z and H.

metaphors, or models. Identity and difference are ontological presuppositions for all of this. But is *similarity* the force that unites things in concepts and representations.

In speaking of similarity here, I am not speaking of a dyadic relationship between objects, available in the world to be used and consumed by science. It is rather a triadic relationship between two objects and an active subject. It is one of those triadic relationships that Peirce talks about. Without a subject there would not be actual similarity.

Both in Aristotle and in Peirce, similarity is understood as a relationship between three poles. From the Platonic point of view as well, the relationship of similarity is triadic: it demands reference to an Idea. Aristotle keeps the triadic scheme but the third pole is no longer an Idea, but a human subject who actualizes a similarity that exists in the objects as a real possibility. Consequently, similarity is not one of those relationships that Peirce calls relationships of "brute force" among pairs, but a triadic relationship. This triadic relationship is removed from its original setting within Platonism and comes to rely, not on immobile Ideas, but precisely on the activity of a subject.

Nevertheless, this triadic character does not strip similarity of its objective basis. If it lacked an objective basis, we could establish any relationships of similarity we wished at whim, between any objects and in any way. We know from experience that this is not so, that sometimes reality simply says *no* to our desires to connect things, that our classifications are sometimes erroneous, that laws do not always predict correctly, that the theories, models and metaphors with which we try to understand reality are not always satisfactory. This is due to the fact that reality also has its say. In fact, similarity has an objective basis. It is rooted as a possibility in physical difference and identity.

In the case of organisms, the objective possibility of similarity derives from genesis, i.e. from differentiation. Differentiation is the physical basis for similarity. But the objective possibility of two things being seen as similar is only actualized because of the activity of a subject. So, thanks to similarity, we can pass from the game of physical identities and differences to the game of concepts and representations, with its logical relationships of conceptual identity and comparative differences. We do this by actualizing the similarities that exist as possibilities in reality. And the connection itself between the logical and the physical level, that is, between thinking and being, should be described not as identity, nor as absolute difference, but as a kind of similarity.

13.4 Conclusion

Robustness is defined as the capacity that a system has to maintain its identity despite the internal and external perturbations that it may suffer. For example, an organism can suffer various perturbations over the course of its differentiation and even so still be able to preserve its identity. This phenomenon is called canalization.

The canalization of development constitutes a paradigmatic case of robustness. Therefore, prior to beginning a philosophical reflection on the robustness of systems in general, I believe it is useful to focus on one paradigmatic case, that of the systems we call organisms, and their particular form of robustness that we call canalization. In this case, there are certain base ontological conditions that can illuminate the phenomenon in question.

The development of organisms is a process of differentiation. It will be, therefore, vitally important to study the ontology of difference, the relationship between difference and identity, and the relationship between difference and intelligibility. I have done so by using the Aristotelian distinction between the *logikos* point of view and the *physikos* point of view. From the physical point of view, identity and difference—in Heidegger's terminology—belong to each other mutually, since the identity of the organism is constituted by differentiation, and the result of a process of differentiation can only be a difference. From the logical point of view, we speak of conceptual identity—based on the similarity amongst differences—and of comparative difference. The play of both allows for the construction of an entire scaffolding of scientific models. On the other hand, the relationship between the two planes, physical and logical, has been characterized as being a relationship of similarity.

The canalization of development of organisms occurs thanks to the mutual belonging of identity and difference in the organism itself, and it is intelligible to us via a certain interweaving of similarities. Although robustness admits of an analogous formal description in diverse types of systems, my conjecture—and it is only that, a conjecture—is that we will encounter diverse ontological foundations. It is most likely the case that we will only encounter mutual belonging between identity and difference, or, stated in other terms, the capacity to canalize differentiation itself, in living organisms. It is likely, instead, that the ontological bases of a system of another kind (a computational one, a building, a city…) are displaced out of the system itself, and will ultimately refer to the identity/difference of a living organism.

References

Amundson, R. (2005). *The changing rule of the embryo in evolutionary biology: Structure and synthesis*. New York: Cambridge University Press.

Aristotle. *Metaphysics*.

Aristotle. *On the Generation of Animals*.

Aristotle. *On the Parts of Animals*.

Aristotle. *On the Soul*.

Aristotle. *Physics*.

Aristotle. *Sophistical Refutations*.

Balme, D. (1987a). Aristotle's use of *divisio* and *differentiae*. In A. Gotthelf & J. Lennox (Eds.), *Philosophical issues in Aristotle's biology* (pp. 69–89). Cambridge: Cambridge University Press.

Balme, D. (1987b). Aristotle's biology was not essentialist. In A. Gotthelf & J. Lennox (Eds.), *Philosophical issues in Aristotle's biology* (pp. 291–312). Cambridge: Cambridge University Press.

Deleuze, G. (1968). *Différence et répétition*. París: P.U.F..

Derrida, J. (1967). *L'Écriture et la différence*. París: Seuil.

Heidegger, M. (2002). *Identity and Difference* (J. Stambaugh, Trans.). Chicago: University of Chicago Press.

Inciarte, F. (1974). *El reto del positivismo lógico*. Madrid: Rialp.

Lennox, J. (2017). An Aristotelian philosophy of biology: Form, Function and Development. In M. Bertolaso (Ed.), *Emerging trends in the philosophy of biology. Acta Philosophica* (Vol 26, pp. 33–52).

Lyotard, J. F. (1983). *Le différend*. París: Minuit.

Marcos, A. (1996). *Aristóteles y otros animales*. Barcelona: PPU.

Marcos, A. (2012). *Postmodern Aristotle*. Newcastle: CSP.

Mitchell, S. (2004). Why integrative pluralism? *E:CO Special Double Issue, 6*(1–2), 81.

Nagel, T. (2012). *Mind and cosmos. Why the materialist Neo-Darwinian conception of nature is almost certainly false*. Oxford: OUP.

Nicholson, D. J. (2014). The return of the organism as a fundamental explanatory concept in biology. *Philosophy Compass, 9*(5), 347–359.

Pellegrin, P. (1982). *La classification des animaux chez Aristote*. Paris: Les Belles Lettres.

Scaltsas, T. (1994). *Substances & Universals in Aristotle's metaphysics*. Ithaca: Cornell University Press.

Zubiri, X. (1980). *La inteligencia sentiente [Sentient Intelligence]*. Madrid, Alianza. English translation from http://www.zubiri.org/works/englishworks/si/SI1C1.htm.

Alfredo Marcos is Full Professor at the University of Valladolid, Spain, and regularly teaches courses and conferences at several universities in Spain, France, Italy, Poland, Mexico, Argentina and Colombia. He has published several books and chapters on philosophy of science, environmental ethics and Aristotelian studies, including: Ciencia y accion (Mexico: FCE, 2010; translated into Italian and Polish); Paths of Creation: Creativity in Science and Art (Bern: Peter Lang, 2011); and Bioinformation as a triadic relation (Information and Living Systems, edited by G. Terzis and R. Arp; MIT Press, 2011). He has also published several papers in journals of prestige, such as Studies in History and Philosophy of Science (Elsevier), Epistemologia: An Italian Journal for the Philosophy of Science (Tilgher) and Information System Frontiers (Kluwer).

Index

© Springer Nature Switzerland AG 2018

M. Bertolaso et al. (eds.), *Biological Robustness*, History, Philosophy and Theory of the Life Sciences 23, https://doi.org/10.1007/978-3-030-01198-7

C